Special Topics in Quantum Optics

Weiping Zhang · Zeng-Bing Chen
Editors

Special Topics in Quantum Optics

上海交通大学出版社
SHANGHAI JIAO TONG UNIVERSITY PRESS

Springer

Editors
Weiping Zhang
Shanghai Jiao Tong University
Shanghai, China

Zeng-Bing Chen
Nanjing University
Nanjing, China

ISBN 978-981-99-8453-4 ISBN 978-981-99-8454-1 (eBook)
https://doi.org/10.1007/978-981-99-8454-1

Jointly published with Shanghai Jiao Tong University Press
The print edition is not for sale in China (Mainland). Customers from China (Mainland) please order the print book from: Shanghai Jiao Tong University Press.

Translation from the Chinese language edition: "Special Topics in Quantum Optics" by Weiping Zhang and Zeng-Bing Chen, © Shanghai Jiao Tong University Press 2023. Published by Shanghai Jiao Tong University Press. All Rights Reserved.

This Springer imprint is published by the registered company Springer Nature Singapore Pte Ltd.
The registered company address is: 152 Beach Road, #21-01/04 Gateway East, Singapore 189721, Singapore

Contents

1 **Optical Quantum Computing** 1
Zeng-Bing Chen

2 **New Progress in Quantum Optics and Atom Optics** 131
Weiping Zhang

Index 225

Chapter 1
Optical Quantum Computing

Zeng-Bing Chen

1.1 Introduction

Since the shift from the passive observation to the active manipulation of quanta (photons, electrons, atoms, molecules, etc.) in the 1980s and onward, the combination of quantum physics and information technology has blazed a completely new trail in information processing and communication that is fundamentally different from the conventional approach—quantum information sciences. A relatively new but quickly emerging discipline, it has become one of the most vibrant frontiers for research in physics and information science in the past few years. Researchers have discovered that quantum control technology can revolutionize the way we encode, store, transmit and handle information, which is envisioned to overcome bottlenecks in classical physics-based information technology with regard to aspects like information transmission capacity increase, computing speed improvement and information security augmentation, providing a new perspective and growth point for the practical applications of quantum theory.

Quantum informatics relies on microscopic particles as carrier, along with some of the fascinating properties unique to quantum mechanics that differ from the physical laws of the macroscopic world, to perform functions that are extremely difficult or impossible to achieve through classical computing and communication. Among the new subfields in this area, quantum computation has been a magnet for attention because of its enormously greater computing power than classical computers and incomparable advantages over classical computers in solving complex mathematical problems. Quantum computation is a form of computation that adopts the fundamental principles of quantum mechanics to perform calculations. It is known

The Chinese version of this chapter was coauthored by Xing-Can Yao, Zeng-Bing Chen, and Jian-Wei Pan.

Z.-B. Chen (✉)
School of Physics, Nanjing University, 22 Hankou Road, Nanjing 210093, China
e-mail: zbchen@nju.edu.cn

W. Zhang and Z.-B. Chen (eds.), *Special Topics in Quantum Optics*,
https://doi.org/10.1007/978-981-99-8454-1_1

that the most basic unit of information processing in a classical computer is the bit, and information is customarily represented in the binary system of "0" and "1," but an information unit can only exist as either one of these two states at a given moment. Meanwhile, the basic information unit of a quantum computer is the quantum bit (qubit), which has more computation power than the bit and is capable of remaining in any superposition state ($\alpha|0\rangle + \beta|1\rangle$) of "0" and "1". This superposition characteristic of qubits results in a dramatic reduction in the time required to process information when dealing with problems of the same complexity (for classical computers, the computation time increases exponentially with computational complexity, whereas for quantum computers it increases only polynomially). To put it differently, an N-bit classical memorizer can store just one of 2^N possible data, while an N-qubit quantum memory is able to store 2^N numbers at the same time, meaning that the latter's capacity to store information increases exponentially as N increases. In information processing, the coherent superposition property of quanta allows mathematical calculations to be performed simultaneously on all data in the quantum memory, that is, parallel calculations of data. Thus, quantum computers can operate in a high degree of parallelism unmatched by classical computers that will result in tremendous reduction in the amount of computing resources (time, storage cells, etc.) required for the same job. Technological advancements in recent years have thrust quantum information to the fore in domains such as coding, transmission, and information processing. Efforts to purposefully control these quantum mechanics effects have tapped into some the powerful capabilities and functions of quantum information: Quantum key distribution (QKD) enables absolutely secured and confidential communications; quantum metrology provides more accurate measurements thanks to the theoretical support of quantum mechanics; and quantum lithography has the ability to fabricate and assemble devices much smaller than the wavelength of light. These applications of quantum information have found widespread uses in various industries, but the quantum computer is perhaps the most stunning and quintessential example of the future potentials of quantum technology, because the massive superiority over classical computers information processing power and speed makes it a beacon of the future no one can ignore.

Of the most intriguing, important and identifying features that distinguishes the quantum world from the classical world is quantum entanglement, which is a manifestation of the hyperconnectivity and uncertainty among particles and deemed a core resource in quantum information processing and quantum computing. The quantum entanglement state is defined as a special type of quantum state describing a complex system (with two or more constitutive systems) that cannot be decomposed into the tensor product of the respective quantum states of the constitutive systems. In a two-particle entangled state, for example, we cannot discern the difference between the two particles without destroying the entanglement, which means the two particles are fully isotropic. Many important schemes and ideas in quantum information processing and quantum computing rely on quantum entanglement, thus a stable and efficient source of quantum entanglement is the cornerstone to quantum computation.

In the quest to fulfill quantum computing tasks of all sorts, a number of experimental systems based on different scientific principles such as linear optics, quantum dots, cold atoms, nuclear magnetic resonance, and ion traps continue to emerge. Among these, linear optical system boasts several unrivalled advantages: Photons as information carriers are easy to propagate, thereby facilitating communications and the building of distributed quantum computing networks; with the help of simple linear optical components, qubits can be prepared and manipulated in high precision; and other strengths such as strong resistance to environmental interference and constantly improving technical means. Linear optical system has always been the physical system preferred by researchers for quantum computing, and it is also the most sophisticated experimental platform for quantum information. Numerous significant advancements in the field of quantum information were realized through linear optics, including the likes of quantum teleportation, quantum entanglement swapping, quantum entanglement purification, CNOT and Shor's algorithm.

The most basic entangled light source available today is produced in a process called Spontaneous Parametric Down-Conversion (SPDC): A pumped laser hitting a nonlinear crystal can generate pairs of photons whose momentum and frequency are tangled up, and similarly, a process known as type II parametric down-conversion can also create photon pairs that are polarized and entangled with each other. Through entanglement swapping, the entangled information of such a basic pair of polarized and entangled photons can be transferred to other independent photons that are not directly interacting with each other, which facilitates the formation of multi-photon entanglement state. In 1999, during a groundbreaking effort in Austria Zeilinger's team became the first to create a three-photon GHZ state using SPDC and entanglement swapping technology [1], and in 2001 the same group found success with generating a four-photon entangled state [2]. In 2004, Jian-Wei Pan's team from the University of Science and Technology of China (USTC) achieved the world's first five-photon entanglement and terminal-open quantum teleportation by using the entanglement of weakly coherent light and a four-photon GHZ state [3], laying a crucial theoretical and experimental foundation for the eventual actualization of distributed quantum communication and fault-tolerant quantum computing. In 2006, Pan's team realized the first manipulation of a six-photon entangled state by increasing the pump light power [4], and in 2007 the same team successfully created a six-photon graph state for unidirectional quantum computation [5], offering a technical basis conducive to further research on quantum computation, quantum error correction and essential questions in quantum mechanics. At the same time, theory and experiments have shown that hyper-entanglement can be achieved using more degrees of freedom of photons, thereby substantially enhancing and accelerating quantum information processing and computing. In 2005, Pan's team was once again launched into the limelight after producing the world's first source of hyper-entangled photon pairs with high brightness and high contrast, and they tested and verified the GHZ theorem for two photons with a fidelity higher than 95%, revealing for the first time and in an emphatic manner the contradiction between quantum mechanics and local realism in the two-particle system [6]. In 2007 and 2008, Pan's team [7] and Martini's team [8] of Italy created two-body four-dimensional cluster states, and upon this basis they demonstrated a simple one-way quantum computing model. In 2009 and 2010, using the path freedom and polarization freedom of single photons, Martini's team

[9] and Jian-Wei Pan's team [10] respectively produced two-photon six-qubit and four-photon six-qubit cluster states, and applied them in quantum mechanics nonlocality test and quantum computation. In 2012, Pan's team relied on innovation of multi-photon manipulation technology and made history by creating the world's brightest and purest pulsed polarized and entangled photon source. Subsequent optimization and fine-tuning of details yielded a low-noise eight-photon interferometer linear optics technology capable of manipulating the eight-photon entangled state [11].

With the progressive development of laboratory techniques for entanglement sources, which are core resources, significant progress has also been made in optical quantum computing. In 2001, Knill, Laflamme, and Milbum demonstrated that pervasive quantum computing could be achieved by using single-photon detectors to perform post-selective measures with the aid of linear optical elements and auxiliary photons. This significant discovery has inaugurated more capacious space for the application of multi-photon entangled states. In 2003, White's team in Australia implemented an all-optical path qubit-based CNOT gate [12]; another CNOT gate based on polarized qubits was implemented in 2005 by the Weinfurter team in Germany [13]; in 2007, Jian-Wei Pan's team realized the Shor's Quantum Factoring Algorithm using an optical quantum computer [14]; in 2001, Raussendorf et al. [15] proposed the concept of a unidirectional quantum computer. The basic unit of such a quantum computer is a highly entangled multi-qubit state known as a cluster state. When quantum computation is performed, only a series of single qubit measurements on the cluster state is required according to the quantum algorithm that you want to implement. However, the cluster states will be destroyed, so this kind of computer is called a unidirectional quantum computer. The key to realizing such quantum computer is to be able to prepare large quantities of cluster states for quantum computation. In 2005, Zeilinger's team prepared four-photon cluster states based on this scheme, and implemented basic quantum logic gates and quantum algorithms [16]. In 2010, Pan's team used four-photon six-qubit hyper-entangled cluster states to achieve deterministic CNOT.

It is well known that there are two particles in three-dimensional space: boson and fermion. When two bosons are transposed, the wave function of the whole system remains unchanged, but if they are two fermions, the wave function of the system will have a negative sign. In two-dimensional space, there is another kind of particle: anyon. If two anyons swap positions, the wave function may show an arbitrary phase change, which is called fractional statistical property. Based on this property, Dennis et al. [17] proposed topological quantum storage using a quantum state model with a certain mathematical topology in 2002. A year later, Kitaev came up with the idea that manipulating topology-derived anyons could achieve fault-tolerant quantum computation [18]. In 2006, Bombin and Harrington [19], Bombin and Martin-Delgado [20] applied this novel result to the color code and cluster state. As such mathematical topological quantum states require a large number of qubits, experimental research is just at the beginning stage. In 2012, Pan's team successfully implemented topological quantum error correction using an eight-photon cluster state with topological properties, thereby increasing the maximum incidence of tolerable logical operation errors in quantum error correction by three magnitudes to 10^{-2} magnitudes [21]. Non-abelian anyon fractional statistics were observed in the laboratory in 2008 [22]. In 2009, Pan's team implemented an abelian anyon demonstration that could be used for quantum storage, testifying that its properties were different from those of fermions and bosons [23].

Despite all the progress that has been made so far, building optical quantum computers on a large scale still requires a lot of work. For instance, it is not yet known which is more promising, the quantum circuit model or the cluster state model. In fact, if the two techniques can be united, respectively cluster-state coding technology to alleviate noise [cluster state has a higher tolerance threshold (about 1%) for errors] and continued use of CNOT for computation, it is possible to make new breakthroughs. What roles nonlinear optical techniques will play in future optical quantum computers depend on their own efficiency and practicality. In addition, experimental implementations have primarily relied on non-scalable single-photon sources, copious optical elements, and highly efficient single-photon detectors up to now. This implies that the expandable quantum computation cannot be achieved overnight, and it requires powerful nanoscale and even atomic-scale manufacturing processes as well as a deeper understanding of quantum entanglement. What we will gain on the road to optical quantum computing is unprecedented control over natural sources and a deeper understanding of them.

This topic is divided into five subsections: Sect. 1.1 is the foreword; Sect. 1.2 introduces some representative quantum algorithms; Sect. 1.3 briefly describes the theoretical background of optical quantum computing; Sect. 1.4 focuses on the preparation of light quantum bits and coherent manipulation; Sect. 1.5 goes into the particulars of most representative progress made in the forefront of optical quantum computing.

1.2 Quantum Algorithm

Quantum algorithms are an important way to demonstrate the advantages of quantum computing and are a key to achieving a variety of quantum information processing tasks. To provide a clear understanding of quantum computing, this subsection will introduce some of the most important current quantum algorithms and their advantages over traditional classical algorithms.

1.2.1 Shor's Algorithm

The security of communications and passwords relies on the complexity of cracking puzzles through mathematics, and the solutions to NP problems are often considered but they are overly complex to be solved in polynomial time. The number and time of operations required for a computer to address NP problems increase exponentially with bits, making the computational efficiency decline quickly until it is impossible to finish computation. By contrast, a quantum computer features high parallelism in computation and the number and time of operations it requires to deal with NP problems will increase polynomially with bits. So, time is no big deal for quantum computers.

1.2.1.1 RSA and Shor's Algorithm

Proposed by Ron Rivest, Adi Shamirh, and Len Adleman at the Massachusetts Institute of Technology (MIT) in 1977, RSA public-key cryptography is the most influential public-key encryption algorithm today and is recommended by ISO as a standard for public-key encryption. The RSA algorithm is based on a quite simple number theory: it is easy to compute the product of two extremely large prime numbers, but its inverse process—the factoring of large numbers—is an NPC problem. According to this simple principle, the product can be made public as an encryption key. This algorithm has been widely used in multiple fields such as e-banking and Internet, as described in Fig. 1.1a.

If an ordinary computer is used to factorize the number N (binary length $\log_2 N$), its computational steps (time) will grow exponentially with the input length ($\log_2 N$). No reliable method was published to break down RSA before 2008. In recent years, the wide application of the RSA public-key system has accelerated the development of cracking technology. It may be completely replaced someday, for European and American mathematicians and cryptographers have stumbled upon a vulnerability in the system where a small number of public keys are not truly randomly generated. In other words, it may be possible to find the secret prime numbers that form public keys. But this was based on computer attacks on the RSA system. As early as 1994, Shor proposed a quantum algorithm called Shor's algorithm [24] to factorize large numbers, as shown in Fig. 1.1b. With Shor's algorithm, quantum computers can

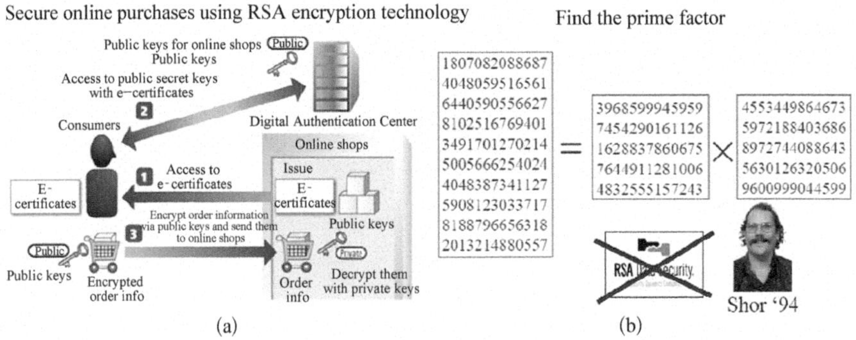

Fig. 1.1 **a** Demonstration of e-commerce using classical RSA passwords; **b** Shor proposed a quantum algorithm for large prime factorization in 1994, meaning that with a quantum computer, RSA passwords will no longer be secure

factorize 1000 bits in a fraction of a second (it takes classical computers 1025 years), and the operation time increases with the lengths of input numbers by the power of three. Shor's algorithm makes things much easier. This however is nothing less than a thunderclap for the RSA cryptosystem, which will have no confidentiality at all, and delivers a devastating blow to the information security of the financial industries and administrative setups across the globe. Shor's pioneering work is an important milestone in the history of quantum information science that has greatly stimulated the development of quantum computers and quantum cryptography.

The power of Shor's algorithm is that it resorts to the parallelism of quantum computing to quickly decompose the prime factors of large numbers. Factorizing large numbers is closely related to password security, although it sounds boring. Let's do some simple math. If you are given a pencil and a blank piece of paper to write out the product of two prime numbers 73 and 137, you can quickly figure out it equals 10001. Let's do it the other way round. If you are given a number 10001, which is the product of two prime numbers, and asked to find the prime factors of the number, it will take you more time to solve the problem. A simpler algorithm traverses the prime numbers from small to large to see if they can be exactly divided by 10001, and this will take roughly 100 attempts to calculate the root of 10001. In fact, even today's most advanced computer algorithm needs $2^{m/3}$ steps to factorize a large *m*-bit number like 2^{m}. It is a terrible exponential issue. If *m* is fairly large, the fastest classical computers in the world can do nothing about it. For instance, if $m = 400$, it would take a classical computer tens of billions of years to finish the calculation, which is already the age of the universe. Seeing that it was difficult to decompose large numbers but easy to test, cryptographers invented an RSA public-key encryption system, which is now widely used to maintain the information security of banks, Internet, and e-commerce, and has an extensive and far-reaching impact on the modern economy, particularly national security.

With Shor's algorithm, it takes only a few minutes to decompose such a large number as $m = 400$ on a quantum computer (the time needed is a polynomial function of *m*). The introduction of Shor's algorithm was undoubtedly a blockbuster, warning people to reexamine the security of RSA ciphers. Arguably, the year of 1994 was a watershed for quantum information research. In the past, people who were only interested in the basic theory of quantum information felt that quantum computers would be powerful, yet no one knew how powerful it would be. After 1994, the study of the physical implementation of quantum computers was meaningful for practical application, and the application was revolutionary. Huge sums of money poured into the research of quantum computation across the world shortly after the theory came into being.

For a detailed description of the Shor's algorithm, see the review by Ekert and Jozsa [25] published in *Rev. Mod. Phys.* in 1996 and the popular scientific article published by Gerjuoy [26] in *Am. J. Phys.* in 2005. There is no doubt that the experimental realization of the Shor's algorithm on a quantum computer is one of the most important goals in the field of quantum information. At the same time, it has been an internationally recognized knotty problem. In 2001, IBM and Stanford University jointly published a paper in *Nature*, revealing the experiment to factorize $15 = 3$

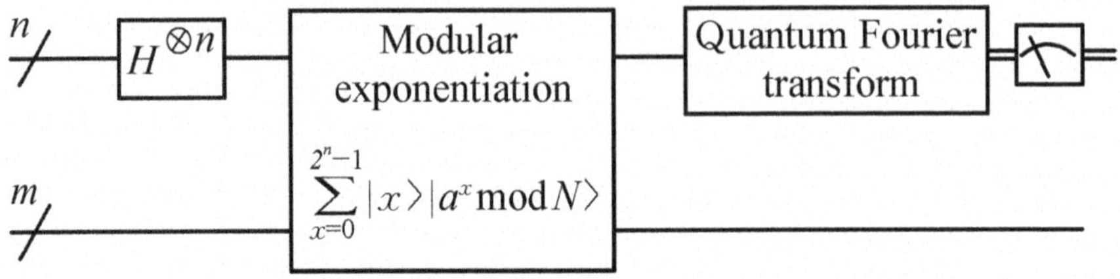

Fig. 1.2 Schematic diagram of Shor's algorithm

× 5 using nuclear magnetic resonance [27]. However, due to the inherent flaw of liquid-state NMR, their experiments could not really prepare quantum pure states, and therefore entanglement could not be observed. N. Linden and S. Popescu said in 2001 that their experiments could not show the quantum essence of the Shor's algorithm. Moreover, it is difficult to expand the liquid-state NMR system to more bits, and this has limited its further upgrading and applications. Photon bits, in comparison, have the advantages of purity, resistance to decoherence, and simple and precise single-qubit manipulation, and so it is helpful to explore and validate Shor's algorithm on this platform.

It should be noted that the Shor's algorithm is actually a derivative of the quantum Fourier transform (see Fig. 1.2), other algorithms like Deutsch-Jozsa and discrete logarithm are of the same kind.

1.2.1.2 Quantum Teleportation

Another highlight is quantum teleportation, which was proposed in 1993 by Bennet et al. [28]. As in Fig. 1.3, its main idea is about the setup of an invisible channel through an EPR entanglement pair shared between the sender and the receiver so that the Bell state measurements are transmitted via the classical channel to the receiver to restore the qubits sent. What is noteworthy is that it contains the classical process of information transmission, so it does not violate the principle that nothing is faster than the speed of light. At the same time, due to measurement collapse, the sender will lose the qubits it has, so the process does not defy the quantum non-cloning theorem either. Quantum teleportation transmits quantum information, which is the most fundamental process of quantum communication. The academic community is currently working on key technologies for quantum communication networks, such as high-brightness entanglement sources, entanglement operations and purification, quantum relay and quantum processors, and has made important progress. It can be asserted that an original and extraordinary way of communication may soon be discovered.

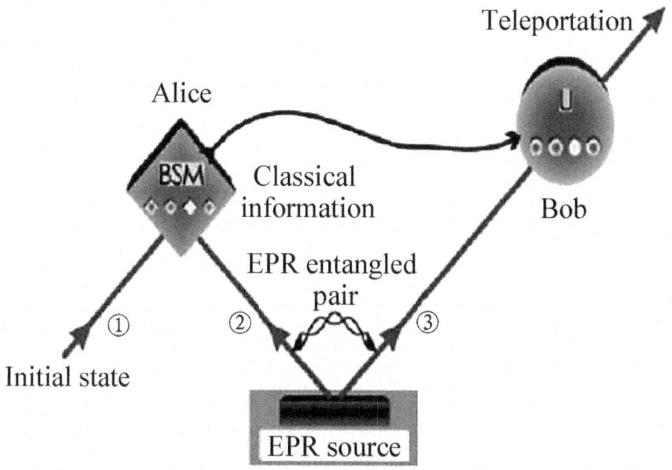

Fig. 1.3 Schematic diagram of quantum teleportation

1.2.2 *Grover's Search Algorithm*

Grover's search algorithm, completely different from the quantum Fourier transform-based algorithm, is suitable for finding a specific target in a set that is disordered and lacks a priori information. For classical algorithms, because there is no a priori knowledge and all the elements are randomly arranged without any pattern, the only method is enumeration. This requires an average of $N/2$ lookups to get the target element for a set with N number of elements. Grover's search algorithm has overcome the dilemma of traditional search algorithms—the difficulty in finding appropriate and efficient solution paths; by taking advantage of quantum parallelism, it searches all the elements at the same time, and increases the amplitude probability of the specified element through continual trials to find the target element with a high probability. Grover's search algorithm can be used in many areas, not only to find an element, but also to look for game strategies and even to attack the cryptosystem. Grover published his algorithm in PRL under the title of "Quantum mechanics can find a needle in a haystack" [29].

The main problem that the Grover algorithm attempts to solve is how to increase the range of solutions. For a search space in a given size N, if there is no a priori information about the search space, how long will it take to find an element that satisfies a particular condition? Under a classical algorithm, when the search space is totally unknown, we can only use the exhaustive method, and the search time is proportional to N. Grover's search algorithm is able to accelerate search with the search time proportional to $\sqrt{N}$. Thus, it can be proved that the Grover's search algorithm is optimal in a completely disordered structure.

However, Grover's search algorithm also has obvious drawbacks, i.e., not using the structural information of the search space N, the need to know the number M

of solutions in advance, and high universality leading to low efficiency. To address these disadvantages, Michel Boyer et al. put forward multiple improvement plans. Researchers further proposed incomplete but more efficient algorithms such as Tad Hog, GSAT, and WALK-SAT.

1.2.3 Quantum Simulation Algorithm

It is possible to simulate quantum systems with classical algorithms, but it is generally resource and time consuming. One of the biggest problems is that the differential equations to be solved are exponential. For a system of n qubits, we need to create 2^n equations to obtain the evolutionary process of the quantum system. Quantum simulation algorithm is similar to classical algorithms, but has fundamental differences. It generally iterates over the Hamiltonian to be calculated, with the new state replacing the old one at each step of the iterative process, and tactfully choosing final measurement to obtain the required information. Some classical algorithms cannot effectively simulate Hamiltonian, which is exponentially difficult for classical algorithms. Thus, quantum simulation may be a workable approach.

1.3 Theoretical Foundation of Optical Quantum Computing

The superiority of photons as qubits is exceptionally long coherence time, so long that people do not usually discuss it (compared to the often headache-inducing coherence time in atomic and quantum dot systems). Photons have long coherence time because it barely interacts with other photons, which makes it difficult to implement control gate operations between optical qubits. As a result, for quite some time, academia considered optical quantum computing impossible to realize. In 2000, Knill, Laflamme, and Milburn demonstrated that it was possible to build universal quantum computers using only linear optics, single-photon sources, and single-photon detectors, which is the famous KLM scheme [30]. In this section, we will first explain the principles of the KLM scheme in detail, then describe another measurement-based quantum computing scheme, and finally relate how to correct the errors in quantum computation.

1.3.1 Basic Quantum Gates

Physically, since there is no interaction between photons, it is impossible to construct a deterministic two-qubit gate with the polarization degrees of freedom or path degrees of freedom of photons. The only way for photons to interact with each other is the Bose commutation relation $\left[\hat{a}, \hat{a}^{\dagger}\right] = 1$. In fact, this property has been used in linear optical quantum computing. To better understand its implications, readers can think of two photons on different paths acting simultaneously on a 50:50 beam splitter, i.e.

$$\begin{aligned} |1,1\rangle_{ab} &= \hat{a}^{\dagger}\hat{b}^{\dagger}|0\rangle \rightarrow \frac{1}{2}\left(\hat{c}^{\dagger} + \hat{d}^{\dagger}\right)\left(\hat{c}^{\dagger} - \hat{d}^{\dagger}\right)|0\rangle_{cd} \\ &= \frac{1}{2}\left(\hat{c}^{\dagger 2} - \hat{d}^{+2}\right)|0\rangle_{cd} = \frac{1}{\sqrt{2}}\left(|2,0\rangle_{cd} - |0,2\rangle_{cd}\right) \\ &|1,1\rangle_{\text{in}} \rightarrow_{\text{trans}} \cos^2\theta|1,1\rangle_{\text{out}}, \\ &|1,1\rangle_{\text{in}} \rightarrow_{\text{refl}} \sin^2\theta e^{i\varphi}e^{-i\varphi}|1,1\rangle_{\text{out}}. \end{aligned} \tag{1.1}$$

It can be seen that the bosonic property of photons gives rise to a clustering effect, specifically, the incident photons will come out in pairs from the same port. But in classical physics, the two photons are supposed to emit with an equal probability of randomness from two different ports of the beam splitter. Obviously, it is a quantum effect. If $|1,1\rangle_{\text{in}}$ can eventually become $|1,1\rangle_{\text{out}}$, then there are only two routes: both photons pass through the beam splitter simultaneously or are reflected by it at the same time. Then, we can write out the following expression:

$$|1,1\rangle_{\text{in}} \rightarrow_{\text{trans}} \cos^2\theta|1,1\rangle_{\text{out}}, |1,1\rangle_{\text{in}} \rightarrow_{\text{refl}} \sin^2\theta e^{i\varphi}e^{-i\varphi}|1,1\rangle_{\text{out}}$$

For a 50:50 beam splitter, $\cos^2\theta = 1/2$, $\sin^2\theta = 1/2$ cancel each other out, and φ can be ignored here. This phenomenon is the Hong–Ou–Mandel effect [31], which is of paramount importance in linear optical quantum computing. In subsequent chapters, we will introduce it in detail. But using the Hong–Ou–Mandel effect alone is not enough to construct a deterministic scheme of linear optical quantum computing. On this account, scientists shifted their focus to probabilistic quantum gates.

In 1995, Lloyd proved [32] that almost all two-qubit gates plus a few single-qubit gates could be used for generalized quantum computation. Nevertheless, in linear optics, controlled phase gates (CZ or C-PHASE gate) and controlled Not gates (abbreviated CNOT or C-NOT) are often applied. First, let us make a truth table to better understand them (see Table 1.1).

Table 1.1 C-PHASE gate and CNOT truth table

Control	Object	CZ (C-PHASE) gate	CNOT
$|0\rangle$	$|0\rangle$	$|0,0\rangle$	$|0,0\rangle$
$|0\rangle$	$|1\rangle$	$|0,1\rangle$	$|0,1\rangle$
$|1\rangle$	$|0\rangle$	$|1,0\rangle$	$|1,1\rangle$
$|1\rangle$	$|1\rangle$	$-|1,1\rangle$	$|1,0\rangle$

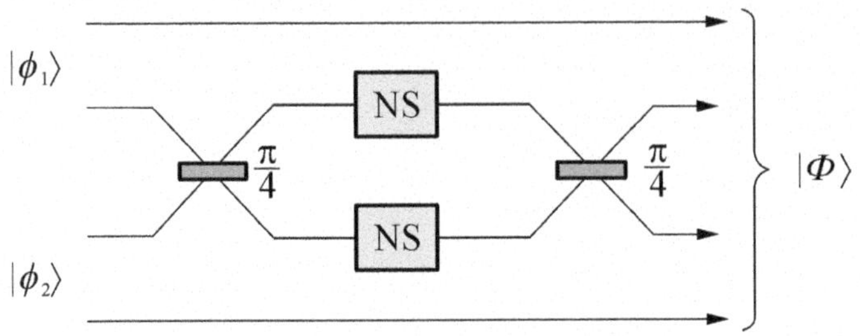

Fig. 1.4 Conditional controlled phase gate (CZ Gate)

It can also be written as $|q_1, q_2\rangle \rightarrow_{CZ} (-1)^{q_1 q_2}|q_1, q_2\rangle, |q_1, q_2\rangle \rightarrow_{\text{CNOT}} |q_1, q_2 \oplus q_1\rangle$. Here, q_k is qubit value 0 or 1 and $q_2 \oplus q_1$ refers to the division of the sum of the two values by two (modulo 2 operation).

A CZ gate can be constructed using two nonlinear sign gates (NS gate). The NS gate acts on the smallest triplet Fock state and can be written as $\alpha|0\rangle + \beta|1\rangle + \gamma|2\rangle \rightarrow \alpha|0\rangle + \beta|1\rangle - \gamma|2\rangle$. This action does not change the range of $|0\rangle$, $|1\rangle$or $|2\rangle$and only changes the phase position of |2. In the circuit diagram given in Fig. 1.4, $|\phi_1\rangle \otimes |\phi_2\rangle = (\alpha|0, 1\rangle + \beta|1, 0\rangle)(\gamma|0, 1\rangle + \delta|1, 0\rangle)$ is an input state, wherein NS gate will bring in a phase position of π. After passing through the second beam splitter, it can get the output state of $|\Phi\rangle = \alpha\gamma|0, 1, 0, 1\rangle + \alpha\delta|0, 1, 1, 0\rangle + \beta\gamma|1, 0, 0, 1\rangle - \beta\delta|1, 0, 1, 0\rangle$. This state has become an inseparable state. Actually, when $\alpha = \beta = \gamma = \delta = \frac{1}{\sqrt{2}}$, the emergent state is a maximally entangled state. The success probability of the CZ gate is $P_{\text{CZ}} = P_{\text{NS}}^2$.

It is clear that a conventional phase shifter cannot be used to construct a NS gate, because only state $|2\rangle$ carries a phase, and moreover a linear optical phase shifter would introduce an overall phase factor i (or − i) on state $|1\rangle$. Knill, Laflamme, and Milburn first proposed the method to use linear optics to construct CZ gates in 2001: such probabilistic NS gates have 3 ports, including the two auxiliary ports measured with the photon number detectors (see Fig. 1.5). The auxiliary input state consists of a vacuum state and a single photon state, and $\eta 1$, $\eta 2$, and $\eta 3$ are the reflectances of the beam splitter. When detectors D1 and D2 measure zero and one photons respectively, it represents a successful NS gate operation. For an arbitrary input state $\alpha|0\rangle + \beta|1\rangle + \gamma|2\rangle$, the probability of success of a nonlinear symbol gate is $P_{\text{NS}} = \frac{1}{4}$, and it is impossible to exceed 1/4 without the support of any feedback system. After numerical calculations and theoretical analysis, Scheel and Lutkenhaus [33] and Eisert [34] have proven that for a NS gate defined to be the success probability is $\frac{1}{N^2}$.

$$\sum_{k=0}^{N} c_k|k\rangle \rightarrow_{NS_N} \sum_{k=0}^{N-1} c_k|k\rangle - c_N|N\rangle \tag{1.2}$$

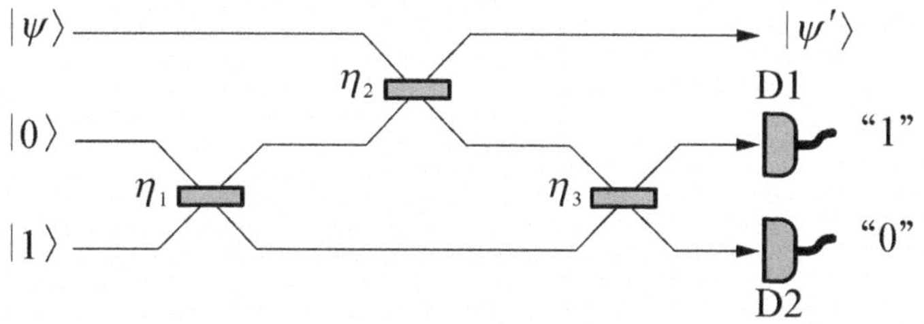

Fig. 1.5 Schematic diagram of the structure of the nonlinear symbol gate proposed by KLM (NS gate)

After the KLM scheme came into being, a few simplified NS gates were reported. The NS gate proposed by Ralph et al. [35] contains only two beam splitters (see Fig. 1.6) and has a success probability of $P_{\mathrm{NS}} = \frac{3-\sqrt{2}}{7}$, which is slightly lower than the theoretical upper limit. Similar schemes were made by Zou et al. [36] and Scheel et al. [37], both using two auxiliary photons, and the success probability is 20 and 25%, respectively.

Rudolph and Pan [38] come up with a more feasible scheme: replacing variable reflectivity beam splitters with polarization rotation method, which made experimental realization more feasible (see Fig. 1.7). This scheme is further elaborated below.

Fig. 1.6 Two equivalent NS gates conceived by Ralph and White

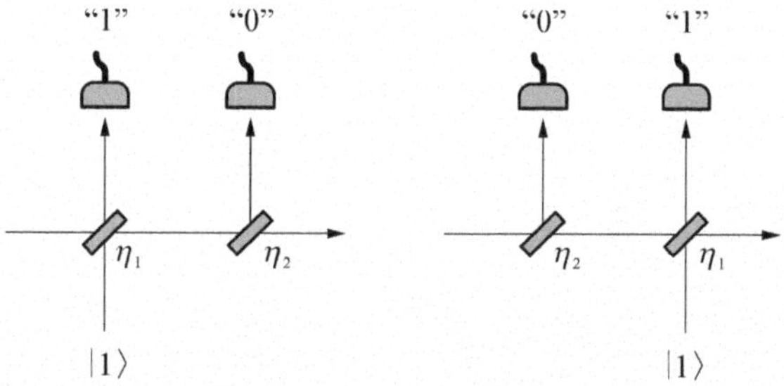

Fig. 1.7 Schematic diagram of the polarized NS gate presumed by Rudolph and Pan

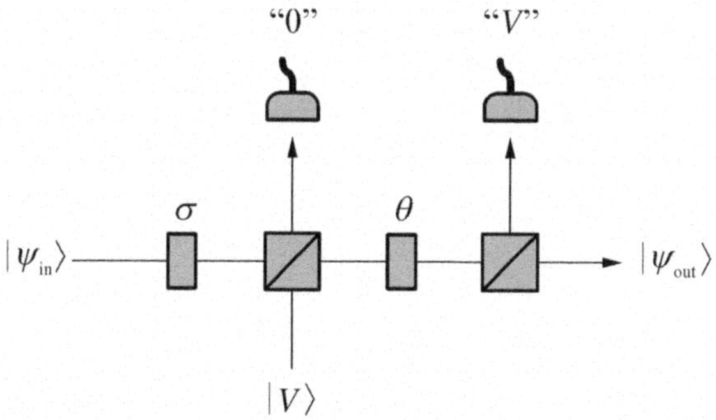

If the input state is horizontally polarized, we first rotate the polarization so that the input state is:

$$\begin{aligned}&\left(\alpha+\beta\hat{a}_H^{\dagger}+\frac{\gamma}{\sqrt{2}}\hat{a}_H^{\dagger 2}\right)\hat{b}_V^{\dagger}|0\rangle\\&\quad\rightarrow[\alpha+\beta\cos\sigma\hat{a}_H^{\dagger}+\beta\sin\sigma\hat{a}_V^{\dagger}\\&\quad+\frac{\gamma}{\sqrt{2}}\left(\cos^2\sigma\hat{a}_H^{\dagger 2}+\sin 2\sigma\hat{a}_H^{\dagger}\hat{a}_V^{\dagger}+\sin^2\sigma\hat{a}_V^{\dagger 2}\right)]\hat{b}_V^{\dagger}|0\rangle\end{aligned}\tag{1.3}$$

The case where photons are not measured at the first port can be written as $\left(\alpha+\beta\cos\sigma\hat{a}_{\mathrm{H}}^{\dagger}+\frac{\gamma}{\sqrt{2}}\cos^2\sigma\hat{a}_{\mathrm{H}}^{+2}\right)\hat{b}_{\mathrm{V}}^{\dagger}|0\rangle$. Immediately after, rotate polarization again: $\hat{a}_{\mathrm{H}}^{\dagger}\rightarrow\cos\theta\hat{a}_{\mathrm{H}}^{\dagger}+\sin\theta\hat{a}_{\mathrm{V}}^{\dagger}$, $\hat{a}_{\mathrm{V}}\rightarrow-\sin\theta\hat{a}_{\mathrm{H}}+\cos\theta\hat{a}_{\mathrm{V}}$. Finally, the output state is changed to be

$$\begin{aligned}&[\alpha+\beta\cos\sigma\left(\cos\theta\hat{a}_{\mathrm{H}}^{\dagger}+\sin\theta\hat{a}_{\mathrm{V}}^{\dagger}\right)+\frac{\gamma}{\sqrt{2}}\cos^2\sigma\left(\cos\theta\hat{a}_{\mathrm{H}}^{\dagger}+\sin\theta\hat{a}_{\mathrm{V}}^{\dagger}\right)^2]\\&\cdot\left(-\sin\theta\hat{a}_{\mathrm{H}}^{\dagger}+\cos\theta\hat{a}_{\mathrm{V}}^{\dagger}\right)|0\rangle\end{aligned}\tag{1.4}$$

When a single vertically polarized photon is detected in the second output channel, it is expressed below:

$$\begin{aligned}|\psi_{\text{out}}\rangle&=\alpha\cos\theta|0\rangle+\beta\cos\sigma\cos 2\theta|1\rangle\\&\quad+\gamma\cos^2\sigma\cos\theta\left(1-3\sin^2\theta\right)|2\rangle\end{aligned}\tag{1.5}$$

When $\sigma\approx 150.5^{\circ}$ and $\theta\approx 61.5^{\circ}$, an NS gate with a success probability of $\cos^2\theta=\frac{(3-\sqrt{2})}{7}$ is obtained.

All of the above are "non-disruptive" quantum gates. In some special applications, such as Bell measurements during quantum teleportation, only "disruptive" quantum gates are needed. A "disruptive" quantum gate will greatly increase the probability of success in gate operations and reduce the resources needed. Ralph et al. [39] stated a realization scheme for "disruptive" CNOT gates, which requires guaranteeing one polarized photon at each output port through post selection in the experiment. The scheme adopts only beam splitters with a reflection coefficient of 1/3 and polarizing beam splitters and its success probability is 1/9. Hofmann and Takeuchi [40] came up with an equivalent CNOT gate at the same time. In 2004, Ralph [41] also demonstrated that the success probability of a set of n CZ gates could reach $P=\left(\frac{1}{3}\right)^{n+1}$ other than $\left(\frac{1}{9}\right)^n$.

Next, we will describe an essential class of gates in optical quantum computing, that is, parity check gate. It contains a polarizing beam splitter and a polarization analyzer. Let the input ports of a polarizing beam splitter to be a and b and its output ports c and d. If the input state of the polarizing beam splitter is $\{|H\rangle,|V\rangle\}$, then

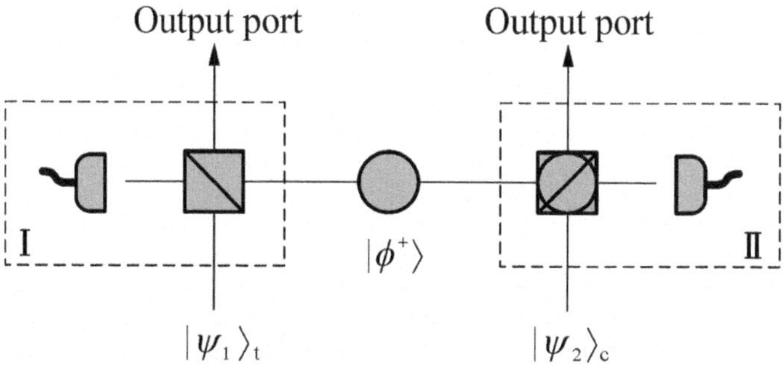

Fig. 1.8 Schematic diagram of the CNOT gate proposed by Pittman

$$
\begin{aligned}
|H, H\rangle_{ab} &\rightarrow |H, H\rangle_{cd} \\
|H, V\rangle_{ab} &\rightarrow |HV, 0\rangle_{cd} \\
|V, H\rangle_{ab} &\rightarrow |0, HV\rangle_{cd} \\
|V, V\rangle_{ab} &\rightarrow |V, V\rangle_{cd}
\end{aligned} \tag{1.6}
$$

In the expression above, $|HV, 0\rangle_{cd}$ means that both a vertically polarized photon and a horizontally polarized photon are output in channel c but none in channel d. The projection measurement $\frac{(|H\rangle \pm |V\rangle)}{\sqrt{2}}$ of channel c yields a parity check: if a photon is detected in channel c, the input qubits have the same logical value. Conversely, if no photon is detected or two photons are detected in channel c, the input qubits are not the same. Parity check gates are very important in optical quantum computing and can be used to build CNOT gates. As shown in Fig. 1.8 [42], two parity check gates and a maximal entangled state $|\phi^+\rangle = \frac{(|H,H\rangle + |V,V\rangle)}{\sqrt{2}}$ as an auxiliary form a CNOT gate whose success probability is 1/4.

1.3.2 Quantum Computing Based on Quantum Teleportation

Assume that the success probability of a single quantum gate in a computational scheme is P. The entire computation requires N quantum gates, and the success probability of the scheme is P^N. For a large N and a small P, the success probability becomes extremely small, and the resources (time or scheme) needed for computation are exponentially multiplied relative to the number of quantum gates. In order to economize probabilistic quantum gates to perform computation, we must solve a puzzle: how can we use quantum gates to do computation without being affected by their probabilistic properties?

In 1999, Gottesman and Chuang [43] suggested a method to remove probabilistic quantum gates from quantum computation circuits: they creatively bypassed this problem by using quantum teleportation to transfer the probabilistic properties of quantum gates to entanglement resources that can be prepared offline.

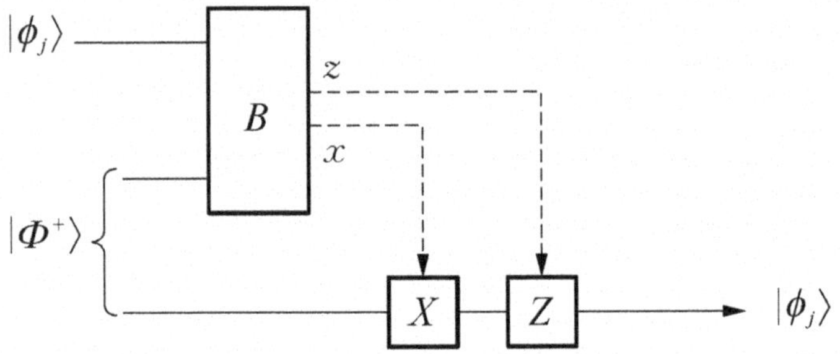

Fig. 1.9 Schematic diagram of $|\phi_j\rangle$ quantum teleportation

Provided a probabilistic CZ gate needs to be acted on the quantum states $|\phi_1\rangle$ and $|\phi_2\rangle$. If the gate is acted directly on the qubit, there is a high probability of failure to destroy the qubit. However, when there is an entanglement resource $|\Phi^+\rangle$ that has been prepared offline, the qubit can be manipulated by means of quantum teleportation (see Fig. 1.9). In Fig. 1.9, B represents Bell measurements, x and z are binary variables that are the results of Bell measurement. In order to realize quantum teleportation of a single qubit state, a σ_x operation (Pauli operator, denoted as X in the figure below) needs to be applied to the qubit after teleportation when x = 1; when z = 1, a σ_z operation (Pauli operator, denoted as Z in the figure below) needs to be applied to the qubit after teleportation; when x and z are equal to 0, no operation needs to be done. In this process, the auxiliary entanglement resources $|\Phi^+\rangle$ do not need to be prepared in real time, and the entangled states prepared offline can be stored in a quantum memory for future computation.

When we want to perform a CZ operation on two quantum states after teleportation, a very intuitive result is that the quantum circuit of manipulation becomes more complex (adding two teleportation circuits) and still there is a high probability of failure that will destroy the quantum state. Gottesman and Chuang de-complicated the process to capture an important property of the CZ gate, which belonged to the Clifford group; CZ gates could be commuted with X and Z with a few more simple Pauli matrices. In other words, by adding some Pauli operators to the quantum circuit, the CZ gate at the far right of the circuit can be moved to the far left of the quantum circuit. After calculation, it was found that a deterministic CZ gate could be achieved simply by changing the entanglement resources prepared offline into $I \otimes U_{cz} \otimes I|\Phi^+\rangle \otimes |\Phi^+\rangle$ (see Fig. 1.10).

1.3.3 KLM Scheme

But unfortunately, Gottesman and Chuang scheme still had a serious problem for linear optics, and Bell state measurement—the core operation of the quantum teleportation—was incomplete and had a 50% probability of success at best at the time.

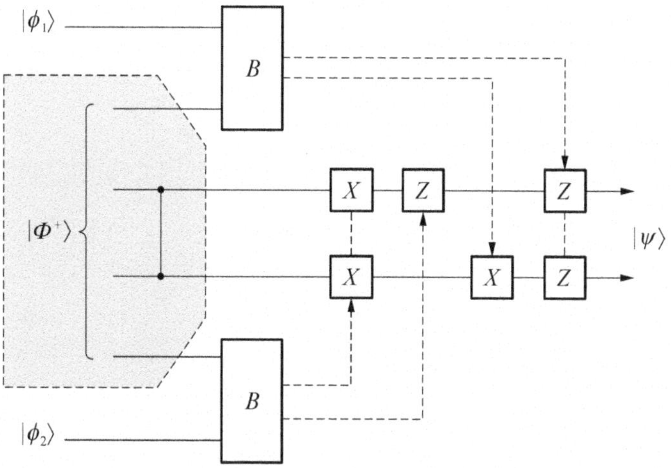

Fig. 1.10 Schematic diagram of the implementation of a CZ gate using invisible quantum teleportation ($|\psi\rangle = U_{CZ}|\phi_1\phi_2\rangle$)

The Probabilistic element which was not easy to remove was again stubbornly in our way, and the prospect of optical quantum computing was clouded. At that time, most people in the academic community believed that scalable optical quantum computing was unachievable. In 2001, Knill, Laflamme and Mlburn published an article with the title of an efficient scheme of linear optical quantum computation in the *Nature* to resolve this issue radically. This groundbreaking work cleared the theoretical hurdles for optical quantum computing, which has since entered a period of vigorous and rapid development. We will be going into the details of the KLM scheme.

In the KLM scheme, the qubits are encoded on a dual mode, and the teleportation acts on a universal single mode. $\alpha|0\rangle + \beta|1\rangle$, $|0\rangle$ is the vacuum state, $|1\rangle$ the single-photon Fock state, and α and β are complex coefficients. An n quantum state that is a specially constructed state with $2n$ modes is chosen as the auxiliary entangled state required for teleportation.

$$|t_n\rangle = \frac{1}{\sqrt{n+1}} \sum_{j=0}^{n} |1\rangle^j |0\rangle^{n-j} |0\rangle^j |1\rangle^{n-j} \tag{1.7}$$

In the equation above, $|k\rangle^j \equiv |k\rangle_1 \otimes \cdots \otimes |k\rangle_j$.

The first $n+1$ discrete quantum Fourier transforms (QFT) are applied to the input and the first n modes of $|t_n\rangle$ states to calculate the total number of photons m detected at the output port. By this process, the input state will then be invisibly transferred to the nth + mth modes of the auxiliary entangled state $|t_n\rangle$ with a probability close to 1, as shown in Fig. 1.11.

Next, we will explain this process by example. The discrete quantum Fourier transform F_n can be written as a matrix element $(F_n)_{jk} = \frac{1}{\sqrt{n}} \exp\left[2\pi i \frac{(j-1)(k-1)}{n}\right]$

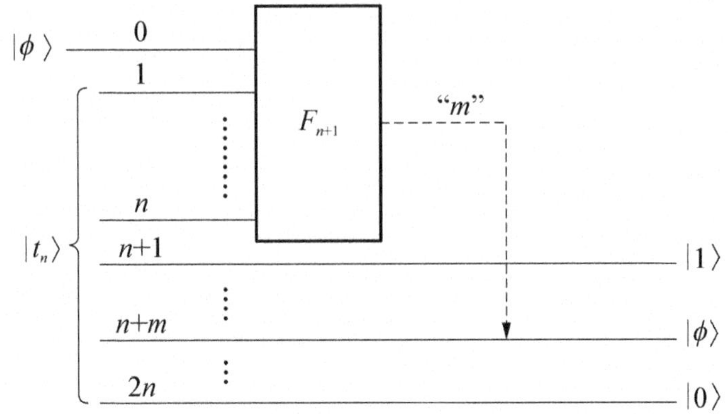

Fig. 1.11 KLM's near-deterministic quantum teleportation scheme

whose main purpose is to erase the mode information of the input state and auxiliary state. Experimentally, a beam splitter can be used to implement transformation. Let $n = 5$, then $|t_n\rangle$ has 10 optical modes. Let's further assume that two photons are measured, i.e., $m = 2$, then there are only two possible outcomes:

① The input state is a vacuum state with amplitude of α, in which case both photons come from the $|t_5\rangle$ state itself; ② the input state contains a photon, in which case the second photon is provided by the $|t_5\rangle$ state. Since the auxiliary entangled state $|t_5\rangle$ is specially constructed, the seventh (here $n + m = 7$) mode of the auxiliary state must contain the same number of photons as the input state. The discrete quantum Fourier transform has erased their mode information, so the two possible outcomes are quantum-mechanically indistinguishable and can therefore be written in the form of coherent superposition. In other words, the input state is invisibly transferred to the seventh mode of the auxiliary state (see Fig. 1.12).

If 0 or $n + 1$ photons are detected at the QFT output port, the input state will collapse to the state of zero or one photon, resulting in a failure of teleportation. Thus, the success probability of a single-qubit teleportation in the KLM scheme is $\frac{n}{n+1}$ (here $|\alpha^2| + |\beta^2| = 1$). As n increases, the success probability of teleportation approaches 1 infinitely.

Using near-deterministic quantum teleportation to construct an assisted entanglement as in Eq. (1.8) can also enable a near-deterministic entanglement-assisted CZ gate:

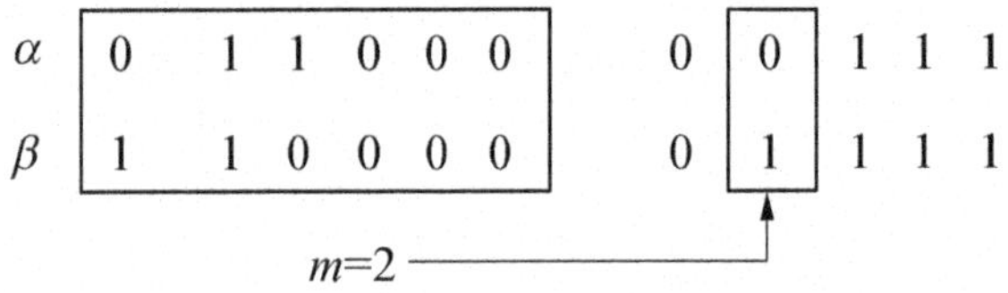

Fig. 1.12 Near-deterministic five-photon assisted quantum teleportation scheme

$$|CZ_n\rangle = \frac{1}{n+1}\sum_{i,j=0}^{n}(-1)^{(n-i)(n-j)}|1\rangle^i|0\rangle^{n-i} \times |0\rangle^i|1\rangle^{n-i}|1\rangle^j|0\rangle^{n-j}|0\rangle^j|1\rangle^{n-j} \tag{1.8}$$

As mentioned above, near-deterministic quantum teleportation can be achieved using the KLM scheme. Readers may get confused as it is previously explained that deterministic Bell measurements cannot be achieved using only linear optics, and the core operation of quantum teleportation is Bell measurement. In fact, there is no contradiction, because Bell measurements in the KLM scheme always have an arbitrarily small probability of errors ε. So, it does not contradict the conclusion. As a matter of fact, the theorem of quantum error correction tells us that arbitrarily long and precise quantum computations can be effectively realized as long as the error rate of each step of computation is less than a threshold. Thus, scalable quantum computation can be achieved as long as the error rate ε is below a fault-tolerant threshold.

1.3.4 One-Way Quantum Computation

Just as a coin has two sides, although the KLM scheme has demonstrated that scalable quantum computing can be achieved via single-photon sources, linear optics, and single-photon detectors, it needs large amounts of resources to realize merely one or two qubits in the LKM scheme. At the current technological level, we are unable to do large-scale quantum computation in real physical systems. To perform scalable quantum computation in real physical systems, researchers have proposed a variety of schemes, of which the most important and most likely to serve as a theoretical framework for future quantum computation is one-way quantum computation. It was first put forward by Raussendorf and Briegel [15]. After nearly 10 years of development, one-way quantum computation has made great progress, both theoretically and experimentally. In this section, we will describe the scheme in detail.

In the circuit quantum computation model, quantum information is encoded as qubits and calculated by coherent manipulation of the qubits (single qubit gates and two-qubit gates). In the one-way quantum computation model, universal quantum computation is performed simply by appropriate single-qubit measurements, which is an important difference from the teleportation-based quantum computation model we described in Sect. 1.3.2. In one-way quantum computation, we first need to prepare a universal highly-entangled two-dimensional cluster state (it has nothing to do with the specific quantum algorithm to be performed) (see Fig. 1.13). Then, we only need to make single-qubit projection measurements of the cluster state in a specific order and measurement basis vector to finish the computation—the quantum algorithm to be executed determines how the measurements are made.

Cluster states are members of a family of highly entangled multiparticle quantum states that can be easily represented by graphs and belong to a special class of graph

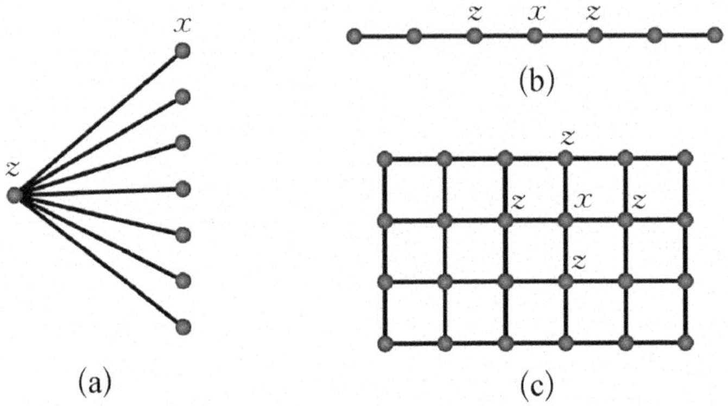

Fig. 1.13 **a** GHZ state; **b** ID cluster state; **c** 2D cluster state

states. In the presentation of a graph state, the graph state (entangled state) $G = (V, E)$ consists of a set of vertices V and edges $E \subseteq [V]^2$ (Fig. 1.13). The edges connect any two vertices, representing their interactions, and each graph corresponds to an N-qubit quantum state $|G\rangle$. The graph state can be made according to a very simple rule: first, N qubits are prepared on the eigenstate of Pauli operator σ_x with a proper value of $+$ 1, i.e. $|+\rangle = \frac{(|0\rangle+|1\rangle)}{\sqrt{2}}$, and then controlled-phase gate operation is done on two qubits connected by an edge, $U_{PG} = \mathrm{diag}\ (1, 1, -1)$, to obtain graph state $|G\rangle = \prod_{(i,j)\in E} U_{\mathrm{PG}}^{(i,j)} |+\rangle^{\otimes N}$. Graph state $|G\rangle$ can also be expressed by a stabilizer code, and each vertex has a stabilizer code $K_j = \sigma_x^{(j)} \otimes_{(i,j)\in E} \sigma_z^{(i)}$. It can be concluded by simple deduction that graph state $|G\rangle$ is the eigenstate of all its stabilizer codes K_j, i.e., $K_j|G\rangle = |G\rangle, \forall j$.

The cluster states required for quantum computation can be constructed by entanglement operations, single qubit operations, and single qubit measurements. For example, when we make a Z measurement on any qubit of a cluster state, we remove it from the cluster state and cut off all the interactions (edges) associated with it. When we do an X measurement on any qubit, we also remove it from the cluster state, but transfer the edges that connect it to neighboring qubits. One-way quantum computation is universal, and although the measurements are random in each step of computation, any quantum computation can be implemented deterministically. Let's see how cluster states are used for quantum computation in Fig. 1.14.

In the quantum circuit model, quantum computation is achieved by unitary operations. In one-way quantum computation it is just the opposite; the information is processed through a series of single-qubit measurements. These measurements are made on a generalized two-dimensional cluster state that is independent of the specific quantum algorithm to be implemented. One-way quantum computation proceeds like this: (i) provide a classical input value to determine the data and program to be processed; (ii) prepare a 2D cluster state large enough to be used as a resource for quantum computation; (iii) perform a series of suitable single-qubit measurements on

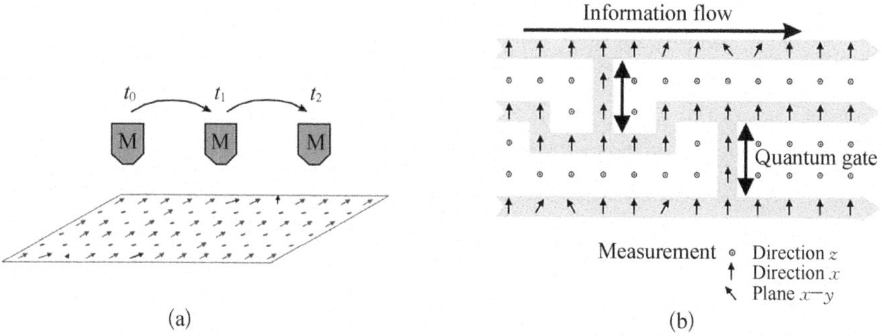

Fig. 1.14 Schematic diagram of a one-way quantum computer

the determined qubits—each measurement to be performed depends on the computational program and previous measurements, and the classical computer is used to compute and select the measurement base vector; and (iv) after the measurements, the quantum state of the system changes to $|\varepsilon^{\alpha}\rangle\left|\Psi_{\text{out}}^{\alpha}\right\rangle$, wherein α represents the measurements of different computation branches. The states $\left|\Psi_{\text{out}}^{\alpha}\right\rangle$ of all branches computed differ from the desired output states by only one local Pauli operator; the measured qubits are on a direct product state $|\varepsilon^{\alpha}\rangle$ that depends on the measurements. The only caveat is that the time sequence of measurement is important.

The proposal of one-way quantum computation aroused wide concern and research fervor in the academic community. Researchers designed many one-way quantum computation schemes based on linear optics, and less-qubit one-way quantum computations based on linear optical systems were implemented one after another. In these experiments, the polarization degrees of freedom or spatial degrees of freedom of photons were usually encoded as qubits, and the operation of qubits could be precisely realized by an interferometer or a polarized wave sheet and then measured using a high-efficiency single-photon detector. In addition, during one-way quantum computation, it does not need to generate cluster states in real time and deterministically. Therefore, scientists worked out many schemes to prepare cluster states through non-deterministic quantum gates, and use the prepared cluster states as an offline preparation resource to perform deterministic one-way quantum computation as well. We will introduce them one by one.

1.3.4.1 Yoran–Reznik Scheme

Yoran and Reznik [44] were the first to bring forward a measurement-based scheme for optical quantum computation. Although this scheme is not rigorously based on the initial cluster-state model, there are still a lot of commonalities. What is worth mentioning is that it employs the quantum teleportation of photonic "entangled chains" to transmit information.

It has been mentioned several times in the preceding paragraphs: it is impossible to achieve deterministic Bell measurements in linear optics by encoding qubits on bimodal degrees of freedom (polarization and space) alone. Therefore, they proposed to encode the qubits simultaneously on the polarization and spatial freedom of a photon, also known as hyper-entangled state (Fig. 1.15). The transformation rule [45] for a single photon with both polarization and path information is listed below:

$$
\begin{aligned}
|H,1\rangle &\rightarrow \frac{1}{\sqrt{2}}(|V,3\rangle + |H,4\rangle) \\
|V,1\rangle &\rightarrow \frac{1}{\sqrt{2}}(|V,4\rangle - |H,3\rangle) \\
|H,2\rangle &\rightarrow \frac{1}{\sqrt{2}}(|H,4\rangle - |V,3\rangle) \\
|V,2\rangle &\rightarrow \frac{1}{\sqrt{2}}(|V,4\rangle + |H,3\rangle)
\end{aligned}
\tag{1.9}
$$

The transformed quantum state looks very similar to the Bell state; the only difference is that this quantum state contains only one photon, and the second qubit exists as path information for the photon. By measuring the polarization of the output modes 3 and 4, the input state can be projected onto the Bell state.

To realize quantum computation (exemplified in Fig. 1.16), we first need to construct vast chained hyper-entangled states.

Fig. 1.15 Schematic diagram of hyper-entanglement, one photon with both polarization and path information

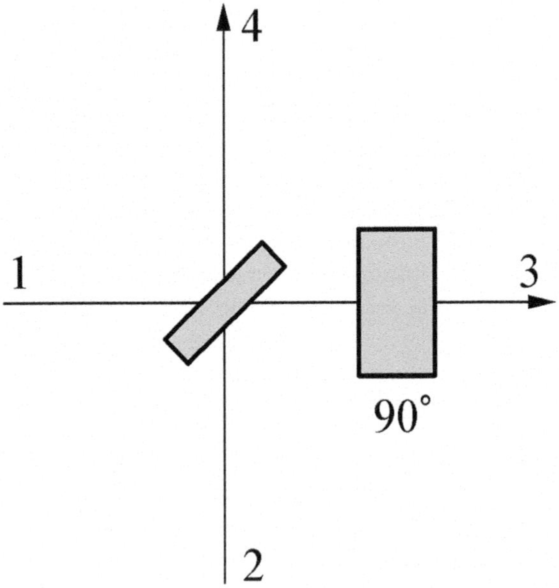

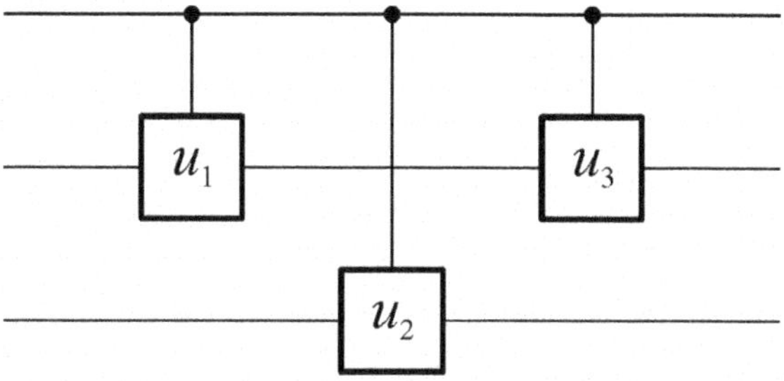

Fig. 1.16 A typical three-qubit quantum computation circuit

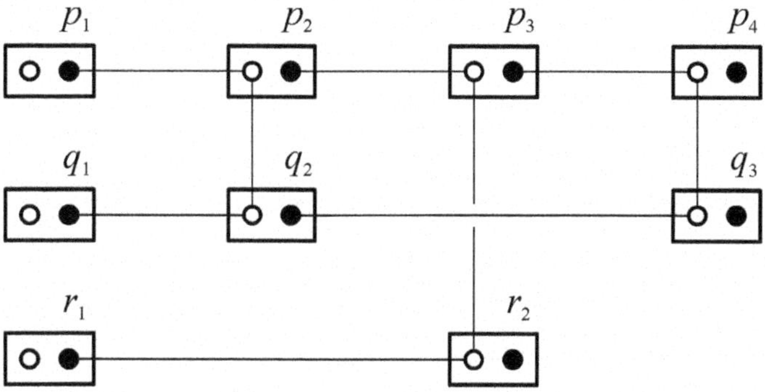

Fig. 1.17 Quantum computation circuit based on hyper-entangled chain states

$$\begin{aligned}&\big(\alpha|H\rangle_{p_1}+\beta|V\rangle_{p_1}\big)\big(|1\rangle_{p_1}|H\rangle_{p_2}+|2\rangle_{p_1}|V\rangle_{p_2}\big)\cdots\\&\quad\big(|2n-1\rangle_{p_n}|H\rangle_{p_{n+1}}+|2n\rangle_{p_n}|V\rangle_{p_{n+1}}\big)|2n+1\rangle_{p_{n+1}}\end{aligned}\tag{1.10}$$

Here, p_j indicates a single photon, and the Bell measurement of it causes the input bits $\alpha|+H\rangle+\beta|V\rangle$ of the first photon p_1 to be transmitted to the second photon p_2. The quantum circuit shown in Fig. 1.16 is then given in a more intuitive image of hyper-entangled states (Fig. 1.17), wherein the hollow dots stand for the photon's polarization degree of freedom while the solid dots represent the photon's path degree of freedom, and there are also vertical connections between these hyper-entangled states at the two-qubit quantum gates u_1, u_2, and u_3.

1.3.4.2 Nielsen Scheme

In 2004, Nielsen [46] came up with a more explicit scheme of linear optical one-way quantum computation, and pointed out that in order to construct a cluster state, the probability of success in adding a chain must be > 1/2, but need not be infinitely close to 1, as in the KLM scheme.

He suggested a CZ gate based on quantum teleportation be used to add a qubit to the cluster state: teleport a single qubit each time other than teleporting two qubits of the CZ gate simultaneously. In this way, the first quantum teleportation can be prepared offline as a resource, thus increasing the success probability of the CZ gate from $CL_n^2/(n+1)^2$ referred to in the KLM scheme to $CZ_{n/n+1}$. We note that when $n = 2$, the success probability of the CZ gate is 2/3, which already satisfies the requirement of adding a chain with a success probability higher than 1/2. Namely, with the Nielsen scheme, only two qubits are required to add one qubit to the cluster state, greatly reducing the resources needed to prepare cluster states.

As mentioned earlier, it is 2D cluster states that are the universal resources for quantum computation. So, we must have the capability to prepare 2D cluster states, which is equivalent to connecting a single qubit between two chained cluster states, and thus requires two successful CZ gates. With the Nielsen scheme, the success probability of this operation is 4/9, lower than 1/2 (the hypothesis is that each CZ gate uses only two auxiliary qubits). This means the process will remove one qubit from the cluster state on average. To solve the problem, Nielsen suggested one qubit be added as a "buffer" before operation. In this way the average effect of the "lost" qubits in the construction of the 2D cluster state would be absorbed by the buffered qubit.

1.3.4.3 Browne-Rudolph Scheme

Browne and Rudolph presented a scheme for constructing cluster states with probabilistic parity check gates in 2005 [47], and the notable merits of this scheme are its ease of implementation and its strong noise immunity. Figure 1.18a is a type I welding gate, including a polarizing beam splitter and a half-wave sheet rotated 45° to polarize. If detector D detects only one photon polarized to H or V, this gate will perform the following projection operation on the input state.

$$
\begin{aligned}
V &: \frac{1}{\sqrt{2}}(|H\rangle\langle HH| + |V\rangle\langle VV|) \\
H &: \frac{1}{\sqrt{2}}(|H\rangle\langle HH| - |V\rangle\langle VV|)
\end{aligned}
\tag{1.11}
$$

The success probability of this gate is $p = 1/2$. We hypothesize that there are two cluster states with lengths n_1 and n_2. When a type I welding gate acts on two photons of these two cluster states, the success of the gate yields $n_1 + n_2 - 1$ photons of the cluster state. If the welding gate fails, the two cluster states will lose the detected photons.

Browne and Rudolph also designed a type II welding gate, shown in Fig. 1.18b, comprising a polarizing beam splitter of 45° rotation. Detectors are mounted on both ports to detect the polarization of photons. When each detector detects the presence of only one photon, it is a successful gate:

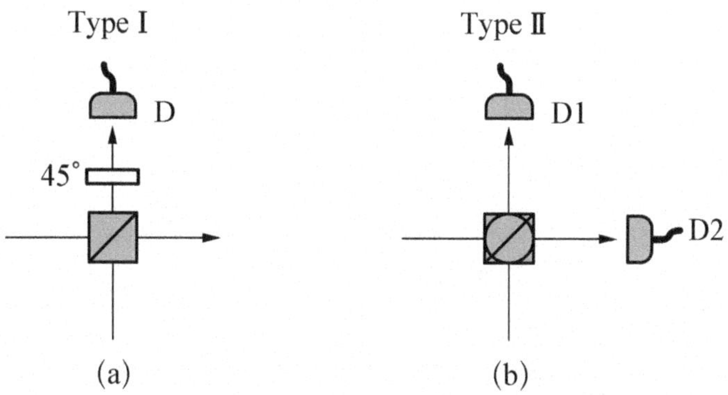

Fig. 1.18 Schematic diagram of type I and II welding gates

$$
\begin{aligned}
H, V \text{ OR } V, H &: \frac{1}{\sqrt{2}}(|H, H\rangle + |V, V\rangle) \\
H, H \text{ OR } V, V &: \frac{1}{\sqrt{2}}(|H, V\rangle + |V, H\rangle)
\end{aligned}
\tag{1.12}
$$

The success probability of the gate is 1/2, and type II welding gate is actually an incomplete Bell measurement [48].

A successful type I or II welding gate will project the qubits of the welded cluster state onto the Bell state. Based on this, we are able to obtain larger cluster states by performing only one Hadamard operation on the qubit adjacent to the detected qubit. Nevertheless, type I and type II weld gates behave completely differently and need to be used on a case-by-case basis. For example, if we assume that only the Bell state is available as an initial resource, only type I welding gate can be adopted when at least three-qubit cluster states are required.

The reason for this is simple: type II welding gates require destructive probing of the photons of two cluster states, and therefore cannot form cluster states of three qubits. We can distinguish between type I and type II welding gates by a simple example: provided only the Bell state is the initial resource, type II welding gate cannot form a cluster state of three qubits because it requires destructive probing of the photons of two cluster states, whereas type I welding gate does not have this restriction; when the welding gate succeeds, it can prepare a cluster state of three qubits. In addition, when type I welding gate fails, it acts as a Z-measurement on the input qubit, but if type II welding gate fails, it acts as an X-measurement on the input qubit. Let's recall the effects of X and Z operations on the cluster state described above: a Z measurement removes the qubit and all the links between it and its nearest neighbor from the cluster state; an I-measurement also destroys the measured qubit, but it transfers the link from the measured qubit to the neighboring one, thus generating a redundant qubit. Therefore, when the measured qubit is at the head or tail of a cluster state chain, type I and type II welding failures lead to the same result: losing qubits. But if the qubit to be welded is in the middle of the cluster state, it is a different story in terms of welding failure. Type I welding will cause a break in the cluster chain, while type II welding only shortens the chain length and produces a redundant qubit adjacent to the measured qubit. As it consumes resources to weld the two broken chains together, type I welding should be avoided if possible.

Nonetheless, we can exploit the properties of a failed type II welding gate to efficiently build a vertical link between cluster state chains to produce two-dimensional cluster states. First, type II welding gate is operated on the qubits of two different chained cluster states. When the welding gate succeeds, the vertical link between the chained cluster states can be effectively formed to construct a two-dimensional cluster state; when it fails, each chained cluster state generates a redundant qubit at the top, and type I welding gate can then act on both of the redundant qubits. If type I welding gate succeeds, a vertical link is constructed; if it fails, the two chained cluster states will lose the two redundant qubits. For sufficiently long chains of clustered states, this process can be repeated until it works.

1.3.5 Errors in Optical Quantum Computation

In the KLM scheme, it is assumed that single-photon sources, linear optics, and single-photon detectors are ideal, and thus scalable quantum computation can be achieved by turning probabilistic quantum gates into near-deterministic ones through quantum teleportation. Still and all, due to decoherence effects, errors will inevitably occur during real optical quantum computing. Fortunately, Shor, Steane, and Gottesman, among others, developed the concept of a quantum error-correcting code (QECC) in a different way, and later proved that arbitrarily long and accurate quantum computation could be realized as long as the error probability of the quantum gate was less than some fault-tolerant threshold. This is the famous threshold theorem in the field of quantum computation, which lays the theoretical foundation for the practical application of quantum computers. This section will focus on common errors in optical quantum computation, followed by a quantum error correction experiment in Sect. 1.5.

1.3.5.1 Detector Errors

In linear optical quantum computation, quantum information encoded on photons is obtained primarily by means of optical detectors. In principle, there are two types of detectors: one will tell us how many photons have been detected, and the other will only tell us whether there are photons or not. The first is called single-photon detector that can distinguish the number of photons, and the second vacuum detector. The original KLM scheme requires exact numbers of photons whereas general linear optical quantum computing just needs a vacuum detector. A real detector usually has two types of errors:

(1) The number of photons detected by the detector is fewer than the actual number of photons, which is generally called photon loss.
(2) The detector finds more photons than the actual number of photons, which is commonly referred to as dark counts. We define the detection efficiency as $\eta \in [0, 1][0, 1]$, indicating the probability of an input photon being converted to a detector count, and let the dark count be the detector count obtained when the input state is the vacuum state. A perfect detector capable of distinguishing the number of photons can be expressed as $|n\rangle\langle n|$ by the projection operator of the Fock state. As a real detector has photon loss, we generally assume that all photons entering the detector have the same probability of being lost, and so they obey Poisson distribution. For a single-photon detector able to distinguish the number of photons, we can write out the following equation [49]:

$$\hat{E}_n = \sum_{k=n}^{\infty} \binom{k}{n} \eta^n (1-\eta)^{k-n} |k\rangle\langle k| \tag{1.13}$$

For a vacuum state detector, on the other hand, the equation is like this [50]:

$$\hat{E}_0 = \sum_{n=0}^{\infty} (1-\eta)^n |n\rangle \left\langle n \right|; \hat{E}_1 = \sum_{n=0}^{\infty} \left[1-(1-\eta)^n\right] \left| n \right\rangle\langle n| \tag{1.14}$$

In linear optical quantum computation, the detectors we often use are vacuum state detectors based on avalanche diodes. The design principle is to convert photons into electrons and eventually probe the electric current by relying on the avalanche effect. These detectors have imperfections in terms of detection efficiency, dark counts, dead time, etc.

1.3.5.2 Light Source Errors

In the KLM scheme, light source is strictly required to be a single-photon source. By single-photon source, we generally mean that merely one photon can be detected in a given detection time window t. So, an ideal way to determine a single-photon source is to directly measure the number of photons received; but as we do not have a perfect single-photon detector that can distinguish the number of photons, we usually measure the second-order autocorrelation function $g^{(2)}(\tau) = \frac{\langle I(t)I(t+\tau)\rangle}{\langle I(t)\rangle\langle I(T+\tau)\rangle}$ of the light source to determine whether it is a single-photon source or not.

If the total detection probability p of a photon is low, the second-order autocorrelation function can be measured using an HBT interferometer [51]. Here, $p = \eta_{\mathrm{red}}$ $T\eta_{\mathrm{det}}$, ηred denotes the probability of emission from a single-photon source, and T indicates the probability of the emitted photons being coupled into the detector, representing the quantum efficiency of the detector. The HBT interferometer comprises a 50:50 beam splitter, two single-photon detectors, and an electronic instrument for measurements (Fig. 1.19).

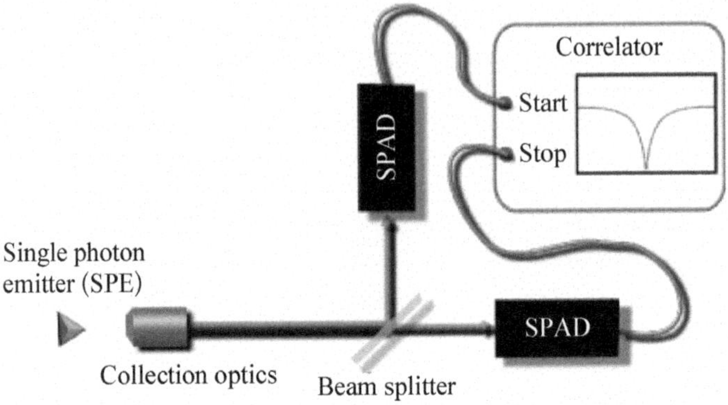

Fig. 1.19 Schematic diagram of Hanbury-Brown and Twiss interferometers

A randomly emitted continuous light source, such as an ideal laser and an LED, $g^{(2)}(\tau) = 1$, indicates that there is no correlation between the emission time of any two photons emitted by the light source. For non-coherent light $g^{(2)}(\tau = 0) > 1$, on the other hand, photons tend to reach the detector simultaneously, to form a bunching effect. In contrast, for quasi-single photon sources $g^{(2)}(\tau = 0) < 1$, the photons tend to be separated and show an anti-bunching effect. It can be understood that because the single-photon source can only emit one photon at a certain time interval in ideal conditions, it is impossible for both detectors to capture the photons simultaneously after the light passes through the 50:50 beam splitter, and so the joint probability is zero, i.e., $g^{(2)}(\tau = 0) = 0$.

For linear optical quantum computing, the single photon sources based on cavities or quantum dots are too complex. As a result, most of theory verification experiments have been conducted around weakly coherent light and SPDC (Spontaneous Parametric Down-Conversion) technology. In the SPDC process, when a high-intensity laser beam threads a nonlinear crystal, the high-frequency photons have a certain probability of splitting into two low-frequency correlated photon pairs. We then select a correct detector event (called post-selection technology) to obtain entangled light sources with high fidelity. SPDC photons are highly correlated in polarization, space, and time due to the momentum and energy conservation. However, the SPDC of light sources is probabilistic, and there are cases where two or more pairs of light sources are generated simultaneously, thus reducing the contrast of the entangled light sources and affecting the accuracy of quantum computation.

1.3.5.3 Circuit Errors and Quantum Storage

In addition to the errors caused by light sources and detection, quantum circuits themselves can also result in errors. The most likely cause is a mode/pattern mismatching brought about by an imperfect wave packet. For instance, the interaction Hamilton of optical devices is not perfect, leading to changes in phase, mode, transmittance, polarization, etc., and further to errors in Hong–Ou–Mandel interference, the action of the CZ quantum gate, etc.; when the light source passes through linear optical elements, such as beam splitters, half-wave plates (HWP), and 1/4 wave plates, there is a quite small probability that photons will be lost, thus causing computational errors; the feedback system can also induce classical errors, a process that includes detector readout, classical post-processing, and optical switching, to name a few, and the loss of optical switching is relatively great. Giacomini et al. [52] and Pittman et al. [53] demonstrated feedback control for linear optical quantum computing in 2002.

Basically, all scalable quantum computations based on linear optics require the technique of quantum storage. When the probability of success of quantum gates and in the preparation of cluster states is low, we need to prepare some entanglements offline and then store them for a period of time for later computation. At this point, it becomes infeasible to use optical fibers for storage; generally speaking, telecommunication optical fibers have an attenuation of 0.17 dB per kilometer at 1550 nm, and the storage of 100 μs will result in the loss of about 54% of photons. On the whole, the errors induced by quantum storage are primarily changes from the input state to the output state, which can be measured with fidelity, i.e., $F_{\mathrm{qm}} = \left[\mathrm{Tr}\left(\sqrt{\sqrt{\rho_{\mathrm{in}}}\rho_{\mathrm{in}}\sqrt{\rho_{\mathrm{in}}}}\right)\right]^2$. In addition to photon loss, the storage process can also fail due to decoherence effects, mode change, and other reasons.

1.4 Preparation and Coherence Manipulation of Optical Qubits

Qubit is the most basic unit of quantum information coding. In classical information theory, the basic unit of information is bit, which can take the value of "0" or "1" in the binary system and represents "yes" or "no" in logic. In the technology of electronics, we can implement it by going high and low. Unlike classical physics, qubits can not only be at either 0 or 1 of two orthogonal quantum states but also at any of their coherent superposition states [54].

Any two-energy system can be used as a qubit, e.g., a particle with spin 1/2, the polarization state of a photon, or the path degree of freedom of a photon. Experimentally, it is difficult to manipulate the system with spin 1/2, whereas it is relatively much easier to manipulate the polarization or the path freedom of photons. Thus, plenty of qubits are currently based on photons. A simple example of coding with the polarization of photons is as follows:

$$\left|\Psi_{\mathrm{qubit}}\right\rangle = \alpha|H\rangle + \beta|V\rangle \quad (1.15)$$

This is the coherent superposition state mentioned above. In the formula, $|H\rangle$ and $|V\rangle$ denote the horizontally polarized and vertically polarized states of the photon, and $|\alpha|^2 + |\beta|^2 = 1$. By means of simple linear optical elements such as polarizing beam splitters, polarizers, and zero-order wave plates, polarized qubits can be prepared and manipulated with very high accuracy (> 99%).

In this section, we will first introduce some basic concepts of optical qubit preparation and then interpret some basic theories and techniques of coherent manipulation of optical qubits. Finally, we will elaborate some basic concepts and techniques of measurement.

1.4.1 Preparation of Qubits

Optical qubits are the basis of optical quantum computation, and in this section, we mainly describe some basic concepts and experimental techniques for the preparation of optical qubits.

1.4.1.1 Weakly Coherent Single Photon Sources

Attenuating laser pulses is an effective way to prepare quasi-single photon sources. Despite the fact that weakly coherent light pulses obey the Poisson distribution and have a certain probability of multiphoton events, they are easy to prepare and manipulate experimentally and close to the properties of single photon sources. Hence, they are widely used.

The number of photons in a laser pulse follows the Poisson distribution [55]. It can be simply deduced that the ratio of the probability of a pulse containing multiple photons to the probability of a pulse containing photons can be approximated as follows:

$$p_{\text{multi}} = \frac{p_{\geq 2}}{p_{\geq 1}} = \frac{1 - p(0) - p(1)}{1 - p(0)} \approx \frac{\langle n \rangle}{2} \tag{1.16}$$

From Eq. (1.16), attenuating laser pulses can obtain weakly coherent light pulses with a very low average photon number $\langle n \rangle$ and thus effectively reduce the probability of multiple photon states in light pulses. But when $\langle n \rangle$ is extremely low, the number of photons in numerous light pulses is zero. In principle, by increasing the repetition frequency of light pulses, a certain communication bit rate can be guaranteed, but the detector has a limit count rate and also a dark count (which will be dwelled on in the section of detectors). Moreover, the detector needs to be turned on at the arrival of each pulse so that the dark count will go up to lessen the signal-to-noise ratio. Theory shows that the signal-to-noise ratio decreases as $\langle n \rangle$ falls. As a result, the average number of photons currently used for quantum communication is usually set to be 0.1 [56].

No matter how small n is, a single pulse of weakly coherent light still has a certain probability of having two or more photonic events. In a lot of optical quantum computing schemes, an exact number of single-photon sources is required, and so weakly coherent single-photon sources are mainly used in quantum communication. In optical quantum computation, it is often taken as the supplement of SPDC entanglement sources.

1.4.1.2 SPDC Entanglement Sources

SPDC is currently the most mature and reliable technique for generating entangled photon pairs [57]. When a high-frequency laser beam penetrates a nonlinear crystal (usually a BBO crystal, i.e., β-type barium borate), the nonlinear interaction causes the high-frequency photon to cleave into two low-frequency photons—signal photon and idle frequency photon. Because this process accords with the conservation of momentum and energy, their correlation allows the preparation of entangled light sources. The main principles of generating entangled photon pairs are simply described below.

The parametric down-conversion process must satisfy the momentum and energy conservation conditions. So, we write out the following:

$$\omega_{\mathrm{p}} = \omega_{\mathrm{i}} + \omega_{\mathrm{s}}, \quad k_{\mathrm{p}} = k_{\mathrm{i}} + k_{\mathrm{s}} \tag{1.17}$$

In the equation, p, i, and s represent pump light, idle light, and signal light respectively, while w and k denote the frequency and momentum of the photons. If the polarizations of the two parametric down-conversion photons are parallel to each other, it is called type I parametric down-conversion; if they are orthogonal to each other, it is a type II parametric down-conversion process $e \to e + \mathrm{o}$, which is the main method used in the laboratory to prepare polarized entanglement light sources. Figure 1.20 is a schematic diagram of type II parametric down-conversion. A beam of pumped lasers passes through the BBO crystal, and the parametric down-conversion photons make up two light cones, respectively horizontal polarization (H) and vertical polarization (V). As the photons on the intersecting lines A and B of the two light cones coincide perfectly in space, we cannot tell which light cone they actually come from. From the perspective of quantum mechanics, the quantum states $|H\rangle_A|V\rangle_B$ and $|H\rangle_B|V\rangle_A$ are in principle indistinguishable so that the parametric down-conversion photons on A and B are on the coherent superposition of these two components, i.e., on the entangled state given follow.

$$|\phi\rangle_{AB} = \frac{1}{\sqrt{2}}\left(|H\rangle_A|V\rangle_B + \mathrm{e}^{\mathrm{i}\varphi}|V\rangle_A|H\rangle_B\right) \tag{1.18}$$

Due to the birefringence of BBO crystals, o and e lights have different group velocities and propagate in different directions. This results in a corresponding time difference $\delta T = L\left(\frac{1}{u_{\mathrm{o}}} - \frac{1}{u_{\mathrm{e}}}\right)$ between o and e, wherein L is the length of the crystal and u_{o} and u_{e} are the group velocities of o and e. When $\delta T \geq \tau_{\mathrm{c}}$ (τ_{c} is the coherence time of the down-converted photons), o and e lights become distinguishable in

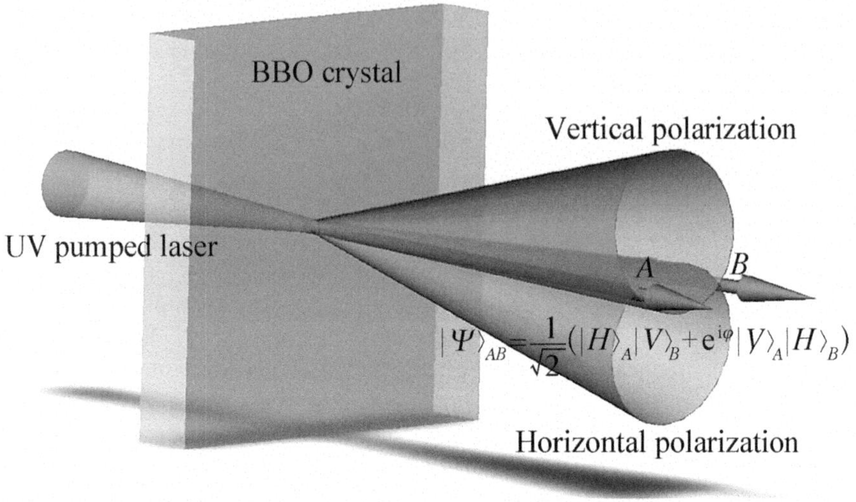

Fig. 1.20 Schematic diagram of an entangled light source generated by parametric down-conversion

theory and we will not be able to observe two-photon interference. In other words, the polarized entangled state will not be obtained. In addition to this, there is a transverse displacement difference $d = L\tan\rho$ between *o* and *e*. Similarly, when this displacement difference is larger than the coherence length of the down-converted photon, we cannot obtain the two-photon entanglement state as well.

Fortunately, since the parametric down-conversion photons are generated coherently over the entire length of the crystal, we are able to fully compensate in principle for both the temporal and spatial walk-off effects. Experimentally, light beams *A* and *B* pass through a half-wave plate to rotate the polarization of the photons by 90°, and with a BBO crystal with the same cutting angle and half the thickness of the main BBO, it can compensate for the walk-off effect between *o* and *e* lights, thus obtaining a two-photon polarization entanglement state with high fidelity.

1.4.1.3 Coherent Optical Conversion

Coherent optical conversion (CPC) [58] is a recently proposed scheme for the preparation of light sources for optical qubit computation and was preliminarily validated by the Zeilinger's team through experiment in Austria in 2011. A great advantage of this scheme is that it allows the preparation of good quality and truly controllable single-photon sources and the realization of all basic operations required for optical quantum computation.

Unlike parametric down-conversion, coherent optical conversion is a deterministic process. The method essentially employs the classical pumping nonlinear effect to produce coherent oscillations between the orthogonal modes of multi-quantum excited states. A typical example of coherent optical conversion is the use of pump-based four-wave mixing interactions to provide a complete set of "tools" for processing optical quantum information. This system fulfils di Vincenzo's requirements for scalable quantum computation, including deterministic multi-qubit entanglement gates (based on an entirely new type of photon-photon interaction), high-quality single- and multi-photon quantum state preparation (independent of higher-order effect), stable and efficient detection, etc.

The basic physical process of coherent optical conversion is the nonlinear interaction between m bosonic modes, such that a single excitation in some modes is coherently converted to the remaining others, presenting coherent oscillations of multiple optical modes and achieving efficient coupling between different quantum modes. A central element of this is the use of a strong classical light field to pump certain modes with higher-order nonlinear effects, which in turn produces this fundamental nonlinear effect. Due to the use of laser pumping, parameters such as coupling strength can be effectively tuned by the classical pump light, and the effective nonlinear coefficient (second order) obtained from the pump light tuning of third-order nonlinear effects can be very strong, even beyond the strength of some natural second-order nonlinear effects.

1. Basic theory

We begin by describing some of the basic physics of the scheme. A standard four-wave mixing process consists of four different frequency modes: *a*, *b*, *c* and *d*. The Hamiltonian of the system is as follows:

$$H = \gamma a b^{\dagger} c^{\dagger} d + \gamma^{*} a^{\dagger} b c d^{\dagger} \tag{1.19}$$

Here γ characterizes coupling strength, introduced by the third-order nonlinear effect $\chi^{(3)}$. $*$ means complex conjugation and $\dagger$ indicates conjugation operator. If pumping mode d is a bright classical light beam with an electric field strength E, an effective second-order interaction will produce, and Hamiltonian can be rewritten in the following:

$$\tilde{H} = \tilde{\gamma}_{ab^{\dagger}c^{\dagger}} + \tilde{\gamma}^{*} a^{\dagger} b c \tag{1.20}$$

In the formula, $\tilde{\gamma} \propto \gamma E$, similar to a nonlinearly coupled augmented, tunable three-wave mixing Hamiltonian. Let the incident state be described by Fock state, set as $|n_a\, n_b\, n_c\rangle$, which will evolve in an integer j} of a well-defined finite Hilbert space $\left\{\tilde{H}^{j}|n_a n_b n_c\rangle|\forall\right.$. It can be determined by conservation of energy that the spatial dimension is $n_a + \min(n_b, n_c) + 1$. During evolution, some elements characterizing coherent quantum processes collapse and reappear. For a two-dimensional subsystem $\{|100\rangle, |011\rangle\}$, $(\tilde{H}|100\rangle \propto |011\}$ and $\tilde{H}^2|100\rangle \propto |100\rangle)$, even though one of the states is a two-photon state, the induced coupling will still drive a Rabi-like oscillation between two ground states. Given incident state $|100\rangle$, in the $\{|100\rangle, |011\rangle\}$ space, the system Hamiltonian can be written as follows:

$$\mathbf{H} = \begin{pmatrix} 0 & \tilde{\gamma}^* \\ \tilde{\gamma} & 0 \end{pmatrix} \tag{1.21}$$

Its eigenvalue is $\{-|\tilde{\gamma}|, |\tilde{\gamma}|\}$, and the eigenvector can be transformed by Nc following expression:

$$\frac{1}{\sqrt{2}}\begin{pmatrix} \frac{-\tilde{\gamma}^*}{|\tilde{\gamma}|} & \frac{\tilde{\gamma}^*}{|\tilde{\gamma}|} \\ 1 & 1 \end{pmatrix} \tag{1.22}$$

From the Schrödinger equation, the incident state ($|\psi(0)\rangle = |100\rangle$) can be simply obtained.

$$|\psi(t)\rangle = \cos(\Gamma t)|100\rangle + i\frac{\tilde{\gamma}}{|\tilde{\gamma}|}\sin(\Gamma t)|011\rangle \tag{1.23}$$

In the above equation, $\Gamma = |\tilde{\gamma}|/h$. Similarly, in a higher-order case, the above analysis can be extrapolated analogously. For example, in the case where the incident state is $|200\rangle$, the system evolves in the middle of $\{|200\rangle, |111\rangle, |022\rangle\}$. Under this basis vector, the Hamiltonian is written as follows:

$$H = \begin{pmatrix} 0 & \sqrt{2}\tilde{\gamma}^* & 0 \\ \sqrt{2}\tilde{\gamma} & 0 & 2\tilde{\gamma}^* \\ 0 & 2\tilde{\gamma} & 0 \end{pmatrix} \tag{1.24}$$

Its eigenvalue is $\{0, -\sqrt{6}|\tilde{\gamma}|, \sqrt{6}|\tilde{\gamma}|\}$, and the eigenvector can be obtained by the transformation of the following basis vector:

$$\frac{1}{\sqrt{6}}\begin{pmatrix} -\frac{\tilde{\gamma}^{*2}}{|\tilde{\gamma}|^2} & \frac{\tilde{\gamma}^{*2}}{|\tilde{\gamma}|^2} & \frac{\tilde{\gamma}^{*2}}{|\tilde{\gamma}|^2} \\ 0 & -\frac{\sqrt{3}\tilde{\gamma}^*}{|\tilde{\gamma}|} & \frac{\sqrt{3}\tilde{\gamma}^*}{|\tilde{\gamma}|} \\ \sqrt{2} & \sqrt{2} & \sqrt{2} \end{pmatrix} \tag{1.25}$$

Let the initial incident state be $|\psi(0)\rangle = |200\rangle$, then the evolution of the Schrödinger equation for the system yields

$$\begin{aligned} |\psi(t)\rangle = &\left[\frac{2}{3} + \frac{1}{3}\cos(\sqrt{6}\Gamma t)\right]|200\rangle \\ &+ \frac{i}{\sqrt{3}}\frac{\tilde{\gamma}}{|\tilde{\gamma}|}\sin(\sqrt{6}\Gamma t)|111\rangle \\ &\left[-\frac{\sqrt{2}}{3}\frac{\tilde{\gamma}^2}{|\tilde{\gamma}|^2} + \frac{\sqrt{2}}{3}\frac{\tilde{\gamma}^2}{|\tilde{\gamma}|^2}\cos(\sqrt{6}\Gamma t)\right]|022\rangle \end{aligned} \tag{1.26}$$

Here $\Gamma = |\tilde{\gamma}|/\hbar$. It can be seen from $|\psi(t)\rangle$ that the initial incident state will always exist in $|\psi(t)\rangle$ and will not be completely eliminated. The result applies to general

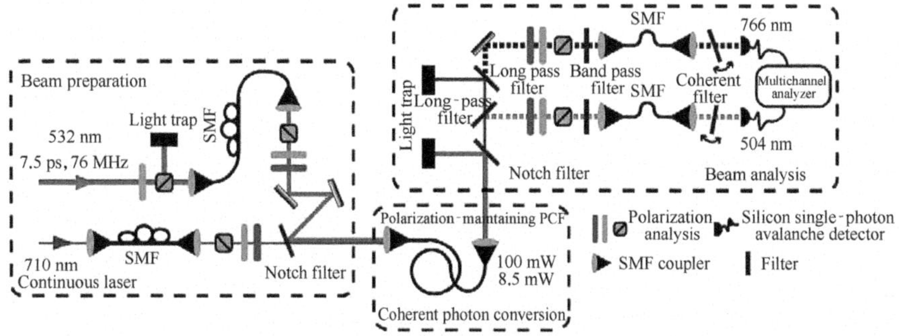

Fig. 1.21 Optical path map of Coherent optical conversion experiment

situations: provided there is a multi-photon incident state of mode a, i.e. $\psi(0)\rangle = |n00\rangle$, the system evolves in state space $\{|j\rangle \equiv |n-j, j, j\rangle\} \cdot (j = 0, \ldots - n)$. Under this basis vector set, the matrix element that is not zero in the Hamiltonian is $H_{j,j+1} = (j+1)\sqrt{n+1}\tilde{\gamma}^*$, $H_{j+1,j} = (j+1)\sqrt{n-j}\tilde{\gamma}^*(j = 0, \ldots, n)$. According to the physical process described above, a series of single-photon operations can be implemented by a specific filtering process.

2. Coherent photon conversion experiment

The afore-mentioned nonlinear process has been implemented in a polarization-maintaining photonic crystal fiber, and the experimental scheme is shown in Fig. 1.21.

The experimental facility consists of 3 parts: beam preparation, coherent optical conversion, and beam analysis. The nonlinear medium is a standard commercial polarization-maintaining photonic crystal fiber (core diameter 1.8 mm, nonlinear characteristics 95 W km). $\chi^{(3)}$ nonlinearities are pumped by a frequency-doubled yttrium vanadate pulsed laser (Nd:YVO_4, wavelength 532 nm, pulse width 7.5 ps, repetition frequency 76 MHz) to produce equivalent tunable effective $\chi^{(2)}$ nonlinearities. A continuous optical external-cavity semiconductor laser (wavelength 710 nm, providing approximately 2×10^5 photons in each Nd:YVO_4 laser pulse width) provides an input state in mode a, which is used experimentally to characterize the interaction strength of coherent optical conversion. By estimating the dispersion characteristics of the photonic crystal fiber, a satisfactory birefringent phase matching 532 nm (H) + 710 nm (H) → 504 nm (V) + 766 nm (V) can be obtained for the four-wave mixing process, wherein H and V represent horizontal and vertical polarizations, respectively. Both the 532 and 710 nm input beams are spatially filtered through a single-mode polarization-maintaining optical fiber, then merged and coupled into a polarization-maintaining photonic crystal fiber through a notch filter. The beam at the output port is first spectrally separated by a series of filters (notch filter, long-pass filter, band-pass filter) and a simple monochromator (rotating coherent filter). The signal is then fed through a fiber-coupled polarization analyzer to a single photon avalanche detector (SPAD), and the measured signal is analyzed by a multichannel coincidence counter.

First, the experiment realizes an adjustable $\chi^{(2)}$ interaction process by pumped $\chi^{(3)}$ interaction. The correlated two-photon sources (wavelengths 504 and 766 nm, respectively) are obtained through the double-pumping process, and the results are in turn used to characterize the interaction strength.

Figure 1.22 shows the measurement results obtained when the average power of the 532-nm pumped laser is 90 mW and the average power of the 710-nm input-state laser is 8.5 mW. During each 532-nm light pulse, the 710-nm laser emits about 2×10^5 photons (an effective power of about 4.8 μW, more than four orders of magnitude smaller than the 532-nm pumped light). The signal light obtained is strong when only the pump light and the input state laser are present independently. When two laser beams exist at the same time, a sharp signal peak can be seen experimentally. When there is only pump light, a periodic background (contributed by other nonlinear effects of the photonic crystal fiber) is experimentally able to be obtained. Figure 1.23 displays the experimental results after removing the background. The coherent optical conversion effect can be clearly seen, i.e., a photon of mode a (710 nm) becomes a photon of mode b (766 nm) and c (504 nm) under the action of pumped light.

3. Photon number multiplication scheme

Through the coherent optical conversion process described above, the photon number multiplication process can be implemented through two different schemes, and Fig. 1.24 shows the first scheme for realizing deterministic photon number multiplication.

As shown in Fig. 1.24, the whole scheme consists of three different coherent optical conversion processes. In the first stage, the input state is $|1\rangle a$, i.e., $|100\rangle$, an evolution process of states as described according to Eq. (1.23). When $t = \pi/2\Gamma$, the

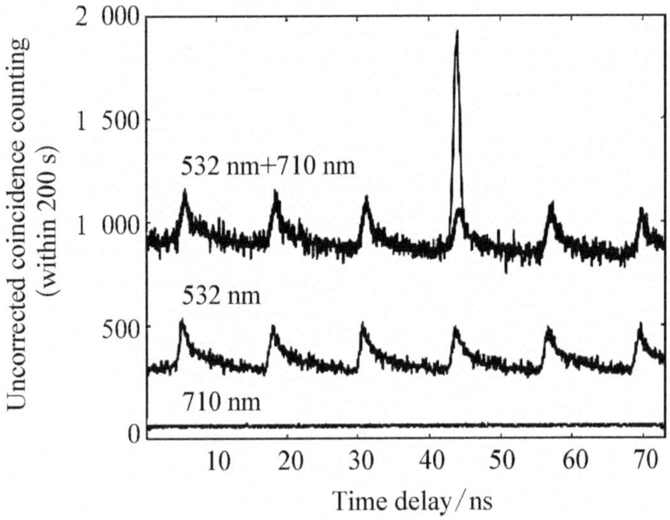

Fig. 1.22 Experimental results of coherent optical conversion

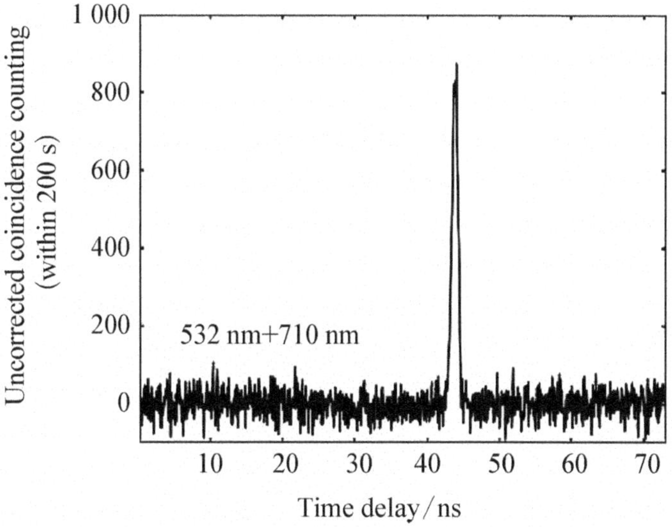

Fig. 1.23 Experimental results after removal of local noise

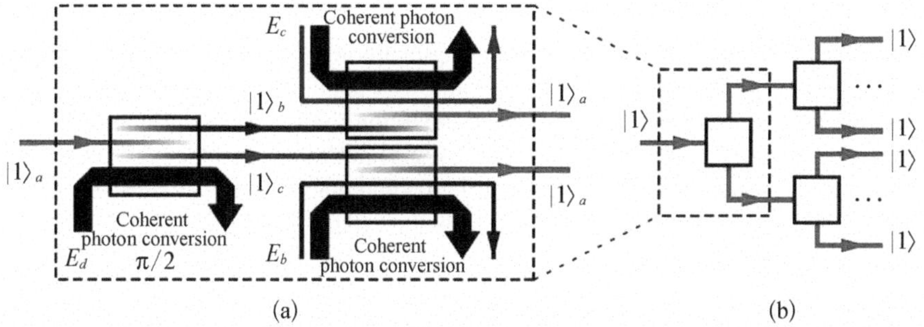

Fig. 1.24 Deterministic photon number multiplication scheme I

output photon state is $|011\rangle$, that is, a deterministic photon number multiplication process is generated, and the output photonic modes are *b* and *c*. In the second stage, the photonic modes *b* and *c* are converted back to photonic mode *a* by two parallel coherent optical conversion processes. It should be noted that the pump light used in this stage is different from that used in stage 1: For the input photon $|1\rangle b$, the pump light is E_c; while for the input photon $|1\rangle c$, the corresponding pump light is E_b. With a specific filtering process, we can obtain two output $|1\rangle a$ photons, thus achieving deterministic multiplication of photon numbers. This process can be adopted to create high-quality scalable multiphoton sources via cascade network (Fig. 1.24b).

Schemes I and II are similar. In scheme I, we only consider a coherent optical conversion process with four different frequencies, but a frequency-degenerate

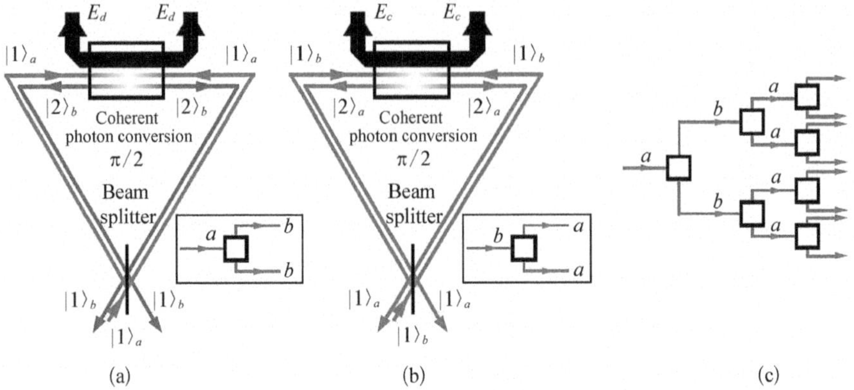

Fig. 1.25 Deterministic photon number multiplication scheme II

coherent optical conversion process is also workable, i.e., b and c are in a degenerate mode, and the Hamiltonian can be written as $H = \gamma ab^\dagger b^\dagger d + \gamma^* a^\dagger bbd^\dagger$. In effect, this is a standard four-wave mixing interaction process used in spontaneous four-wave mixing sources (although it is the degenerate mode that is pumped in the spontaneous four-wave mixing process). When photonic mode d is a classical pump field with an electric vector of E_d, the effective Hamiltonian becomes $H = \gamma ab^\dagger b^\dagger + \gamma^* a^\dagger bb$. At this Hamiltonian, a photon of frequency ωa will produce two photons of frequency ωb. Two degenerated and indistinguishable photons can then be separated into different paths using a reverse Hong–Ou–Mandel interferometer ($|2, 0\rangle + |0, 2\rangle \rightarrow |1, 1\rangle$). Photon b is converted back to photon by the separated photon b through two identical coherent optical conversion processes in parallel. See Fig. 1.25 for details.

4. Single-photon source solution

Coherent optical conversion processes can also be used to prepare efficient and predictable single photon sources. The preparation process needs to take advantage of the fact that different photon number states (Fock states) have different cycles in the coherent optical conversion process. Figure 1.26a reveals the evolution of the input states $|100\rangle$, $|200\rangle$, and $|300\rangle$ with time. It can be easily seen from the diagram that when $n_a = 2$, $|200\rangle$ does not disappear completely; while for $|300\rangle$, the change process is no longer single frequency but a superposition of different base frequencies. By taking advantage of the fact that different Fock states evolve with different time parameters, the system can be filtered at specific time points to gradually attenuate other Fock state evolution processes while retaining the required photon number states.

Provided that the incident state is a coherent state $|\psi(0)\rangle_a = |\alpha\rangle_a$, this input state can be expanded by the Fock state. For the state $|100\rangle$, it remains unchanged after coherent photon conversion and the time $t = \frac{\pi}{\Gamma}$; but for the other $|n_a 00\rangle (n_a \neq 1)$, there is always a certain probability that the photon numbers of modes b and c are not

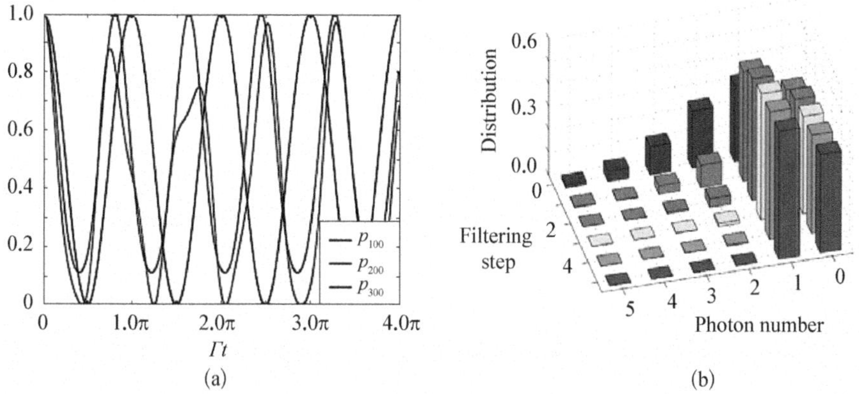

Fig. 1.26 Fock state oscillation process

zero. The filtering process of mode a is retained by filtering modes b and c and there is a certain probability to attenuate the $|n_a 00\rangle (n_a \neq 1)$ component. By repeating the above process, the $|n_a 00\rangle (n_a \neq 1)$ component can be well suppressed, eventually obtaining single photon sources. As in Fig. 1.26b, the weakly coherent optical state $|\alpha|^2 = 1.5$ of mode a, after the above 5-step filtering process, is able to produce predictable single photons by 56% efficiency, while the proportion of multi-photon states is < 0.3%. Using the same process to filter the properties of the system at different time allows the preparation of the Fock state of small photon numbers.

Reviewing the whole process, a very strong nonlinear coefficient is required for enough single photon state or multiple photon states to yield efficiency. In the present scheme, the nonlinear coefficient is related to the pump laser power (theoretically, a linear relationship). Figure 1.27 displays the relationship between the coincidence counting and the pumping power under modes b and c obtained by coherent optical conversion, in which the solid line is theoretical curve, presenting a linear relationship, and the dots are experimental measurements; it can be seen that at low power, it is identical with the theoretical curve, which varies linearly. As the pumping power increases, the coincidence counting is gradually saturated. The saturation derives from two mechanisms: saturation of the detector's counting rate and the poor performance of single mode fiber with power boost, which can be solved by improving experimental setups.

5. Controlled-phase gate solution

Controlled-phase NOT gate, a core element of quantum computation, can also be achieved effectively through coherent photon conversion.

Figure 1.28 gives a scheme for achieving a two-qubit controllable Z-gate directly on photons of modes b and c through coherent photon conversion. A key point in the examination of the coherent optical conversion process is that, like any coherent process that oscillates periodically between two orthogonal states, the oscillation period of a photon number produces a geometric phase. It can be seen in the figure

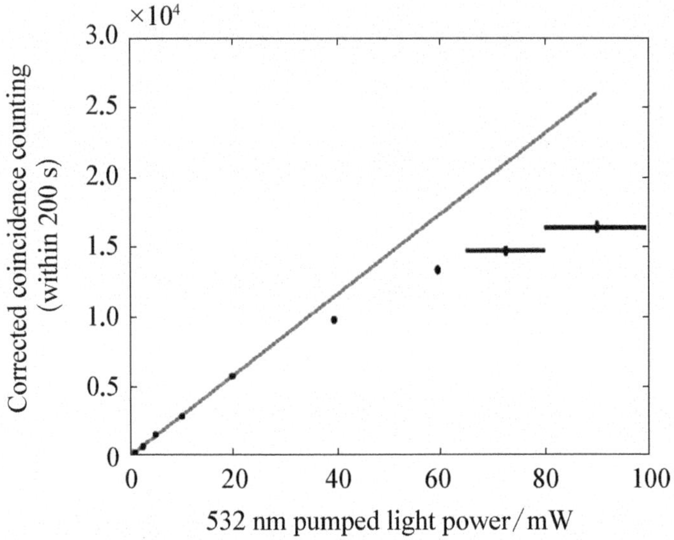

Fig. 1.27 Curve of nonlinear effect changing with pump light power

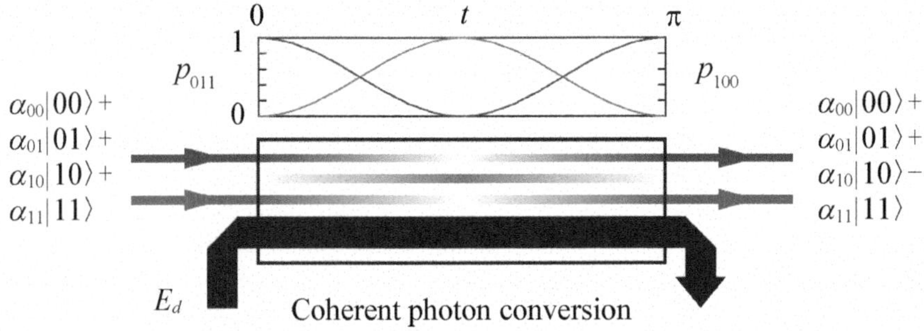

Fig. 1.28 Defined controlled-phase gate scheme

that at the moment $t = 0$, the photon state is input state $|011\rangle$; when $t = \pi/\Gamma$, the input state will finish a complete oscillation while phase of π changes as a whole, and so it becomes $-|011\rangle$. This process directly achieves a controlled-NOT gate between two photons with 100% efficiency. We can switch this gate on and off by modulating the pump light, quickly controlling the connection of the quantum network. This control means is very useful in adaptive quantum algorithms, such as quantum phase estimation.

6. Multi-photon entanglement state preparation scheme

Using the photon number multiplication scheme described in (3) can effectively produce multi-photon entangled states, such as Bell entangled state and GHZ entangled state. The specific experimental scheme is shown in Fig. 1.29a, which describes how the Bell entangled states are generated. The photon of mode a is a coherent superposition state $|\psi(0)\rangle = \alpha|H\rangle + \beta|V\rangle$, in which $\alpha^2 + \beta^2 = 1$. The input photon first passes through a polarization displacer, and then the transmission and reflection paths undergo coherent photon conversion with time $t = \pi/2\Gamma$ respectively, followed by four parallel three-wave mixing processes. Eventually, the outgoing photons are interfered again on the polarization displacer through a half-wave plate to realize the preparation of the Bell state. Figure 1.29b shows the scheme for the preparation of the GHZ state, Fig. 1.29c is a schematic diagram for the preparation of the with predictable nonlocal maximum entanglement state, and Fig. 1.29d gives the schematic of the quantum bits required to achieve the nine-qubit error correction scheme by this method.

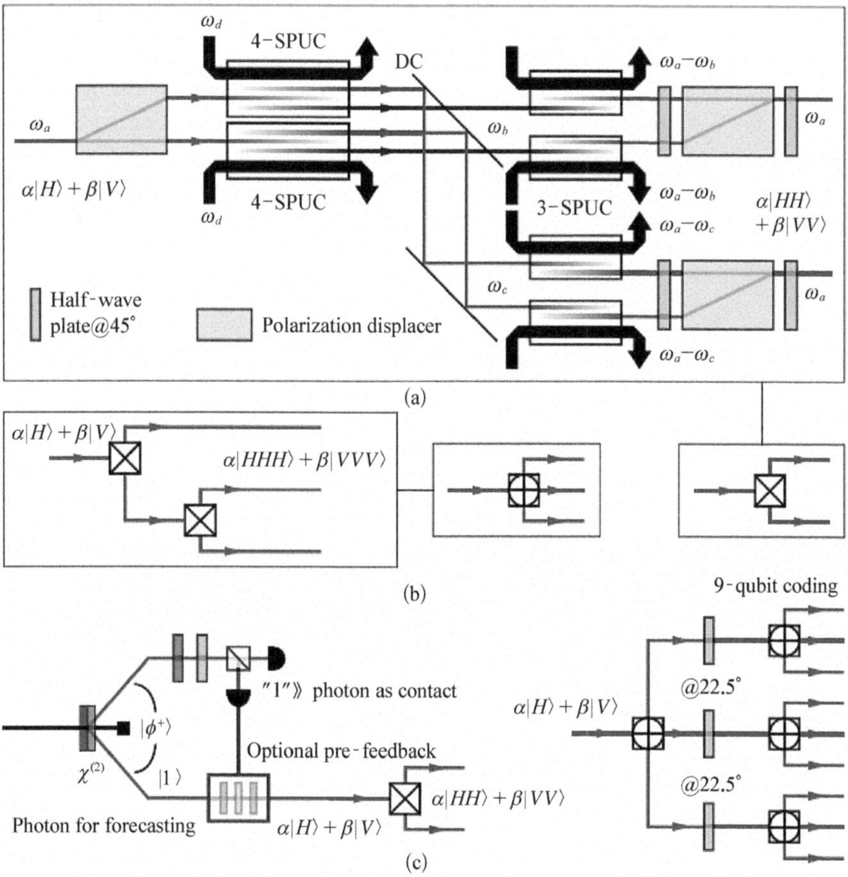

Fig. 1.29 Defined controlled-phase gate scheme

1.4.1.4 Basic Experimental Devices for Optical Quantum Computation

Basic experimental devices required for optical quantum computation include beam splitters, polarizing beam splitters, wave plates, etc. This section will give a brief description of these basic optical components.

1. Beam splitter

Beam splitters are one of the most important and widely used devices in optical quantum computation and play an important role in qubit preparation, manipulation, and measurement. The main function of an ordinary beam splitter (BS) is to transmit or reflect an incident photon with a certain probability (this probability can be 0.5 or some other value), as shown in Fig. 1.30.

When a photon is reflected from the beam splitter, the reflected photon will have an additional phase relative to the transmitted photon. Experimentally, the MZ interferometer can be implemented with two beam splitters.

As shown in Fig. 1.31, both A and B are 50% transmitting and 50% reflecting beam splitters. A photon with a quantum state of $|\phi\rangle$, incident from port 1 of beam splitter A, will have a 50% probability of direct transmission from output port 3 and a 50% probability of reflection from output port 4. After passing through this beam splitter, the quantum state of the photon changes to

$$|\phi\rangle \Rightarrow |\phi\rangle_{a1} = \frac{1}{\sqrt{2}}|\phi\rangle(|3\rangle + i|4\rangle) \tag{1.27}$$

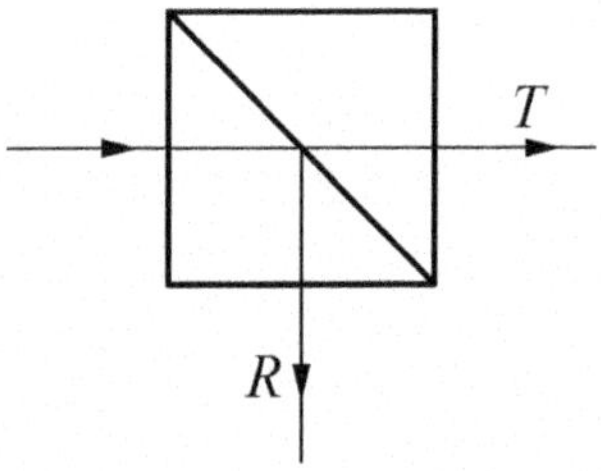

Fig. 1.30 Beam splitter diagram

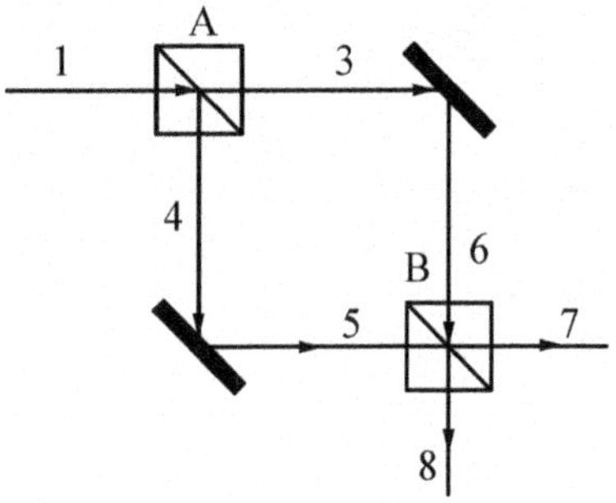

Fig. 1.31 Diagram of MZ interferometer realization using beam splitter

If a beam splitter B is placed behind beam splitter A, the photons will reach the two incident ports 5 and 6 of beam splitter B at the same distance. After the action of beam splitter B, the above formula is written like this:

$$|\phi\rangle_{a1} \Rightarrow |\phi\rangle_{b1} = \frac{1}{\sqrt{2}}|\phi\rangle\left[\frac{1}{\sqrt{2}}(|7\rangle + \mathrm{i}|8\rangle) + \mathrm{i}\frac{1}{\sqrt{2}}(\mathrm{i}|7\rangle + |8\rangle)\right] = \mathrm{i}|\phi\rangle|8\rangle \quad (1.28)$$

That is, the photon will be determined to emanate from output port 8 of beam splitter B. If we add a phase of $\frac{\pi}{2}$ to either optical path between the two beam splitters, the result of the above equation will be $i|\phi\rangle|7\rangle$, and the photon will be determined to emanate from port 7 of beam splitter B.

Now consider the case where there are two photons entering each of the two incident ports of beam splitter A. The states of the two photons are $|\varphi_1\rangle$ and $|\varphi_2\rangle$. After passing through beam splitter A, the input states of the two photons are as follows:

$$|\phi_{\mathrm{f}}\rangle_{12} = \frac{1}{\sqrt{2}}|\varphi_1\rangle \otimes (|3\rangle + \mathrm{i}|4\rangle) \cdot \frac{1}{\sqrt{2}}|\varphi_2\rangle \otimes (\mathrm{i}|3\rangle + |4\rangle) \quad (1.29)$$

If the two photons arrive at beam splitter A at the same time, there is an overlap of spatial modes in the emergent state. If the two incident photons are in the same photon state, the crossover interference between the two photons by the identity principle must be taken into account, denoted by $|\varphi\rangle$, and the result of the above equation is written like this:

$$|\phi_{\mathrm{f}}\rangle = \frac{i}{\sqrt{2}}|\varphi\rangle \cdot |\varphi\rangle \otimes (|3\rangle|3\rangle + |4\rangle|4\rangle) \quad (1.30)$$

2. Polarizing beam splitter

The polarizing beam splitter allows almost all of the horizontally polarized incident photons to penetrate and almost all of the vertically polarized photons to reflect, with the entire changing process drawn in Fig. 1.32a.

The relationship between the incident modes (a and b) and the emergent modes (c and d) is as follows. Note that, as with the beam splitter, there is a phase jump $\frac{\pi}{2}$ in the component of the reflected light.

$$\begin{aligned} \hat{a}_{\mathrm{H}} \to \hat{c}_{\mathrm{H}} \cdot \hat{a}_{\mathrm{V}} \to i\hat{d}_{\mathrm{V}} \\ \hat{b}_{\mathrm{H}} \to \hat{d}_{\mathrm{H}} \cdot \hat{b}_{\mathrm{V}} \to i\hat{c}_{\mathrm{V}} \end{aligned} \quad (1.31)$$

Figure 1.32b and c describe the basic function of the polarizing beam splitter in more details. When the two incident modes are one of H and V and have the same polarization, two photons are emitted from different paths; when the two incident modes are polarized differently, the two photons are emitted from the same output port. Polarizing beam splitters are commonly used in various polarization-related

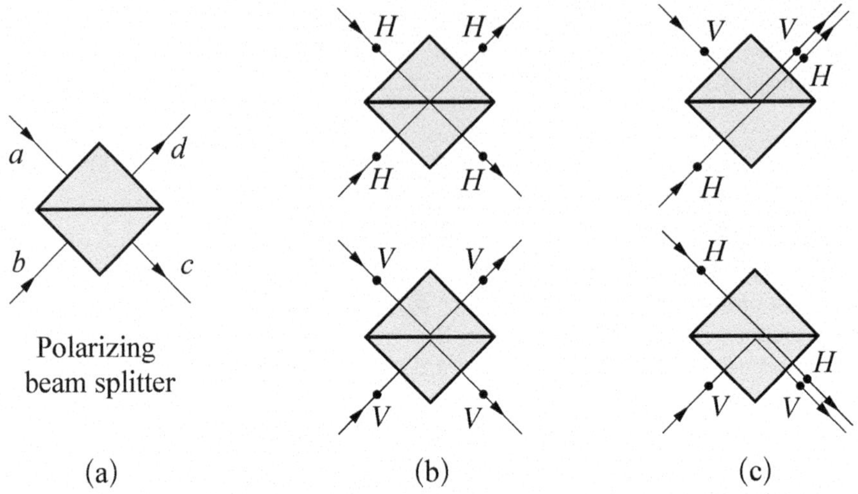

Fig. 1.32 Schematic diagram of the operating principle of the polarizing beam splitter [59]

interferometers, such as the Sagnac interference ring [60] applied to generate PPKTP entanglement, as illustrated in Fig. 1.33. In the following paragraphs, we will further describe another important application of polarizing beam splitters in optical quantum computing experiments—as a core element of the polarization analyzer.

3. Wave plate

Optical wave plates [61] are typically cut from quartz crystals of precise thickness, with surfaces parallel to the direction of the crystal's optical axis. Thus, when a parallel beam of light is incident perpendicularly, the incident light with the electric vector oscillation perpendicular to the optical axis is called *o* light and *e* light if

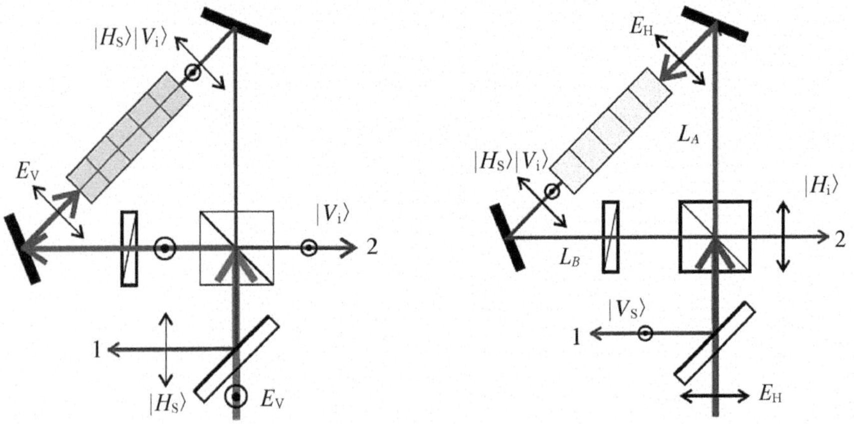

Fig. 1.33 Schematic diagram of colinear ppKTP entanglement source

parallel to it. The two lights propagate at different speeds in the birefringent crystal, accordingly creating a relative phase difference.

$$\Delta = \varphi_o - \varphi_e = \frac{2\pi}{\lambda}(n_o - n_e)d \tag{1.32}$$

If the phase difference is equal to half a wavelength, it is called a half-wave plate; if the phase difference is 1/4 wavelength, it is a 1/4 wave plate. When the optical axis of the half-wave plate has an angle of δ with the vertical direction, if the photon is incident to the e light component parallel to the optical axis, the o light component perpendicular to the optical axis will be delayed by half a wavelength. It causes a phase difference of π, and o light needs an opposite sign. Therefore, both H and V, after passing through the half-wave sheet, are the inversion of the optical axis.

$$\begin{aligned} |H\rangle &\Rightarrow \sin\delta|e\rangle + \cos\delta|o\rangle \\ |V\rangle &\Rightarrow \cos\delta|e\rangle - \sin\delta|o\rangle \end{aligned} \tag{1.33}$$

The Jones matrix for the half-wave plate can be derived from Eq. (1.33):

$$\mathbf{HWP}(\delta) = \begin{pmatrix} -\cos 2\delta & \sin 2\delta \\ \sin 2\delta & \cos 2\delta \end{pmatrix} \tag{1.34}$$

In Eq. (1.34), we can see that if the optical axis of the half-wave plate is placed at an angle of 67.5° to the vertical direction, the half-wave plate performs Hadamard gate transformation on the incident photon:

$$\begin{aligned} |H\rangle &\Rightarrow \frac{1}{\sqrt{2}}(|H\rangle + |V\rangle) \\ |V\rangle &\Rightarrow \frac{1}{\sqrt{2}}(|H\rangle - |V\rangle) \end{aligned} \tag{1.35}$$

Similarly, the Jones matrix for 1/4 wave plate is written out as follows:

$$Q\mathbf{WP}(\delta) = \begin{pmatrix} 1 + i\cos 2\delta & -i\sin 2\delta \\ -i\sin 2\delta & 1 - i\cos 2\delta \end{pmatrix} \tag{1.36}$$

The wave plate is used in the system to regulate the polarization state of the light. It can be proved that for any θ in the polarization state $|H\rangle + e^{i\theta}|V\rangle$, a 1/4 wave plate is able to transform it into linearly polarized light. Three wave plates (two 1/4 wave plates and one half-wave plates) can realize any unitary transformation of the polarization state of the light.

4. Polarized qubit coding

In experiments, both polarizing beam splitters and wave plates are often used to encode the desired polarized qubits, as shown in Fig. 1.34.

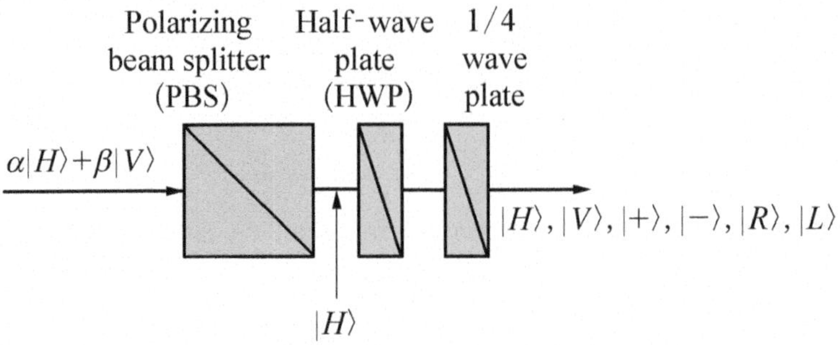

Fig. 1.34 Polarized qubit coding

Assuming that the input state is $\alpha|H\rangle + \beta|V\rangle$, the corners of HWP and QWP can be adjusted to easily prepare commonly-used coding states $|H\rangle$, $|V\rangle$, $|+\rangle$, $|-\rangle$, $|R\rangle$, and $|L\rangle$; if we use the same device and change the direction of the incident light, we can analyze any polarization state; it is also known as polarization analyzer.

1.4.2 Coherent Manipulation of Optical Qubits

In this section, we are going to relate the coherent manipulation technique in linear optical quantum computation. There are two major research directions at present: platform-based linear optical quantum computation (where photons are mainly manipulated in free space). Due to its mature technology, the vast majority of current quantum computation schemes are first validated in this system, but it has the disadvantage of being poorly scalable; there is also waveguide-based optical quantum computation, which, although still in its infancy, is promising because of its good scalability. In the following paragraphs, we will go into the details of both experimental systems.

1.4.2.1 Basic Optical Quantum Interference

All in all, the coherent manipulation of photons is an essential tool in optical quantum computation. The most basic interferometer is Mach–Zehnder interferometer, which was mentioned above when describing the quantum properties of beam splitters. Our focus is on the Hong–Ou–Mandel interferometer, GHZ interferometer, and multi-photon interferometer that are frequently used in optical quantum computation. Next, we will briefly introduce the Bell-state synthesizer.

1. Hong–Ou–Mandel interference

Quantum interference can be entirely caused by interparticle indistinguishability. The most fundamental optical effect arising from inter-photon indistinguishability will

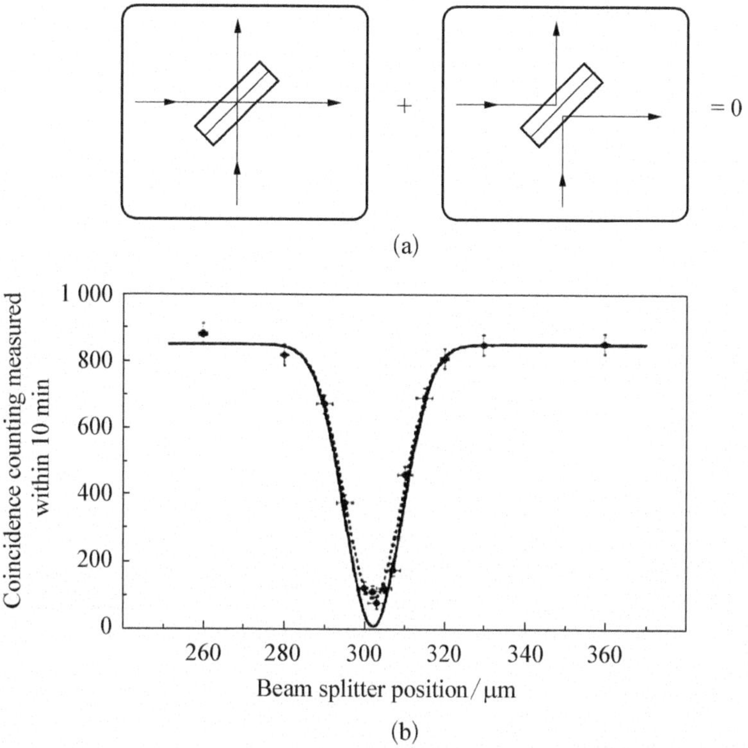

Fig. 1.35 Schematic diagram of HOM interference [59]

be presented here, i.e., Hong–Ou–Mandel interference (HOM interference) behind a 50:50 beam splitter [62]. The overall schematic is shown in Fig. 1.35.

As mentioned earlier, a photon acquires a phase shift i after being reflected by a beam splitter. Assuming that the modes of the input photon are a and b and the modes of the output photon is c and d. If we have two identical photons (both with the same frequency and polarization) entering from each of the two inputs of the beam splitter at the same time, the initial state $\hat{a}^\dagger\hat{b}^\dagger|\Omega\rangle$ is transformed to the following:

$$\frac{1}{\sqrt{2}}(\hat{c}^\dagger + \mathrm{i}\hat{d}^\dagger)\frac{1}{\sqrt{2}}(\hat{d}^\dagger + \mathrm{i}\hat{c}^\dagger)|\Omega\rangle \to \mathrm{i}/2(\hat{c}^{\dagger 2} + \hat{d}^{\dagger 2})|\Omega\rangle \tag{1.37}$$

Two members are canceled out in the above equation, namely $\hat{c}\hat{d} - \hat{d}\hat{c} = 0$, representing the case where photons go out at separate outputs. If two photons are completely indistinguishable, cancellation will happen, in other words, we cannot observe an event where photons appear at both ports at the same time. They will leave either output port of the beam splitter (randomly) in pairs, which is a bunching effect caused by the bosonic properties of photons, as plotted in Fig. 1.35a.

If the photons are at least path-distinguishable (in which case a different notation like $\hat{a}_1$ and $\hat{b}_1$ is needed to represent the annihilation operator of the photon), the initial state $\hat{a}_1^\dagger \hat{b}_2^\dagger |\Omega\rangle$ is converted to the following:

$$\frac{1}{\sqrt{2}}\left(\hat{c}_1^\dagger + i\hat{d}_1^\dagger\right)\frac{1}{\sqrt{2}}\left(\hat{d}_2^\dagger + i\hat{c}_2^\dagger\right)|\Omega\rangle \tag{1.38}$$

As $\hat{c}_1 \neq \hat{c}_2, \hat{d}_1 \neq \hat{d}_2$, the member $\frac{1}{2}\left(\hat{c}_1^\dagger \hat{d}_2^\dagger - \hat{d}_1^\dagger \hat{c}_2^\dagger\right)|\Omega\rangle$ that contributes to the case where the photon leaves the beam splitter from different output ports cannot be canceled out. If β denotes the resolution between photon 1 and photon 2, the probability $\frac{1}{2}\left(\hat{c}_1^\dagger \hat{d}_2^\dagger - \hat{d}_1^\dagger \hat{c}_2^\dagger\right)|\Omega\rangle$ of obtaining coincidence counting at outlets c and d becomes $\frac{1}{2}|\beta|^2$. Thus, if $|\beta|$=1 (the photons $\hat{a}^\dagger$, $\hat{b}^\dagger$ are perfectly distinguishable), this probability is $\frac{1}{2}$; if $\beta = 0$ (the photons are totally indistinguishable), the probability is zero.

The Hong–Ou–Mandel effect therefore depends rigorously on the indistinguishability of photons. In the initial experiment, a pair of photons hit the beam splitter are produced by SPDC, and the photons have the same polarization and frequency distribution. The photon's indistinguishability is given by the delay difference between two paths, and the coincidence counting as a function of the delay difference between the two paths of the photon shows a Hong–Ou–Mandel Dip where the delay is zero, as drawn in Fig. 1.35b.

2. GHZ interference

After many years of research on two-particle entanglement, Greenberger et al. imagined a three-photon interferometer in 1989 [59]. The step from two-photon interference to three-photon interference looks small although, it still has far-reaching implications, one of which is the ensuing famous GHZ theory.

As can be seen in Fig. 1.36, the center of the interferometer is a three-photon source, and the resulting three-photon entanglement state is the GHZ state:

$$|\mathrm{GHZ}\rangle_{123} = \frac{1}{\sqrt{2}}\left(|a\rangle_1 \left|a'\right\rangle_2 \left|a''\right\rangle_3 + |b\rangle_1 \left|b'\right\rangle_2 \left|b''\right\rangle_3\right) \tag{1.39}$$

Here, each photon has two different spatial modes. E.g., photon 1 has states a and b. After relative phase shift and BS action are considered, some new properties of three-particle interference can be obtained. Firstly, it corresponds to the probability of the triplet coincidence technology, e.g., $p_{1c,2c,3c}(\phi_1, \phi_2, \phi_3) = \frac{1}{8}[1 + \sin(\phi_1 + \phi_2 + \phi_3)]$, $p_{1d,2c,3c}(\phi_1, \phi_2, \phi_3) = \frac{1}{8}[1 - \sin(\phi_1 + \phi_2 + \phi_3)]$. The three-body coincidence counting rate given by the above equation takes $\phi_1 + \phi_2 + \phi_3$ as an independent variable to exhibit sinusoidal variation. Secondly, the three-photon interferometer does not yield any two-particle interference fringes. For example, if only two particles are detected to meet $2c$ and $3c$ whereas the $1c$ and $1d$ modes are ignored, the observed coincidence rate will be a constant, $p_{1c,2c,3c}(\phi_1, \phi_2, \phi_3) + p_{1d,2c,3c}(\phi_1, \phi_2, \phi_3) = \frac{1}{4}$, i.e., the two-body coincidence rate

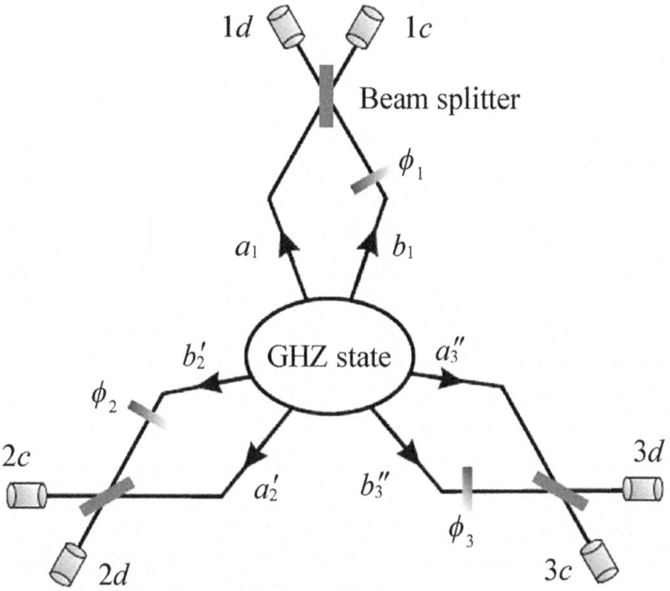

Fig. 1.36 Phase-adjustable GHZ interferometer [59]

is totally independent of the phase. Finally, similar to the argument above, no single-particle interference fringes will of course be observed. In fact, the n-particle interferometer realized according to the above theory yields n-particle interference fringes other than $(n-1)$, $(n-2)$, and other single-particle fringes. Although the aforementioned three-photon interference is described in terms of the photon's path degree of freedom, similar interference phenomena can also be observed using any of the photon's degrees of freedom, such as polarization.

3. Star multiphoton interferometer

The larger the number of photons, the more difficult multiphoton interferometry becomes. One of the difficulties with multiphoton interferometer is that multiphoton coincidence counting depends on random conformity of independent entanglement pairs, and usually the count rate is extremely low. Assuming an entanglement pair coincidence counting of 1000 Hz and a laser pulse frequency of 76 MHz (the frequency of the currently used femtosecond laser), the four-photon coincidence counting is $\frac{1000^2}{76\times10^6} = 0.013$ Hz. As can be seen, it is vital to improve the coincidence counting rate of entangled pairs.

The second key point is quantum erasure. In multiphoton interference, the indistinguishability between different photons greatly determines the final interference contrast. To achieve perfect indistinguishability, the time information, polarization information, and spatial modes of the interfering photons are required to be indistinguishable. In previous experiments, only the time information (no delay between interfering photons) and polarization information (equal polarization of interfering photons) were generally indistinguishable. Because multi-photon interference photons are generated during the SPDC process, a beam of photons contain both o-light and e-light. In order to make the spectra equal, only narrow-band filtering can be used. In principle, the spectral information is indistinguishable only when the narrow-band filter is made narrow enough. But then the photons are lost in large numbers, a suitable half-height narrow-band filter (*FWMH* ~ 3 nm) is experimentally used. In addition, during multiphoton interference, the narrow-band filter also plays a role in increasing the photon coherence time and removing the frequency correlation between different entangled photon pairs. To tackle this problem, our team has designed a unique star multiphoton interferometer [11] in the latest multiphoton quantum computation experiment, allowing only e-light to interfere, as shown in Fig. 1.37. As only e-light interferes, the spectral information of the photons is, in principle, indistinguishable. Furthermore, as there is still a frequency correlation between different entangled photon pairs, it is still necessary to insert a narrow band filter in front of the interfering photon. Nevertheless, due to the narrow spectrum of the e-light, just few photons are lost in the filtering process, and the interference contrast is much higher than in previous experiments.

In addition, while analyzing the noise source of the multiphoton interferometer in previous experiments, we find that the traditional chain structure interferometer introduced a lot of unwanted noise, which will be effectively suppressed by the star interferometer. We can see from Fig. 1.38 that the multi-pair noise in the chain structure interferometer is 6p5, while the multi-pair noise in the star interferometer is only 4p5, down by 50%. In the figure, p is the probability of producing entangled photon pairs per pulse during the SPDC process.

In specific experiments, to ensure that the best coherent contrast is obtained, the interfering optical path needs to be carefully modulated to achieve the best spatial

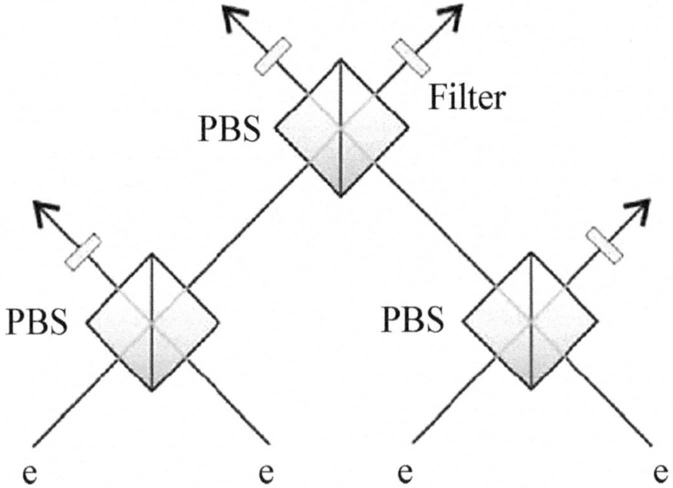

Fig. 1.37 Schematic diagram of multiphoton interferometer structure [11]

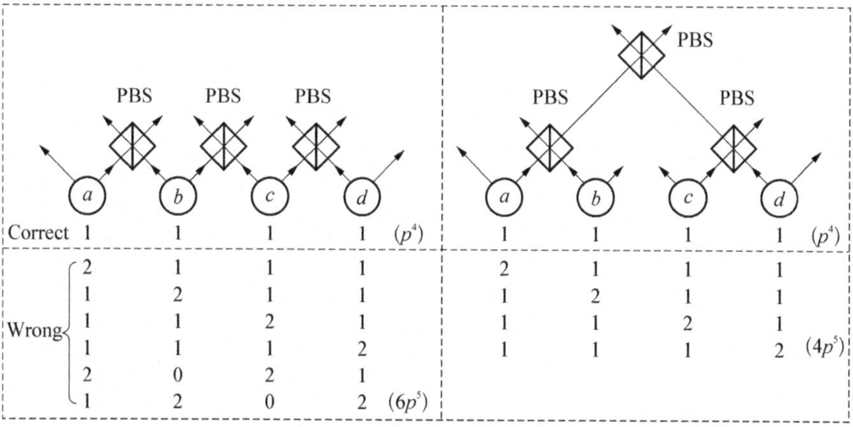

Fig. 1.38 Noise analysis of a chain interferometer and a star interferometer [11]

pattern matching. First of all, an optical-cement polarizing beam splitter with an extremely high polarization extinction ratio is required. As the angular deviation of the optical-cement polarizing beam splitter is quite small, the incident beam is virtually free of deflection, while the very high polarization extinction ratio minimizes the noise introduced by the imperfection of the beam splitter. Second, the two beams of interfering photons incident from different input ports must be perfectly pattern-matched: the Gaussian beam's waist size and position must be equal (deviation < 10 μm). Finally, the two laser beams must be perfectly overlapping in space.

4. Bell state synthesizer

In order to obtain high-brightness high-fidelity pulsed entanglement sources, our team has developed a set of Bell-state synthesizer techniques in recent multiphoton entanglement experiments [21, 63], specifically:

As shown in Fig. 1.39, a beam of femtosecond pulsed violet light passing through a BBO crystal will produce entangled photon pairs with some probability. However, there are two types of undesired temporal information will be correlated with the polarization information during SPDC to reduce the fidelity of the entangled state: (i) the group velocities of different polarized photons are different, which can be removed by a pair of birefringence-compensated crystals; (ii) the spectra of the o- and e-light are different during SPDC of the ultrafast pulsed pump. Most previous multiphoton experiments removed this frequency correlation by introducing a narrow band filter. However, this passive filtering process is very inefficient and a large number of photons are unnecessarily wasted.

The Bell-State synthesizer technique can successfully bypass this frequency correlation. First, photons a and b pass through a half-wave plate and a 1-mm-thick BBO crystal respectively, to compensate for the temporal and spatial walk-off effects. Then, they are coherently superimposed on the polarizing beam splitter by means of a reflector as well as a prism. We inserted a half-wave plate (HWP) in the path

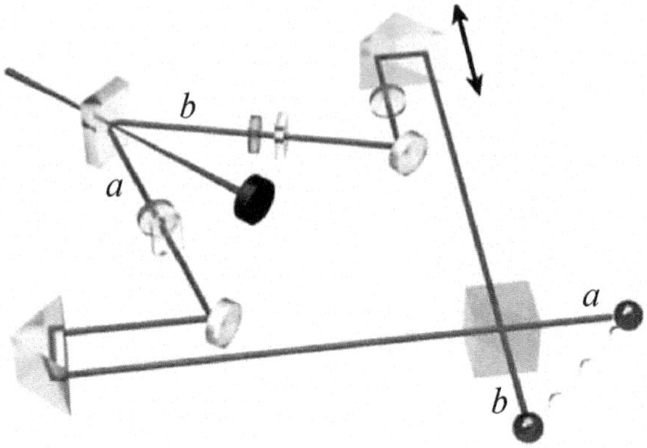

Fig. 1.39 Schematic diagram of the Bell-State synthesizer [21]

of photon b to rotate its direction of polarization by 90°. Since the polarizing beam splitter transmits H light and reflects V light, there are two possible outcomes: both photons transmit (*t*–*t*) or both reflect (*r*–*r*); in this way, the e-light (with a smaller spectral bandwidth) and o-light (with a larger spectral bandwidth) will leave the different output ports of the polarizing beam splitter and the e-photon will be captured by detector a, while the o-photon will be found by detector b. The process can be expressed by the following equation:

$$\begin{aligned} &|\psi\rangle = \tfrac{1}{\sqrt{2}}(|H^o\rangle_a|V^e\rangle_b + |V^e\rangle_a|H^o\rangle_b) \\ &\text{Compensating HWP, rotation of polarization by } 90^\circ \\ &\quad \tfrac{1}{\sqrt{2}}(|V^o\rangle_a|H^e\rangle_b + |H^e\rangle_a|V^o\rangle_b) \\ &b \text{ - path HWP, rotation of polarization by } 90^\circ \\ &\quad \tfrac{1}{\sqrt{2}}(|V^o\rangle_a|V^e\rangle_b + |H^e\rangle_a|H^o\rangle_b) \\ &\text{Coherent superposition on PBS} \\ &\quad \tfrac{1}{\sqrt{2}}(|H^e\rangle_a|H^o\rangle_b + |V^e\rangle_a|V^o\rangle_b) \end{aligned} \tag{1.40}$$

Thereupon, the Bell-State synthesizer is able to effectively separate the time information from the polarization information of the photon pair to form a coherent superposition state. It is also important to note that since there is no case of two photons emanating from the same port of the polarizing beam splitter, post-selection is generally not required. In specific experiments, one of the prisms needs to be mounted on a one-dimensional micron translation stage, and by moving the position of the prism in the optical path, the delay between photon a and photon b becomes zero, truly realizing indistinguishability.

1.4.2.2 Platform-Based Multiphoton Coherent Manipulation

In this section, two specific experiments are taken as examples to introduce the platform-based multiphoton coherent manipulation technique.

1. Preparation of eight-photon entangled states

The development of practical quantum computation necessarily relies on the coherent manipulation of multiple quantum states. A vivid metaphor is that a multi-particle entangled experimental platform is like driving an aircraft carrier for quantum computation, based on which a variety of quantum computation programs are possible to be explored. The experimental preparation of multi-particle Schrödinger cat states epitomizes the core technology of quantum manipulation, which has always been an internationally competitive field. The preparation of multi-particle entangled states with higher particle counts is not only a fundamental scientific research but also boosts the development of experimental techniques of quantum information. From the first experiment [1] of three-photon entangled states in 1999, scientists have made unremitting efforts to prepare multi-particle entangled states for more and more quantum bits. In 2012, our team first realized the preparation of eight-photon entangled states based on the SPDC process in the world [11].

The goal of the experiment was to generate an eight-photon Schrödinger cat state, or Grcenberger–Home–Zeilinger (GHZ) state in the following form (i.e., eight-photon Schrödinger cat state):

$$|SC_8\rangle = \frac{1}{\sqrt{2}}\left(|H\rangle^{\otimes 8} + |V\rangle^{\otimes 8}\right) \tag{1.41}$$

In the formula, H and V denote the horizontal and vertical polarization of a single photon. It contains an equal-weighted superposition of two polarized orthogonal states of eight independent photons. To achieve this, we first employed the SPDC procedure to prepare four entangled photon pairs in state $|\phi^+\rangle = \frac{1}{\sqrt{2}}(|H\rangle|H\rangle + |V\rangle|V\rangle)$ and then combined them into a Schrödinger cat state using an eight-photon interferometer.

The first challenge encountered in the experiment is the extremely low eight-photon coincidence counting rate. To detect eight photons requires that all four independent pairs of entangled photons arrive at the detector at the same time, so the magnitude of the eight-photon coincidence event is $(p \times \xi)^4$, wherein p is the probability of SPDC and ξ the total efficiency with which individual photons are collected and detected. For very small values of p and ξ, this value goes down drastically. If we refer to the data from previous six-photon experiments, the eight-photon event coincidence counting rate we are able to obtain is 2.8×10^{-5} Hz. At such a low counting rate, it will take us more than 10 years to verify whether an eight-photon entangled state has been prepared, and no experimental facility is stable enough to maintain this level of stability for such a long period of time. Hence, it is a completely impossible task. To solve this problem, we must greatly increase the brightness of the entangled photons. The second experimental challenge is noise control during the generation of multi-photon entangled states. Though a higher-power pump laser can be adopted to increase the probability p of generating single-pair photons directly (and thus improve both the two-photon count rate), it will inevitably enhance the probability of generating two pairs of photons simultaneously ($\sim p^2$), which is considered the most important source of noise in multiphoton experiments. It is therefore necessary to keep the power of the pumped light at an appropriate level (in fact, the highest pumped laser power we are able to achieve in the experiment is 1.7 W, twice the final power used).

By employing the Bell-State synthesizer technique described above, we were able to generate 4 pairs of polarized entangled photons with rather high luminance and fidelity. We then controlled these 4 pairs to form an eight-photon Schrödinger cat state. Four e-polarized photons, from each of four entangled photon pairs, were combined through a linear optical network comprising three polarizing beam splitters (Fig. 1.40b). It can be demonstrated that only four incident photons with the same polarization direction can be transmitted ($|HHH\rangle$ or reflected $|VVV\rangle$) by these three polarizing beam splitters simultaneously, and then a coincidence counting is obtained (a single photon at each output port at the same time). It is worth noting that if all other information (e.g., time, frequency, spatial mode) is erased, then in principle there is no way to distinguish between $|HHH\rangle$ and $|VVV\rangle$. Thus, this interferometer effectively projected (projection operators $|HHH\rangle\langle HHH|+|VVV\rangle\langle VVV|$) four pairs of entangled photon pairs to the eight-photon Schrödinger cat state. There are some other interesting features observed from the eight-photon interferometer in the experiment. In the graphical representation, it visualizes the process of fusing the optical qubits into a star topology (Fig. 1.40b, right); we also find that an interferometer with this structure can also be used as a projector for the four-photon GHZ state to be directly applied in the experimental studies of multi-particle entanglement swapping.

In the experiment, a pulsed UV laser with 880 mW power continuously pumped four BBO crystals with a thickness of 2 mm. Different lens combinations were used to ensure that the waist of the laser focused on each of the BBO crystals was the same in size. Using a Bell-State synthesizer, the average coincidence counting rate of two photons that could be observed in the experiment was about 1 MHz, with about 90% fidelity if no narrow-band filter was inserted. With a spectrometer, it is possible to measure the spectral width of PDC photons to obtain the FWHM (λ_{FWHM}) of e-light and o-light, respectively 6 and 12 nm.

After obtaining a high-brightness polarized entanglement source, we coherently cross-linked e-light 1–4–5–8 on three polarizing beam splitters (with well-matched modes between coherent beams), as shown in Fig. 1.41. The delay between different photon paths needed to be precisely adjusted to ensure that all photons reach the polarizing beam splitter at the same time. We filtered the e-light with a λFWHM = 2.8 nm filter (peak transmittance > 98%) in an attempt to remove the frequency correlation between entangled photons and improve the contrast of two-photon interference. In addition, to filter out the noise and increase the photon collection efficiency, we also filtered the o-light (2–3–6–7) with a 8 nm (λFWHM = 8 nm) filter. After filtering the o- and e-light separately, we measured the average coincidence counting

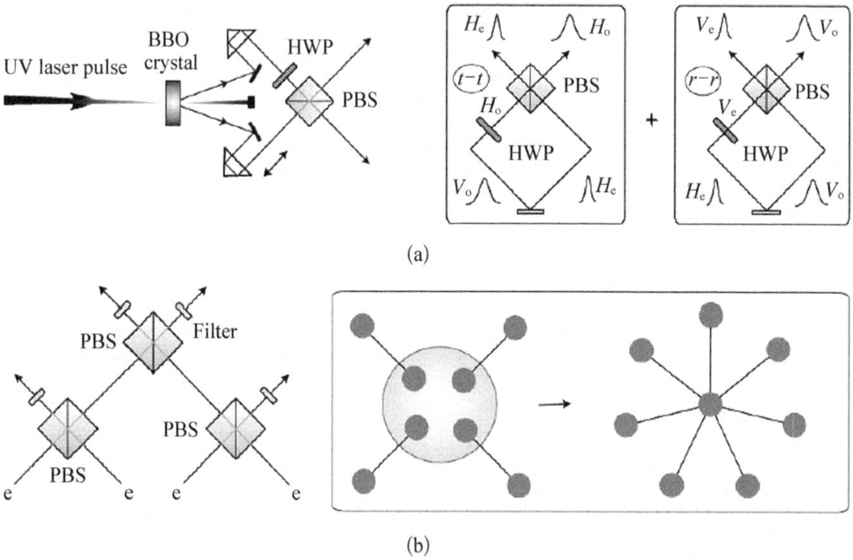

Fig. 1.40 Experimental scheme of eight-photon Schrödinger cat state [11]

count rate of two photons, which was about 310,000 Hz under the basis vector of $|\pm\rangle = \frac{1}{\sqrt{2}}(|H\rangle \pm |V\rangle)$, and a visibility of about 94%. At this brightness, we deduced the probability $p = 0.058$ of an entangled photon pair generated by single pulse and the total collection and measurement efficiency = 0.265. To measure the indistinguishability between independent photons, we experimentally observed Hong–Ou–Mandel-type interference of the photons on three polarizing beam splitters, and the interference visibility was about 76%.

In order to analyze the polarization state of each photon, the experiment used a combination of a ¼ wave plate (QWP), a half wave pate (HWP), and a PBS. A single-mode fiber was used to collect photons and spatially filter the photons, and the collected photons were fed into a single-photon detector. Sixteen single-photon detectors were adopted simultaneously in the experiment to measure the data, and the signals detected were sent into a home-made FPGA-based programmable multichannel coincidence counter that was capable of simultaneously recording all 256 possible combinations of eight-photon coincidence events (when and only when eight detectors at different output ports responded simultaneously) and automatically discarding nine-photon coincidence events, so as to lessen the noise introduced by higher-order PDC events.

In the experiment, it was able to obtain about nine 8-photon coincidence counts per hour. To prove that the eight-photon Schrödinger cat state has been successfully prepared, it can be demonstrated as follows: first of all, among the measured eight-photon coincidence counts, it is theoretically only $|H\rangle^{\otimes 8}$ and $|V\rangle^{\otimes 8}$ can be observed and no other case. By comparing all 256 possible polarization combinations under the H/V basis vector, it is found that the experimental data are perfectly accord

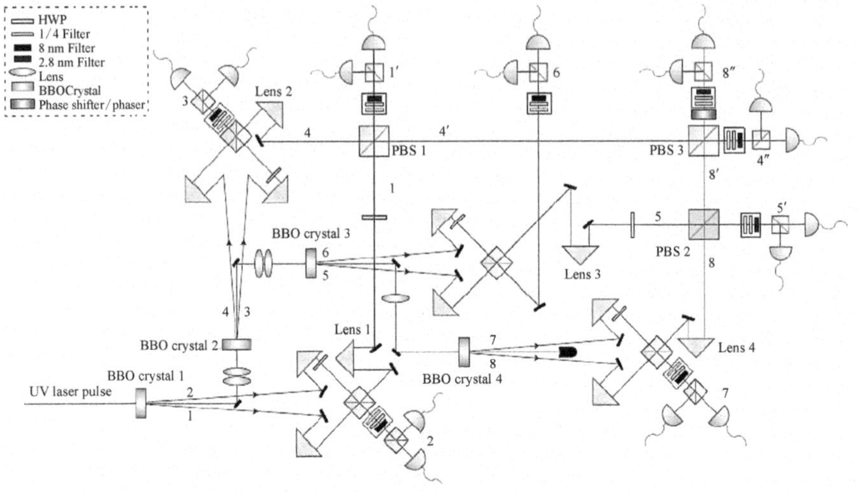

Fig. 1.41 Experimental beam path of eight-photon Schrödinger cat state [11]

with this inference. The experimental data are shown in Fig. 1.41. In all measured eight-photon coincidence events, $|H\rangle^{\otimes 8}$ and $|V\rangle^{\otimes 8}$ dominate.

The signal-to-noise (SNR) ratio (defined as the mean of the desired component divided by the mean of the unwanted component) of the experimental results was 530:1. Next, to prove that $|H\rangle^{\otimes 8}$ and $|V\rangle^{\otimes 8}$ are indeed coherently superimposed, it was also measured under the basis vector $\left(|H\rangle \pm \mathrm{e}^{\mathrm{i}\theta}|V\rangle\right)/\sqrt{2}$, wherein $\theta = \frac{k\pi}{8}$. For example, when $k = 0$, the prepared Schrödinger cat-state could be rewritten in the basis vector $+\rangle/|-\rangle$ state as an expression containing 128 members, with items containing an even number of $|-\rangle$ (e.g. $|+\rangle^{\otimes 8}$, $|+\rangle^{\otimes 6}|-\rangle^{\otimes 2}$) appearing and those containing an odd number not showing up. The measurements bear this out, with a SNR ratio of about 4:1. From the measured experimental data, the expected value of the observation $\left\langle M_k^{\otimes 8}\right\rangle = \left\langle\left(\cos\theta\sigma_x + \sin\theta\sigma_y\right)^{\otimes 8}\right\rangle$ can be calculated as 0.610 ± 0.026 (Fig. 1.42b). Using the entanglement witness operator, the fidelity of the cat states can be calculated to determine whether a true multiparticle entangled state has been prepared. The fidelity is a definition of the ideal degree of the prepared entangled state and can be obtained by calculating the inner product between the experimentally prepared state and the ideal state. For the eight-photon Schrödinger cat state, its fidelity can be calculated (0.708 ± 0.016) based on the experimental data shown in Fig. 1.42. The theory proves that if the fidelity of the prepared Schrödinger cat state exceeds the threshold value of 0.5, it is sufficient to demonstrate that the prepared entangled state is an actual entangled state. In brief, we did successfully prepare the eight-photon Schrödinger cat state and the experimental results were > 14 standard deviations.

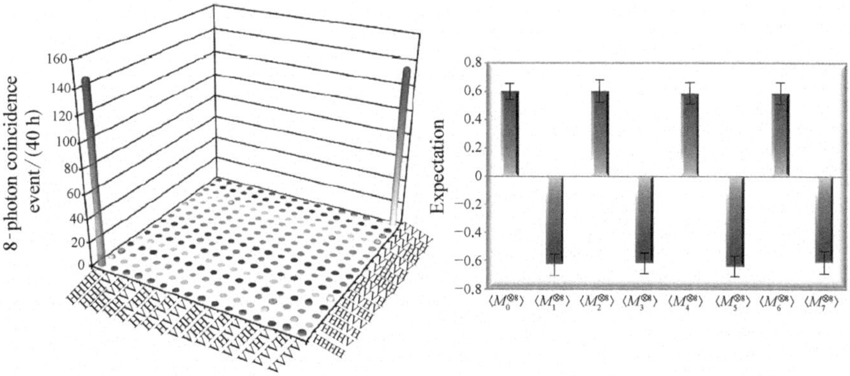

(a) (b)

Fig. 1.42 Results of the eight-photon Schrödinger cat-state experiment [11]

2. Preparation of 10-qubit hyper-entangled states

In this section we will present methods for the preparation of ten-qubit Schrödinger cat states using the polarization and path degrees of freedom of photons [64]. As we know, photons have multiple degrees of freedom, and when they are entangled with each other not only in polarization degrees of freedom but also in other degrees of freedom simultaneously, they can form a special entangled state namely a hyper-entangled state. Figure 1.43 has demonstrated a simple preparation of a polarization-path hyperentangled photon state: when a single photon hits PBS in an arbitrary quantum state $|\Psi_{\text{qubit}}\rangle = \alpha|H\rangle + \beta|V\rangle$, because of the PBS transmitting H polarization and reflecting V polarization, the path of this photon will be divided into two parts, which will penetrate PBS along the original path with the probability of $|\alpha|^2$ and be reflected by it with the probability of $|\beta|^2$ in a direction perpendicular to the original path. If these two paths are also encoded as quantum bits, with the transmission path encoded as $|H'\rangle$ and the reflection path as $|V'\rangle$, we will get the polarization-path hyper-entangled photon state.

$$|\Psi_{\text{qubit}}\rangle = \alpha|HH'\rangle + \beta|VV'\rangle \tag{1.42}$$

The preparation process seems to be easy, but the experimental realization of the polarization-path hyper-entangled state measurement process is not easy. Figure 1.44 has illustrated the process: to measure the path quantum bits, two different paths need to be superimposed on the beam splitter, and x and y can be measured by changing the phase of either path (x/y are Pauli operators); to make z measurements for path quantum bits, only a light barrier is needed to block the transmitted or reflected light path. The polarization qubit measurement is relatively simple, and it only requires

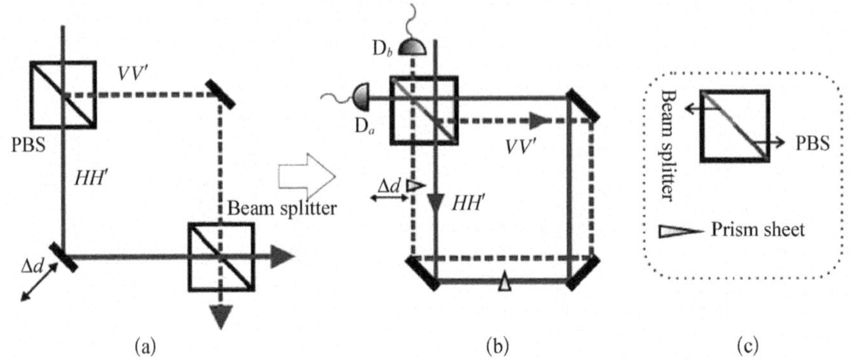

Fig. 1.43 Demonstration of the preparation of polarization-path hyper-entanglement states

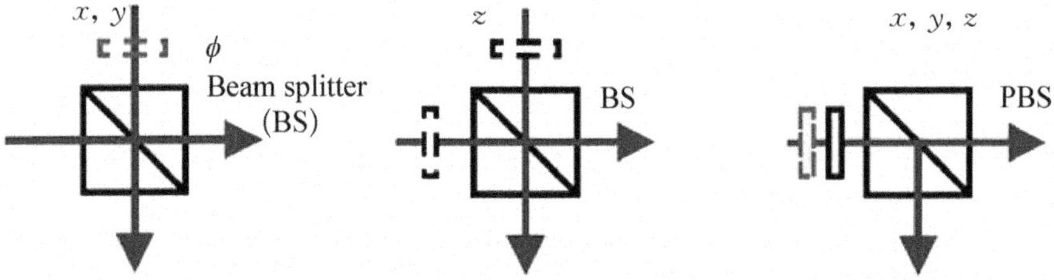

Fig. 1.44 Demonstration of polarization-path hyper-entangled state measurement process

placing a 1/4 wave plate and a HWP before the PBS and changing the number of degrees of the two plates to complete the x, y, z measurements of polarization qubits.

Next, we applied this polarization-path hyper-entangled state to the preparation experiments of ten-qubit Schrödinger cat states, as illustrated in Fig. 1.45. Two pairs of polarized entangled photon pairs $1-2,\ \ 3-4\ :\ \left|\Phi^{+}\right\rangle_{ij} = \frac{1}{\sqrt{2}}\left(|H\rangle_i|H\rangle_j + |V\rangle_i|V\rangle_j\right)$ were first prepared, wherein *H/V* denotes horizontal and vertical polarization whereas *i* and *j* represent the spatial modes of photons. Photon 5 is a weakly coherent single-photon source as described previously, and photons 2, 3 and photons 3, 5 were overlapped on the two polarizing beam splitters so that they could arrive at the polarizing beam splitters at the same time by adjusting their optical distances. After post-selection measurement was conducted at the 5 output ports, five-photon GHZ state $|\mathrm{Cat}_5\rangle = \frac{1}{\sqrt{2}}\left(|H\rangle^{\otimes 5} + |V\rangle^{\otimes 5}\right)$ was obtained.

We placed a PBS at each output port and successfully prepared a ten-qubit entangled state, $|Cat_{10}\rangle = \frac{1}{\sqrt{2}}\left(|H\rangle^{\otimes 5}\left|H'\right\rangle^{\otimes 5} + |V\rangle^{\otimes 5}\left|V'\right\rangle^{\otimes 5}\right)$, as described above.

The measurement of quantum bits becomes extremely difficult due to the use of two degrees of freedom per photon, requiring the use of a single-photon interferometer. In the experiment, the two spatial modes of a single photon interfering on the BS were employed to measure the spatial qubits, and the polarization qubits were

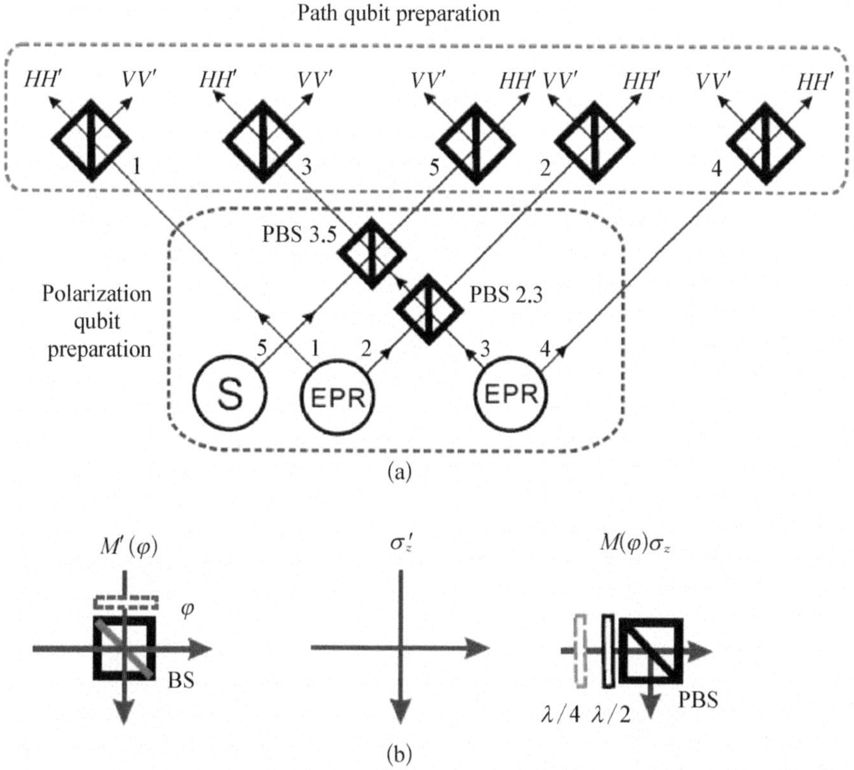

Fig. 1.45 Experimental scheme for ten-qubit hyper-entangled states

measured using a polarization analyzer. Conventional single-photon interferometers are more susceptible to air flow, temperature, etc., and are difficult to maintain stability. In the experiment, an interferometer automatically stabilizing the Sagnac structure was used. As displayed in Fig. 1.46c, after the photons have passed through the PBS, photons H are transmitted counterclockwise and photons V clockwise; after one circle, they interfere on the BS. The phase in the Sagnac interferometer is adjusted by the slender prism placed in it. In experiments, this single-photon interferometer is able to remain stable for up to 10 h.

The experimentally obtained ten-qubit cat-state count is 0.021 Hz with a fidelity of 0.56 ± 0.019, exceeding by 0.5 and more than three standard deviations, thus demonstrating the entanglement properties of the ten qubits. In addition to the methods described here for the preparation of hyper-entangled states, more degrees of freedom of photons, such as angular momentum, can also be exploited to prepare more qubit entangled states, eventually moving towards the goal of scalable optical quantum computation.

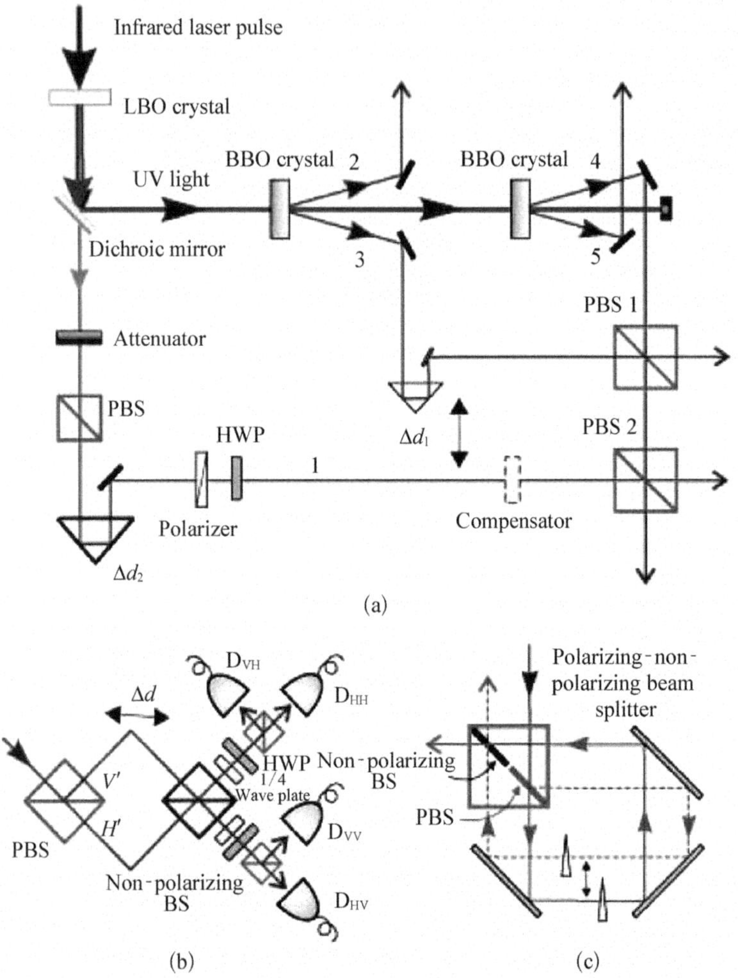

Fig. 1.46 Experimental apparatus for the preparation of ten-qubit hyper-entangled states

1.4.2.3 Coherent Manipulation of Waveguide-Based Optical Quantum

Platform-based experimental system is currently the main way to conduct optical quantum computation demonstration experiments. The technology of this method is relatively mature at present, but due to its hugeness, it is poor in scalability and has reached a bottleneck in technological development. Waveguide-based quantum computation is a new type of optical quantum computing; although it is still in its infancy, but because of its high miniaturization potential and scalability, it has the promise to be used for the realization of scalable optical quantum computing.

1. Basic waveguide manufacturing techniques

To achieve the coherent manipulation of optical quantum states in the waveguide, the corresponding waveguide structure is firstly designed according to the operating wavelength, so that the transmission of the operating wavelength in the waveguide is a single transverse mode. Detailed parameters mainly include the size and shape of the waveguide aperture and the selection of core materials and the refractive index of the cladding. At present, there are two ways to prepare "silica-silicon" chips. We are chiefly introducing related technology of O'Brien experimental group in the University of Bristol.

The first is growth method. The entire waveguide is designed for 78 nm. In Fig. 1.47, I is a silicon substrate fabricated by first generating an undoped silica layer (16 μm thick as shown in II of Fig. 1.47) on the silicon substrate in the way of thermal growth and then by depositing it on a silicon wafer as the lower cladding layer of the waveguide; after that, a 3.5-μm thick germanium- and boron dioxide-doped silica is deposited on II using flame hydrolysis. Immediately thereafter, a 3.5-μm wide region is retained using standard optical printing while the rest part is removed, as shown in III of Fig. 1.47. Next, a layer of phosphorus- and boron-doped silica with a thickness of 16 μm continues to be grown on top of III as an upper cladding for the waveguide; in the meantime, its refractive index is designed to match that of the lower cladding. Finally, a resistive layer and a corresponding metal electrode are grown on the top layer. The waveguide is theoretically simulated to enable single mode transmission of 780 nm light in it. The distribution of the single-mode light intensity transmitted in this theoretically simulated waveguide is displayed in Fig. 1.48.

Another approach is the use of laser writing techniques and femtosecond lasers to write waveguide networks directly into the fused silica, as shown in Fig. 1.49.

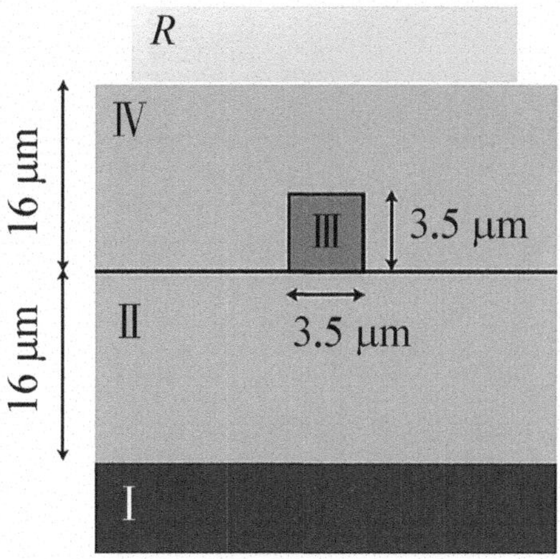

Fig. 1.47 Schematic diagram of the silica-silica waveguide structure prepared by the growth method [65]

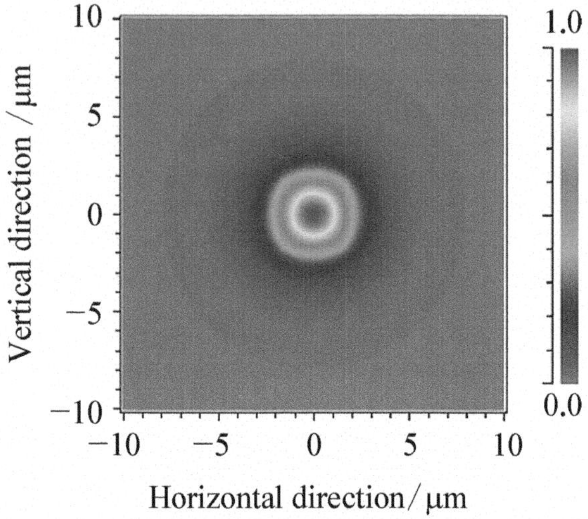

Fig. 1.48 Simulation results of silicon-silica waveguide transmission mode (operating wavelength 780 nm) [66]

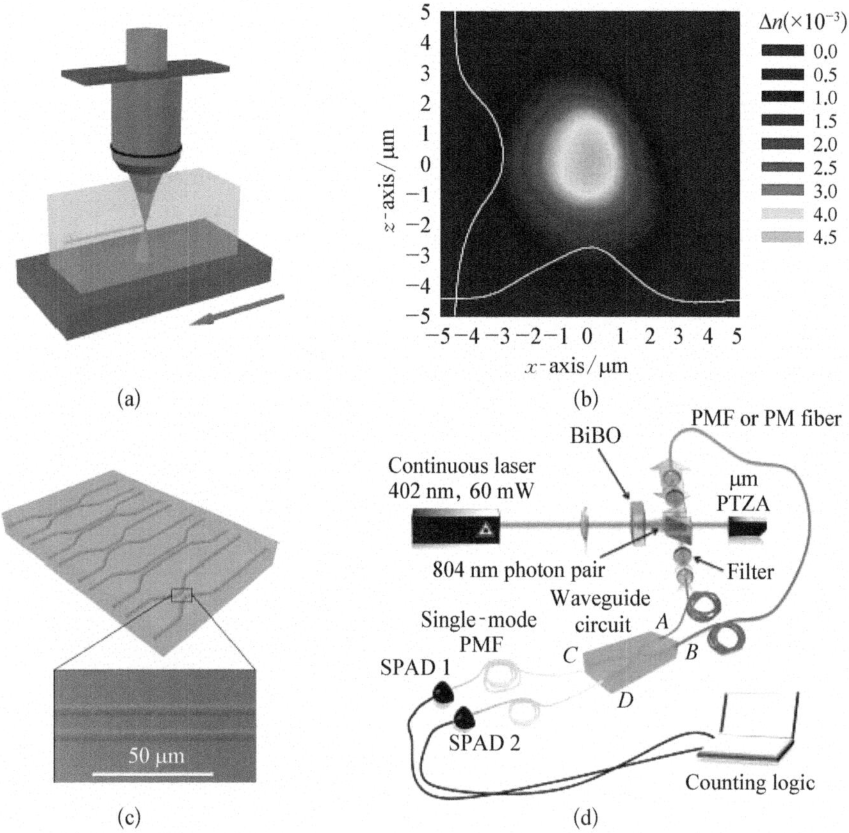

Fig. 1.49 Schematic diagram of fused quartz waveguides prepared by photographic writing [67]

The groundwork of its operation is to focalize a femtosecond laser to make it reaches a sufficient power density at the beam waist and then irradiate the fused quartz. The high-power femtosecond laser will change the refractive index of the fused quartz, thereby achieving a waveguide structure in the fused quartz. Figure 1.49a shows the basic way in which the laser is written into the waveguide; Fig. 1.49b is a distribution plot of the experimentally measured refractive index coefficients of waveguide; Fig. 1.49c displays a schematic representation of the sequence of directional couplers brought about by the waveguide; and Fig. 1.49d gives a map of experimental setups for two-photon interference realized on the waveguide to verify the performance of the prepared waveguide directional coupler. On this setup, HOM interference experiment described earlier was performed with good results, as shown in Fig. 1.50.

2. Waveguide-based "basic optical elements"

First, we describe how a beam splitter-like function can be implemented in a waveguide. In the experiment of O'Brien's team, a directional coupler approach was used: they lithographically made the space interval between two waveguides small enough so that two evanescent fields of waveguide overlap to achieve interference. By precisely accommodating the distance and interaction length between the two waveguides at the directional coupler, the coupling efficiency $1 - \eta$ (η corresponding to the reflectance of the beam splitter) can be regulated. As we can see in Fig. 1.51, a segment where two waveguides close together is a directional coupler.

Next, we introduce how to achieve the phase control of optical qubits based on the waveguide (see Fig. 1.52): Grow a resistive layer and a corresponding metal electrode on top of the waveguide. When a voltage is added between the two electrodes *p1* and

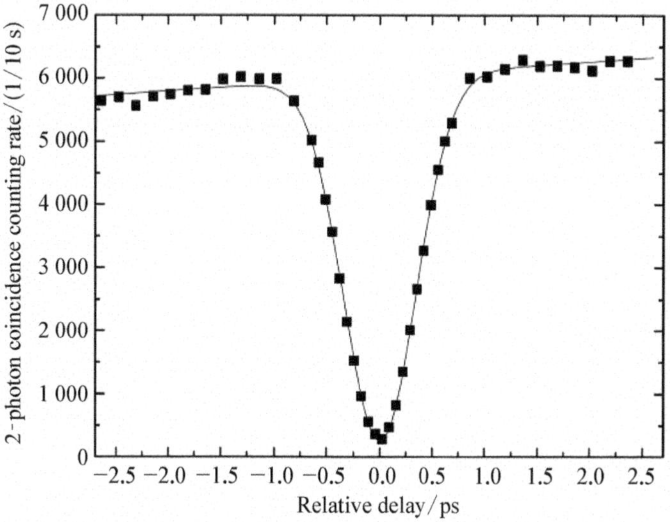

Fig. 1.50 HOM interference curve based on waveguide directional coupler [66]

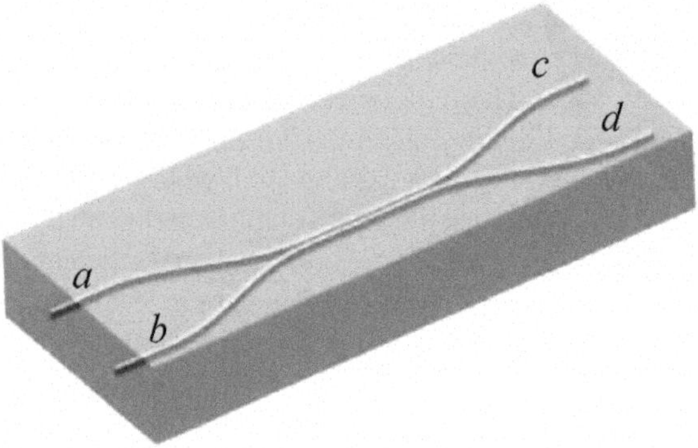

Fig. 1.51 Schematic diagram of a waveguide-based directional coupler [66]

p2 to heat up the resistor passing by, the quantity of heat will be dissipated directly in the waveguide, causing the waveguide region corresponding to the resistor to warm up T. For silica, the change rate $\frac{dn}{dT} = 1 \times 10^{-5}/\text{K}$ of the refractive index with the temperature will result in a change in the group velocity of light in the waveguide in the resistive region. In design, the voltage can be changed by designing a suitable resistor so that a continuous phase shift can be obtained, i.e., $\phi(V) \in \left[-\frac{\pi}{2}, \frac{\pi}{2}\right]$. The thermal effect is used to change the refractive index and transmittance of this path, enabling quantum operation (modulation of phase and intensity) for this path. The special structure of the chip makes it unnecessary to strictly control the external temperature.

We know that the change relationship between phase and control voltage is approximately $\phi(V) = \alpha + \beta V + \gamma V^2 + \delta V^3$. To accurately modulate the phase, we need to calibrate this function experimentally. By testing the two-photon interference curve at different control voltages, we can obtain the relationship between phase $\phi(V)$ and control voltage with the aid of numerical fitting. The following is a description of the specific calibration process, as shown in Fig. 1.52: Suppose the incoming photon state is $|1\rangle_a|1\rangle_b$. After the maximum path entanglement state $\frac{1}{\sqrt{2}}(|2\rangle_c|0\rangle_d + |0\rangle_c|2\rangle_d)$ generated via the directional coupler in the waveguide passes through the phase

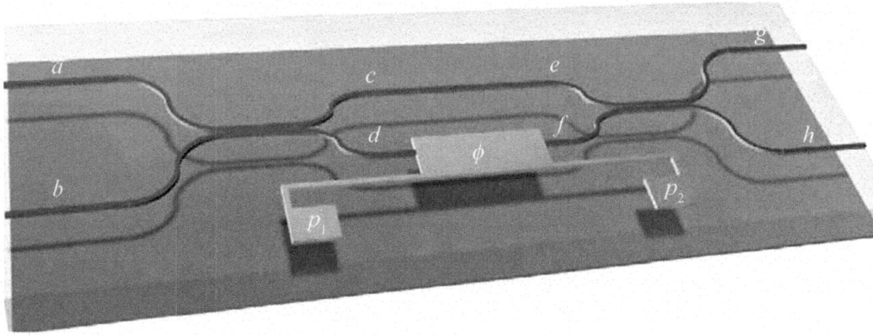

Fig. 1.52 Waveguide-based phase modulation

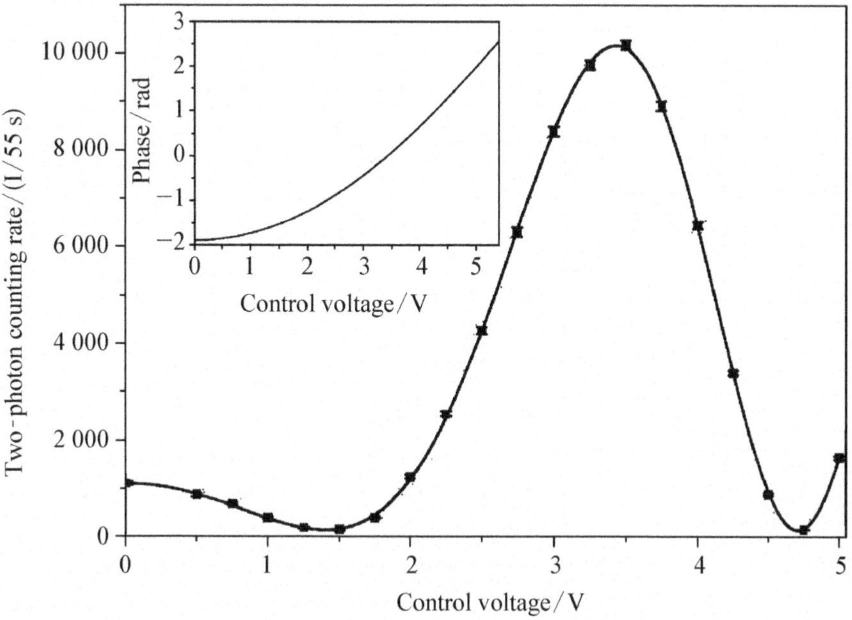

Fig. 1.53 Relationship curve between control voltage and phase modulation

shifter, the entanglement state $\frac{1}{\sqrt{2}}\left(|2\rangle_e|0\rangle_f + e^{i2\phi(V)}|0\rangle_e|2\rangle_f\right)$ is produced. Finally, two-photon coincidence measurement is performed between the output ports g and h of the waveguide to obtain the experimental results plotted in Fig. 1.53.

Through the numerical fitting of the experimental results in Fig. 1.53, we can infer the specific functional relationship of $\phi(V)$, and the fitting results are listed in Table 1.2. It can be seen that by changing the control voltage the phase can be accurately modulated.

3. Waveguide-based quantum state preparation and arbitrary rotation manipulation

A single-photon *MZ* interferometer can be implemented using the directional coupler and voltage-controlled phase modulator described above. As we have introduced previously, by modulating the phase in any one path of the *MZ* interferometer, an interference curve can be observed at the output port g or h. Figure 1.54 is the experimentally measured single-photon interference curve with a fidelity of 0.982,

Table 1.2 Fitting results of Φ(V) [65]

Parameter	Numerical value	Error
α	− 1.887	0.006
β	0.157	0.005
γ	0.0045	0.002
δ	− 0.001	0.0002

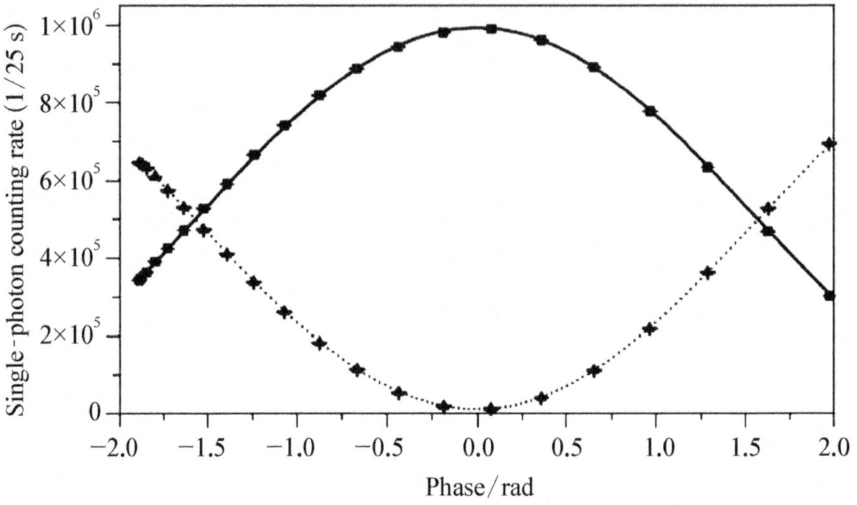

Fig. 1.54 Single-photon interference curves observed at output ports *g* and *h* [65]

which proves the excellent performance of the "beam splitter" and "phase shifter" in the waveguide.

We know that for real phases $\Phi 1$, $\Phi 2$, and $\Phi 3$, arbitrary single-qubit rotation can be decomposed into $U_{arb} = e^{i\phi_3\sigma_z/2}e^{i\phi_2\sigma_y/2}e^{i\phi_1\sigma_x/2}$. Note that $\sigma_x e^{i\phi\sigma_z/2} = e^{-\phi\sigma_z/2}\sigma_x$ and $U_{Mz} = ie^{i\phi\sigma_y/2}\sigma_x$. Therefore, adding a phase modulator before and behind this *MZ* interferometer enables an arbitrary rotation of single qubits. The operation to obtain a qubit from a logic bit preparation is expressed by $U_{\text{prep}} = e^{i\phi_3\sigma_z/2}e^{i\phi_z\sigma_y/2}$, while $U_{\text{prep}}^{\dagger}$ provides an arbitrary projection measurement. So, if it is possible to concatenate *N* waveguides in a similar way, it is possible to achieve the unitary operation of *N* modes.

4. Waveguide-based C-NOT gate

Figure 1.55 provides the waveguide structure diagram used to implement the C-NOT gate, wherein the control bit *C* and the target bit *T* are encoded by the photons in two waveguides respectively. If a signal is detected at the output ports of both the control bit and the target bit (with probability 1/9), it means C-NOT gate operation is successful. The waveguide was designed entirely based on the linear optical C-NOT gate scheme made by Ralpf et al. in 2002 and prepared by the method described above. As can be seen in the figure, the entire waveguide chip contains two directional couplers with 1/2 reflectance and three with 1/3 reflectance.

5. Waveguide-based multiphoton entanglement

The same applies to realize multiphoton entanglement manipulation in waveguides using the directional coupler and voltage-controlled phase shifter described previously. Here, we only introduce the two-photon and four-photon entanglements achieved by the O'Brien team. In the experiments, the first thing was to prepare

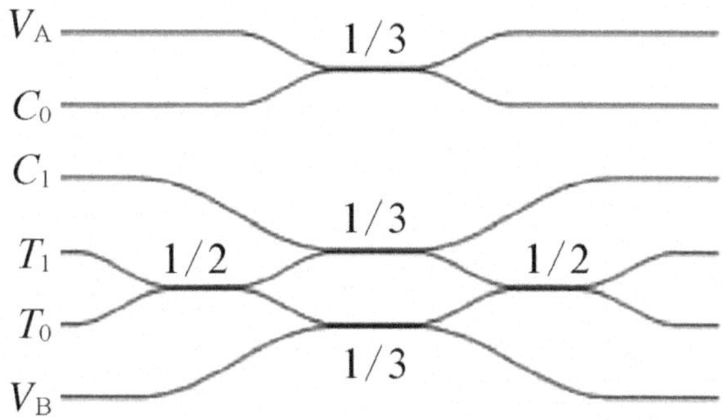

Fig. 1.55 C-NOT waveguide structure diagram [66]

the light source of the incident waveguide, as shown in Fig. 1.56: a 390 nm pulsed laser beam pumps a type-I BiBO crystal, and by controlling the average light intensity of the pumped light, a degenerate 780-nm-central wavelength two-photon and four-photon states are generated in two paths, all of them coupled to a polarization-maintaining optical fiber (PMF) and further into the waveguide chip drawn in Fig. 1.51.

First, the situation where the light source produces a pair of degenerate photons is considered. At this point, the photon state entering the waveguide chip is $|1\rangle_a|1\rangle_b$. After passing the first directional coupler the two photon states are $\frac{1}{\sqrt{2}}(|2\rangle_c|0\rangle_d + |0\rangle_c|2\rangle_d)$, and by phase modulation, it will become $\frac{1}{\sqrt{2}}\left(|2\rangle_e|0\rangle_f + e^{2i\phi}|0\rangle_e|2\rangle_f\right)$. If we measure the coincidence counting of the photons at exits g and h, it can be predicted that the coincidence counting is a function of the phase shift Φ. The experimental measurements are given in Fig. 1.57. The period in which the two-photon coincidence counting varies with the phase is π. By comparing

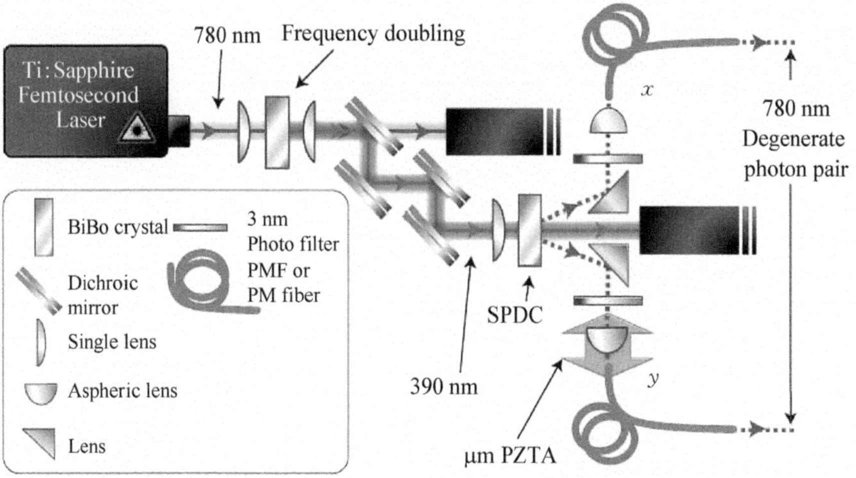

Fig. 1.56 Experimental setup for light source preparation of waveguides [65]

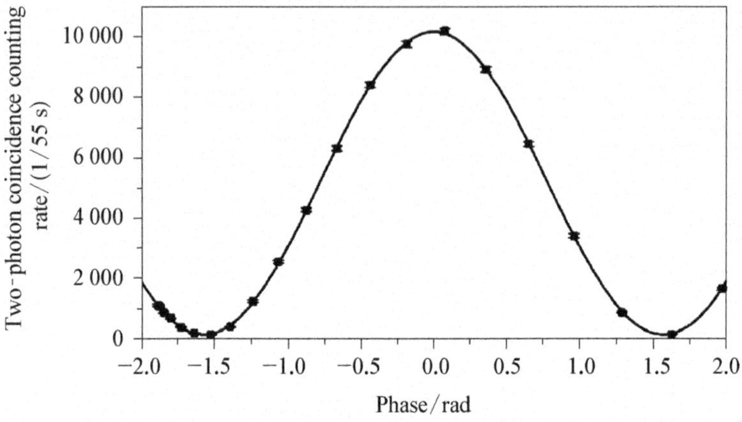

Fig. 1.57 Two-photon entanglement coincidence curve [65]

it with the single-photon interference curve in Fig. 1.54, we found that the period of the two-photon interference curve is half the period of the single-photon interference curve.

$$P_{g,h} = \frac{1}{2}(1 + \cos 2\phi) \tag{1.43}$$

Next is about the case where light source produces two pairs of degenerate photons. At this point, the incoming photon state is $|2\rangle_a|2\rangle_b$; after the first directional coupler, the photon state becomes $\frac{\sqrt{\frac{3}{4}}(|4\rangle_c|0\rangle_d+|0\rangle_c|4\rangle_d)}{\sqrt{2}} + \frac{1}{2}|2\rangle_c|2\rangle_d$. Here, the primary concern is the coincidence counting corresponding to $|3\rangle_g|1\rangle_h$ or $|1\rangle_g|3\rangle_h$. At the second directional coupler, quantum interference makes only $|4\rangle|0\rangle+|0\rangle|4\rangle$ contribute to the items $|3\rangle_g|1\rangle_h$ and $|1\rangle_g|3\rangle_h$. So, it can be predicted that three photons are measured at one of the g and h ports, while the coincidence counting rate for one photon measured at the other port is

$$P_{3g,h} = P_{g,3h} = \frac{3}{8}(1 - \cos 4\phi) \tag{1.44}$$

Three-photon counting at port g (implemented using a fiber-optic beam splitter and three detectors) and single-photon coincidence counting at port h are experimentally measured. Figure 1.8 shows the change of the four-photon counting rate with phase ϕ, and the change period is $\frac{\pi}{2}$ (Fig. 1.58).

It is worth noting that in the above analysis the maximum entangled state $|N\rangle|0\rangle+|0\rangle|N\rangle$ of the photon's path is considered. After the phase shifter the state changes to $|N\rangle|0\rangle + e^{iN\phi}|0\rangle|N\rangle$, and at this point the overall phase sensitivity is $\Delta\phi = \frac{1}{N}$, which excels the classical limit $\Delta\phi = \frac{1}{\sqrt{N}}$.

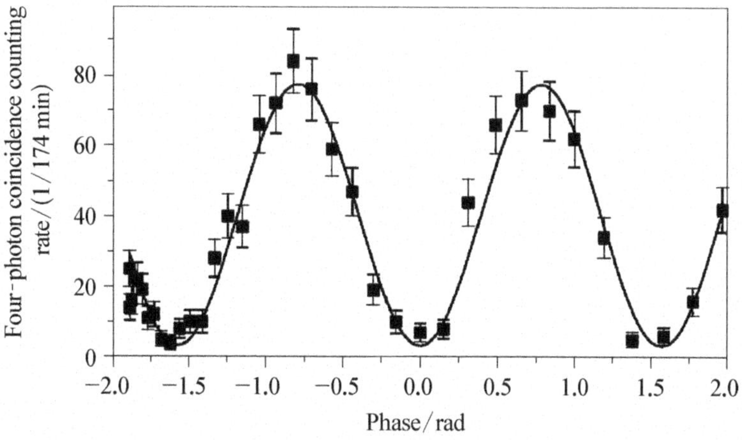

Fig. 1.58 Four-photon interference coincidence curve [65]

1.4.3 Measurement of Optical Quantum States

Quantum measurement is the soul of quantum physics and the cornerstone of all quantum physics-based technologies and applications. Quantum measurement is not simply about detection; it is inherently rich in physical contents, including many fundamental physical problems that have not been solved. It is difficult to understand modern physical experiments without being familiar with quantum measurement theory. In this section, we will describe specifically the measurement devices and methods commonly used in optical quantum computing.

1.4.3.1 Underlying Assumptions of Measurement [68]

The mechanical quantity A of the normalized wave function $\psi(x)$ is always measured according to the family of normalized eigenfunctions $\{\varphi_i(x), \forall i\}$ of operator $\hat{A}$ corresponding to A, as follows:

$$\Psi(x) = \sum_i c_i \varphi_i(x), \left\{\varphi_i(x) | \hat{A}\varphi_i(x) = a_i \varphi_i(x), \forall i\right\} \tag{1.45}$$

The result obtained after a single measurement of the mechanical quantity A will necessarily belong randomly to one of the proper values $\{a_i\}$; at the same time, $\psi(x)$ collapses onto the eigenvector $\varphi_i(x)$ of the corresponding proper value, with the probability of collapse being lc^il^2. All measurements are carried out following this underlying assumption. Specific measurement theories, such as orthogonal projection measurements and general measurements, are elaborated in quite a lot of textbooks and will not be described here anymore. This section mainly considers some measurement devices and means frequently used in optical quantum computation.

1.4.3.2 Single Photon Detector [69]

Single photon detectors, which are experimental devices for photon detection, can usually only detect the presence or absence of photons (a photon number solvable single-photon detector can detect a specific number of photons) and cannot directly give information about the photon quantum state. A combination of other theoretical and experimental means is required to reconstruct the information of light quantum state from the detection results.

Many schemes for optical quantum computing have demanding requirements for single photon detectors. Currently in the visible and near-infrared wavelengths, the energy of a single photon is about 10^{-19} J. Achieving accurate detection of particles of such low energy is a challenging task. Early single-photon detectors were primarily photomultiplier tubes (PMT), but with the development of materials science and quantum information science, the models of single photon detectors have become increasingly diverse. Before introducing particular single photon detectors, we will first describe the main indicators of single photon detector performance: characteristic wavelength range, dead time, dark count, detection efficiency, time jitter, and photon number resolving power.

The characteristic wavelength range of a detector refers to the range of spectral frequencies to which the detector can respond. Current single-photon detectors are all sensitive to only a certain wave band of photons, which is determined by the detector production materials and processing technology, and the spectral response characteristics of the detector also determines its application range. For example, for quantum communication in free space, the photon wavelengths used are mainly concentrated in the visible or near-infrared spectrum, requiring a detector sensitive to this band; for optical communication, on the other hand, as the optical fiber in the 1550 nm wavelength has the minimum loss, the efficiency of the detector must be high enough to detect 1550 nm photons in optical fiber-based quantum information networks.

When a detector discovers a photon, it cannot respond to a new photon for a certain period of time, which is called the detector's dead time. For optical quantum computing, the detector's dead time should be as short as possible. In the current technical conditions, the dead time depends on the detector's electronic post-processing system rather than its photosensitive material. For example, with regard to a SPAD, when the detector finds a photon, it needs to suppress the post-pulse signal brought by the signal detected. The detector must then be turned off for a period of time, and cannot be turned on until the post-pulse signal from the previous detector has been almost eliminated, which is the dead time of the SPAD. The dead time of the detector determines the maximum counting rate of the detector.

When no photons enter the detector, the detector still has a counting rate, which is called dark count. Dark counts arise from defects in the photographic material, voltage bias, and external environmental disturbances (e.g., thermal noise). They have a direct effect on the SNR ratio of an experiment, and so reducing dark counts is a vital target in the development of single photon detectors.

Detection efficiency refers to the probability that when a photon enters the detector it can be detected. Improving the detection efficiency is of great importance in almost all quantum information research, especially in the field of optical quantum computing, which in principle requires perfect single photon detectors.

Time jitter means the time wobble during the process when different photons enter the detector and are converted into an output signal. If the detector's time jitter is obvious while the measured clock period is short, it is possible that the photons detected in the previous cycle end up being counted in the counts of the next cycle, which can cause errors. Therefore, alleviating time jitter is also an important goal for single photon detectors.

Photon number resolving power indicates the ability of the detector to give the exact number of photons arriving at the same time. Presently, most single photon detectors are vacuum state detectors that can only tell the existence and absence of photons other than the number of arriving photons. Currently, for linear optical quantum computing, as the entangled light source under the SPDC process is probabilistic, there are cases where multiple pairs of photons are generated simultaneously, leading to the reduction in the SNR ratio. Optical quantum computing would be greatly developed if single photon detectors with efficient photon number solvability were available. A couple of single photon detectors of different types are presented below.

1. Photomultiplier tube [70]

Photomultiplier tubes are the first generation of single photon detectors that contain a vacuum tube as well as a series of electrodes in the tube. It works by using the photoelectric effect, which means that when a photon enters the detector, there is a certain probability that an electron will be excited on the photocathode. Due to the existence of a series of electrodes between the cathode and anode of the detector, there is a strong potential difference between the adjacent electrodes; the closer to the anode, the higher the potential. As a result, the electrons are accelerated in the electric field to hit the electrodes, thereby exciting more electrons and multiplying the electron numbers. This allows a certain amount of current to form at the anode, ultimately enabling the detection of photons. This type of detector is still in widespread use today. The photomultiplier tube covers the band of about 115 ~ 1700 nm, the maximum detection efficiency can reach 40%; the minimum dark count is around 100/s, while the time jitter is 300 ps or so, which can be reduced to 20 ps through special design; the maximum counting rate is up to 10 MHz, which is able to distinguish single photons in theory, but the resolution is limited.

2. Single photon detector based on avalanche diode (SPAD) [71]

SPADs operate in Geiger mode. When photons arrive at the detector, they excite an avalanche of carriers to form a current that can be processed, thus enabling single photon detection. SPADs are the most widely used detectors for both visible and near-infrared bands. Most of these detectors have a relatively high detection efficiency in the response wave band (currently the maximum detection efficiency of commercial detectors is about 70%, such as PerkinElmer's SPCM), with a minimum dark count of about 25/s, a counting rate of roughly 30 MHz, and a time jitter of 400 ps or so. These detectors are used in the majority of the current experiments of optical quantum computing.

3. Superconducting single photon detector (SSPD)

Superconducting single photon detectors include superconducting transition edge sensors (TES) [72] and superconducting nanowire single photon detectors (SNSPD) [73].

The photosensitive element of the TES is a superconducting thin film at the superconducting transition edge. The detector is normally in a biased state; when a photon enters the detector, the local area of the detector is heated due to the absorption of the photon energy, causing the heating region to change from superconducting to non-superconducting so that the detector current gets a jump signal, which is received and amplified by the SQUID amplifier. At a determined wavelength, the size of the current variation is proportional to the number of photons, and so TES is photon number solvable. Current research findings show that TES can clearly resolve the signals of up to 8 photons, and what's more, it has very high detection efficiency (as high as 95%@1550 nm) and very low dark counts. Therefore, TES is expected to be widely used in the research of linear optical quantum computing. Superconducting nanowire single photon detectors are made to be superconducting by growing a series of bent and aligned nanowires on a substrate and then lowering their temperature. The principle of detecting photons is also similar to that of TES: When a photon enters the detector, there is a certain probability that it will be absorbed by the nanowires, thus creating a local heat source in the detector and destroying the superconducting properties at that point. Since a constant current is maintained in the superconducting nanowire at all times, when the nanowire is out of the superconducting state, a small resistance is produced to heat up the area, enabling a larger area of nanowires to get out of the superconducting state. The detection of a single photon is completed when the resistance in the nanowire increases to such an extent that an observable voltage is generated at both ends of the nanowire. Current relevant developments have demonstrated that up to 57% of detection efficiency, 30 pS time jitter, 1G counting rates, and 10 Hz dark counts can be achieved.

4. Single photon detector based on quantum dot and semiconductor defect [74, 75]

Currently, single photon detectors based on quantum dots or semiconductor defects mainly achieve single photon detection by capturing the charge of quantum dots or defects in the heterojunction of semiconductors of family III–V compounds. The quantum dots that can be controllably distributed in the heterojunction to maximize the detector's internal efficiency and guarantee the best output signal consistency are the main research direction of single photon detectors based on quantum dots or semiconductor defects.

There are two main physical mechanisms of quantum dot detectors: one uses a light-induced photoconductive gain mechanism to change the conductivity of a field-effect transistor by the charge captured by the defect or quantum dot. This type of detector is currently able to achieve a counting rate of 400 kHz, has an intrinsic efficiency of about 68% in the 805 nm band [75], and can resolve three photons at most. Another type is using the absorption of quantum dot light to change the tunneling probability in resonant tunneling diodes so as to realize single photon detection. This class of detectors features 12% detection efficiency at 550 nm, a time wobble of 150 ns, and only extremely low dark counts. Quantum dot detectors are still in the early stages of development; although the dark counts are quite small, both the detection efficiency and the maximum counting rate are low while the time jitter is large. Moreover, they need to operate at low temperature (usually below 4 K) and still lag behind other types of detectors.

5. Frequency up-conversion detector [7]

Due to the limitations of photographic materials, the detection efficiency of single photon detectors based on indium gallium arsenic (InGaAs) in the 1550 nm band is very low, while fiber-optic quantum communications have a strong demand for 1550 nm detectors. Therefore, scientists put forward a new solution: apply frequency up-conversion effect to convert the 1550 nm photons into photons near 775 nm, and then adopt SPAD for detection. As long as the efficiency of the frequency up-conversion is high enough, the detection efficiency of 1550 nm photons can be greatly improved (Table 1.3).

1.4.3.3 Single-Photon Projection Measurement

The principle of single-photon projection measurement is very simple. Take the measurement devices used by our team for example: As shown in Fig. 1.59, the single-photon projection measurement system consists of a polarization analyzer, a fiber optic coupler, a single photon detector, and a multi-channel coincidence counter. The polarization analyzer comprises a 1/4 wave plate (WP), a HWP, and a PBS. In principle, the polarization analyzer made up by the above optical elements is capable of projecting the polarization quantum state of any single photon for measurement.

Experimentally, we collected photons at both outputs of the PBS using couplers connected to a single mode fiber (two detectors are used to "probe" individual photons).

The single mode fiber allows spatial filtering of photons to ensure the indistinguishability of the spatial patterns of photons, which is also a post-selection process. The collected photons are directly detected by a single photon detector via optical fiber transmission. The single photon detector we adopted is an SPCM manufactured by PerkinElmer, with a detection efficiency of up to 60% at 780 nm. In addition, our measurement system could automatically filter out the noise of multiple pairs of photons, tremendously improving the contrast ratio. Take the measurement of entangled photon pairs produced by parametric down-conversion as an example: assuming that the number of SPDC photons obeys a Poisson distribution and that a probability of a single pulse generating a pair of entangled photons is p, the probability of a single pulse producing two pairs of intertwined photons simultaneously is $\frac{1}{2}p$. Then, suppose that the probability of a photon being detected is η (collection efficiency

Table 1.3 Comparison of the performances of different types of single photon detectors [69]

Detector type	Working temp./K	Detection efficiency (η)	Time jitter Δ(FWHM)	Dark count D (no gate-control)	Figure of merit	Max counting rate	Distinguish photo numbers?
Photomultiplier tube (visible-near-infrared bands)	300	40%@500 nm	300 ps	100 Hz	1.33×10^7	10 MHz	Yes
Photomultiplier tube (infrared band)	200	2%@1550 nm	300 ps	200 kHz	3.33×10^2	10 MHz	Yes
Silicon based single photon detector (thick junction)	250	65%@650 nm	400 ps	25 Hz	6.5×10^7	10 MHz	No
Silicon based single photon detector (shallow junction)	250	49%@550 nm	35 ps	25 Hz	5.6×10^8	10 MHz	No
Indium-gallium-arsenic single photon detector (gated mode)	200	10%@1550 nm	370 ps	91 Hz	2.97×10^5	10 kHz	No
Indium-gallium-arsenic single photon detector (self-differential mode)	240	10%@1550 nm	55 ps	16 kHz	1.14×10^5	100 MHz	Yes
Frequency up-conversion detector	300	9%@1550 nm	400 ps	13 kHz	1.7×10^4	10 MHz	No
Frequency up-conversion detector	300	2%@1550 nm	40 ps	20 kHz	2.5×10^4	10 MHz	No
Visible light photon detector	6	88%@694 nm	–	20 kHz	–	–	Yes
Visible light photon detector	6	34%@633 nm	270 ps	7 kHz	1.83×10^5	–	Yes

(continued)

Table 1.3 (continued)

Detector type	Working temp./K	Detection efficiency (η)	Time jitter Δ(FWHM)	Dark count D (no gate-control)	Figure of merit	Max counting rate	Distinguish photo numbers?
Superconducting transition edge sensor (STES)	0.1	50%@1550 nm	100 ns	3 Hz	1.67×10^6	100 kHz	Yes
STES	0.1	95%@1550 nm	100 ns	–	–	100 kHz	Yes
Superconducting nanowire detector	3	0.7%@1550 nm	60 ps	10 Hz	1.16×10^7	100 MHz	No
Superconducting nanowire detector	1.5	57%@1550 nm	30 ps	–	–	1 GHz	No
Quantum dot detector (resonant-tunneling diode)	4	12%@550 nm	150 ns	2×10^{-3} Hz	4×10^9	250 kHz	No
Quantum dot detectors (field effect transistor)	4	68%@805 nm	–	–	–	1 Hz	Yes

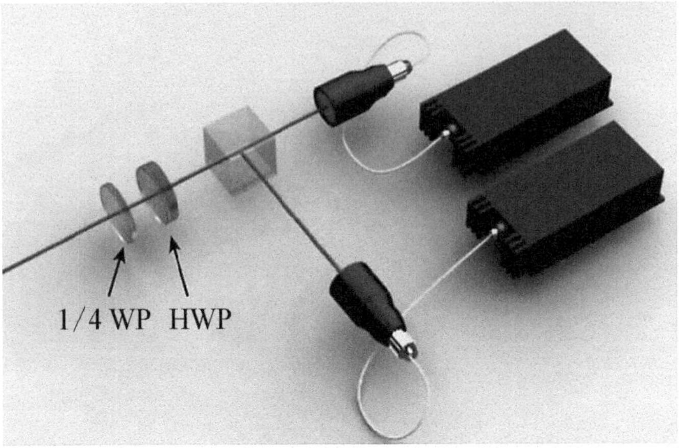

Fig. 1.59 Schematic diagram of single-photon projection measurement devices

multiplied by the quantum efficiency of the detector). If the previous measurement system is adopted, the visibility of a single pair of entangled photons that can be obtained by a simple calculation is approximately 1-p on the condition that only second-order term is considered. If the measurement system we designed is used, the three- and four-photon events will be automatically filtered out as the coincidence counter records one and only one photon at the two output ports of the PBS, ultimately improving the visibility of a single pair of entangled photons to $1 - p(1 - \eta)^2$. When the total detection efficiency reaches 100%, the noise of multiple pairs of photons will be completely filtered out.

1.4.3.4 Null Method of Measurement

Null method of measurement [76] is an efficient optical measurement method with single photon resolution power: the measured signal photon interferes with a local oscillating laser beam on a beam splitter. First, measurement is performed with a linear photodetector at the two output ports of the beam splitter, and then the output signal of the detector is differenced to obtain the final measurement result. This method can make full use of the characteristics of the linear photodetector to achieve near 100% high-efficiency single-photon resolution measurement, and have been extensively applied in the fields of optical quantum computing based on continuous variables and quantum communications [9]. Among countless zero-difference measurement schemes, the balanced homodyne method has obvious advantages in reducing experimental device errors and the jitter of the laser field. In this section, we will introduce the specific working principle of the balanced homodyne method from a simplified model.

Balanced homodyne measurement devices are plotted in Fig. 1.60. A signal and a coherent laser beam acting as a local osillator interfere on a strictly balanced beam splitter (50:50). Two identical linear photodetectors are placed at the two output ports of the beam splitter to detect the light signal and to difference the detector output signal. The local oscillator plays a very important role in heterodyne measurements: it amplifies the signal light and provides a sufficiently accurate reference phase. Therefore, it must be strong enough relative to the signal light to allow us to treat it in the same way as a classical laser field while ignoring its quantum jitter. Next, the relative phase between the local osillator and the signal light must remain stable and may be changed for measurement. Although this condition can be fulfilled by phase locking, its application is somewhat difficult. Thus, in experiments, both the local oscillator and the signal light come from the same laser beam.

Here, we assume that the photocurrent signals exiting from the linear photodetector are I_l and I_2, respectively, and that the current difference $I_{2l} = I_2 - I_l$ including the interference components of the signal light and the local oscillator is the desired measurement. Assuming that the output currents I_l and I_2 are proportional to the number of photons n_l and n_2 reaching the detector, there are $\hat{n}_1 = \hat{a}_1'^{\dagger}\hat{a}_1', \hat{n}_2 = \hat{a}_2'^{\dagger}\hat{a}_2'$, wherein $\hat{a}_1' = \frac{1}{\sqrt{2}}(\hat{a} - \alpha_{LO})$, $\hat{a}_2' = \frac{1}{\sqrt{2}}(\hat{a} + \alpha_{LO})$. As a result, the current difference I_{2l} is proportional to the photon number difference $\hat{n}_{21} = \hat{n}_2 - \hat{n}_1 = \alpha_{LO}^*\hat{a} + \alpha_{LO}\hat{a}^{\dagger}$, wherein a_{LO} is the complex amplitude of the local oscillator field. We know that the generation or annihilation operator corresponds to orthogonal parameters $\hat{p}, \hat{q}$, in phase space, i.e., $\hat{q} = \frac{1}{\sqrt{2}}(\hat{a}^{\dagger} + \hat{a})$, $\hat{p} = \frac{i}{\sqrt{2}}(\hat{a}^{\dagger} - \hat{a})$ and $\hat{q}_{\theta} = \hat{q}\cos\theta + \hat{p}\sin\theta$, $\hat{p}_{\theta} = -\hat{q}\sin\theta + \hat{p}\cos\theta$. Assuming that θ is the phase

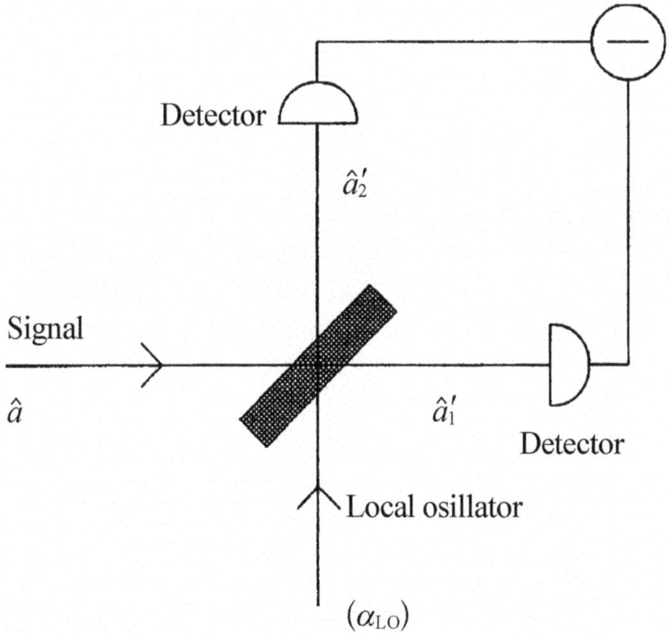

Fig. 1.60 Diagram of balanced homodyne measurement devices [76]

of the local oscillator filed, a very important conclusion can be drawn by simple derivation:

$$\hat{n}_{21} = \sqrt{2}|\alpha_{\mathrm{LO}}|\hat{q}_\theta \quad (1.46)$$

It can be seen in Eq. 1.46 that balanced homodyne method actually measures the orthogonal component $\hat{q}_\theta$ in the phase space of signal photons. In the experiment, the reference phase θ is provided by the local oscillator, and it is possible to change θ by changing the optical length to obtain the distribution of the signal photon $\hat{q}_\theta$. For instance, in the case where the signal photon is a vacuum state, the distribution of the measured $\hat{q}_\theta$ should obey a Gaussian distribution. It is important to note that the sum of the output signals namely $I_l + I_2$ of two photodetectors must be preserved for better analysis and processing of data. In addition, a "balanced homodyne detector" performs the same signal amplification function as other quantum state measurement instruments. For example, because the local oscillator is so strong that it is larger than the electrical noise of the photodetector, the signal can still be observed when the signal photons are single photons. In addition to signal amplification, balanced homodyne measurement method also selects the temporal and spatial modes of signal photons—the local oscillators only "picks" the temporal and spatial modes that match it in a continuous signal quantum light field. In this way, the observer can isolate specific quantum models from the whole system. An in-depth analysis of this application requires a more comprehensive consideration of the theory of quantum electrodynamics. Interested readers are encouraged to consult relevant literature.

1.5 Recent Advances in Optical Quantum Computing

1.5.1 Boson Sampling Experiment

Quantum computers are expected to solve computational problems that classical computers are unable to solve, and although scientists have made great strides on the road to making quantum computers, preparing a fully scalable quantum computer is still an impossible task under current technological conditions. Recently, scientists have proposed a new optical quantum computing scheme—boson sampling [77]. It is a localized quantum computing approach that uses quantum optics to study the interference behavior of n input photons in a m × rn optical network unitary matrix so as to know the final distribution of output states. It is worth mentioning that, unlike quantum computers, boson sampling requires only an indistinguishable single-photon source, linear optical elements, and a single photon detector to implement sampling that is relatively intractable or unattainable for classical computers.

Boson sampling can be described like this: n homogeneous bosons (photons), after passing through a complex linear optical network of several "*MZ* interferometers" (the network can be characterized as the linear orthonormal transformation U of n input photons, and the number of its input/output ports m must be greater than n), performs coincidence measurement on the output ports of this network to obtain the probability distribution of n photons simultaneously appearing at different output ports. It has been theoretically proven that this probability distribution corresponds to the permanent of the matrix U, which for classical computers is an #P problem overriding the *NP* complexity. The proposal of boson sampling has opened a new chapter in the research of optical quantum computing, and scientists from various countries are currently working on the realization of boson sampling of multiple photons. In 2012, four independent experimental groups in Europe and Australia experimentally demonstrated the principle of boson sampling for the first time [78–81]. We are going to introduce these four experiments in this section.

1.5.1.1 Basic Theory [77]

If you assume that n photons are sent to a linear optical network ($m > n$) with m input/output ports and define its computational basis vector as $|i_1, i_2, \ldots i_m\rangle$, there will be a formula of $i_1 + i_2 + \ldots + i_m = n$. Assume that the input state is $|i\rangle = |i_1, i_2, \ldots, i_m\rangle$, which, after being transformed by the $m \times m$ linear optical network U, will have a certain probability to get the output state $|O\rangle = |j_1, j_2, \ldots, j_m\rangle$. This probability is $P_{\mathrm{I,o}} = |\langle O|U|I\rangle|^2 = \frac{|\mathrm{Per}(U_{\mathrm{I,o}})|^2}{i_1!i_2!\cdots i_m!j_1!j_2!\cdots j_m!}$, wherein $\mathrm{Per}(\mathrm{U_{I,O}})$ represents the permanent of the matrix.

We take as an example the interference behavior of two photons in a 3×3 linear optical network. Assume that the network U takes the form $\begin{pmatrix} a & b & c \\ d & e & f \\ g & h & i \end{pmatrix}$: if photons enters the first two input ports of the network, and the input state is $|I\rangle = |1, 1, 0\rangle$; if the photons exiting the last two output ports of the network, the output state is $|0\rangle = |0, 1, 1\rangle$. The columns of the matrix U correspond to the input ports and the rows the output ports. Since there are no photons at the port 3 of the input state and the port 1 of the output state, we can eliminate the elements of the third column and first row of matrix U and consider only the submatrix $U(I, O) = \begin{pmatrix} d & e \\ g & h \end{pmatrix}$, then the distribution probability of this output state is $\mathrm{P_{I,O}} = \frac{|\mathrm{Per}(\mathrm{U_{I,o}})|^2}{1!1!0!0!1!1!} = |dh + eg|^2$. By traversing all the combined forms of the input/output ports, it is possible to obtain the permanent of all the submatrices and completely reconstruct this matrix.

1.5.1.2 Experimental Realization of Boson Sampling

In 2012, Philip Walther's team in Austria implemented the boson sampling of three independent photons in a 5×5 linear optical network [78], and reconstructed this 5

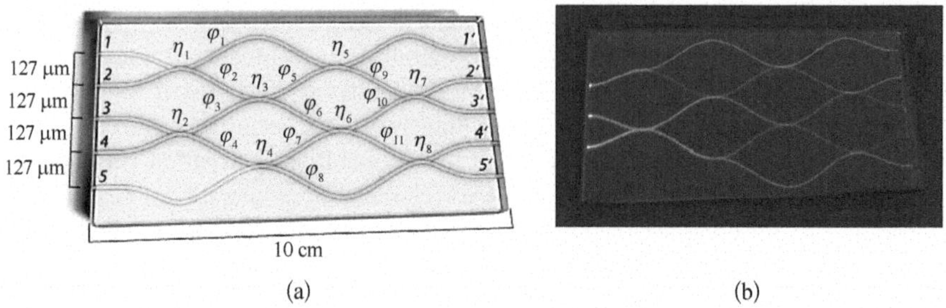

Fig. 1.61 Waveguide chip used in the experiment [78]

× 5 matrix according to the photon interference results. The 5 × 5 optical network in the experiment was completed using a waveguide structure. With the laser writing technique described previously, eight directional couplers and 11 phase shifters with different reflectivity identical to each other were implemented on a waveguide chip about 10 cm long, with an adjacent input/output port at a distance of 127 μm, and the output port was directly connected to a single photon detector, as shown in Fig. 1.61a. In the experiment, the reflectance of the directional coupler was characterized by the light and shade of the fluorescence imaging of the chip, as plotted in Fig. 1.61b.

The photons input to the waveguide chip were prepared by the parametric down-conversion process. The entire initial-state preparation process was divided into two steps: a beam of pump light with a central wavelength of 394.5 nm, a pulse width of 150 fs, and an average power of 700 mW passed through a 2-mm thick non-collinear BBO crystal to produce four-photon states via SPDC process. A 3 nm filter was used to ensure the indistinguishability of photons in the experiment.

$$|\Psi\rangle_{a,b} = (|HH\rangle_a|HH\rangle_b + |HV\rangle_a|HV\rangle_b + |VV\rangle_a|VV\rangle_b)/\sqrt{3} \tag{1.47}$$

In the equation above, a and b are the path modes of photons. As shown in Fig. 1.62, the prepared four-photon state is divided by a PBS, and the four photons are distributed in four different paths a′, a″, b′, and b″. Then, they directly measured "a" as the trigger signal, and the other three photons would collapse to a state $|H\rangle_{a'}|H\rangle_{b'}|V\rangle_{b''}, b''$. Photon b″ then hit a half-wave plate, and the polarization was rotated to H to ensure that a′, b′, b″ photons could be simultaneously input to the waveguide chip with the same polarization state. After photons entered the waveguide chip shown in Fig. 1.61a and transformed by this optical interference network, the output photons would be measured directly by the single photon detector.

The experimental results are displayed in Fig. 1.63. In the experiment, they prepared two different waveguide chips to represent two different U matrices. The input ports of three photons were fixed at 3, 4, and 5, and the U matrices were rebuilt by traversing the coincidence probability distributions of all the output ports. Figure 1.63a and b present the experimental results of the first waveguide chip (U matrix), a showing the probability distribution of three photons at the output ports

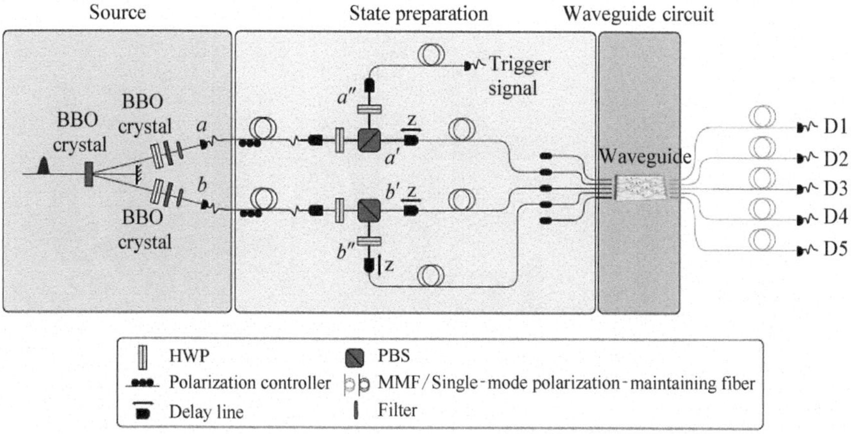

Fig. 1.62 Boson sampling experimental facility [78]

while b referring to the reconstructed U matrix. Figure 1.63c and d exhibit similar results with the only difference being the waveguide chip (U-matrix).

1.5.1.3 Boson Sampling on Optical Quantum Chips

Ian Walmsley's team at the University of Cambridge has also implemented boson sampling on optical quantum chips [79]. Figure 1.64 is a schematic representation of the quantum boson sampler (QBSM) used in their experiment. Its primary role is to obtain the probability distribution of the output state $|\psi_{\mathrm{OUT}}\rangle$ after the evolution of the linear unitary matrix A after an initial state $|T\rangle$ is given. The initial state $|T\rangle$ can be written as $|T\rangle = |T_1, \ldots, T_M\rangle \propto \prod_{i=1}^{M} \left(\hat{a}_i^\dagger\right)^{T_i}$, which describes the case where $N = \sum_{i=1}^{M} T_i$ particles distributed over M input ports.

The quantum devices mainly consisted of an indistinguishable single-photon source, an optical quantum interference network containing six input/output ports (a waveguide chip with 10 directional couplers, as shown in Fig. 1.65a), and a single photon detector. The experiment considered only the case where the numbers of output and input photons were equal. In order to rebuild element $\Lambda_{ij} = \tau_{ij}e^{i\phi_{ij}}$ of the matrix Λ, several single-photon and two-photon transmission measurements need be made experimentally. First, to determine the amplitude τ_{ij}, a single photon should be emitted at input port i, and the probability of detecting this photon at output port j is $P_1(j, i) = \left|\Lambda_{ij}\right|^2 = \tau_{ij}^2$; to determine the phase ϕ_{ij}, two-photon interference need be measured. If a single photon was emitted at input port i_1, i_2, then the probability of observing one and only one photon at output ports j_1 and j_2 was $P_1(j_1, j_2, i_1, i_2) = \left|\Lambda_{i_1 j_1}\Lambda_{i_2 j_2} + \Lambda_{i_1 j_2}\Lambda_{i_2 j_1}\right|^2$. Putting the previously obtained

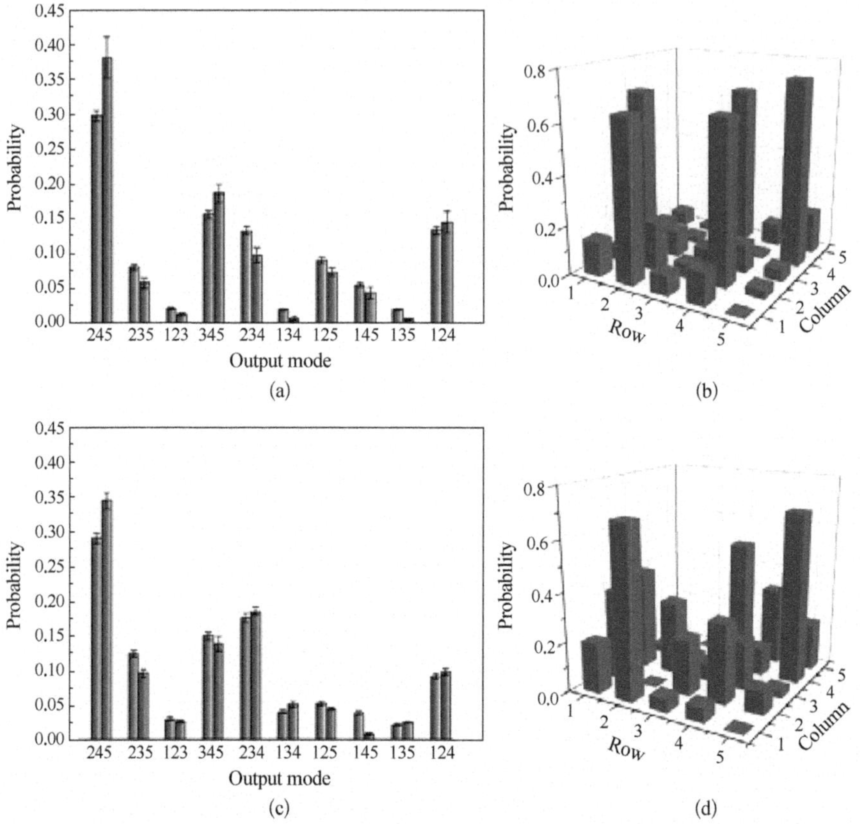

Fig. 1.63 Plot of the results of boson sampling experiment [78]

amplitude in the equation would get a corresponding phase ϕ_{ij}. Figure 1.65b and c are submatrices reconstructed from the experimental results.

They completed the three-photon and four-photon boson sampling on two identical waveguide chips. For the three-photon boson sampling, the input ports were 2, 3, and 5, and the quantum state was $|T\rangle = |011010\rangle$, repeatedly injected into the waveguide chip. All possible three-photon coincidence counts were collected at the output port. As in Fig. 1.66a, the gray line is the experimentally obtained distribution probability P^{exp}, and each set of output states are collected for 160 h; light grey shading is the error bar for the experimental data; the black line is the result Pth calculated from the permanent of the reconstructed matrix, while the dark grey shading indicates errors and comes from the uncertainty of the reconstructed matrix. For the four-photon boson sampling, they entered two photons at ports 1 and 3, with a quantum state of $|T\rangle = |202000\rangle$. Similarly, they collected all possible four-photon coincidence counts at the output ports, as listed in Fig. 1.66b.

Finally, to analyze the performance of this boson sampling quantizer, they compared the difference between the ideal device and the experimental results. This

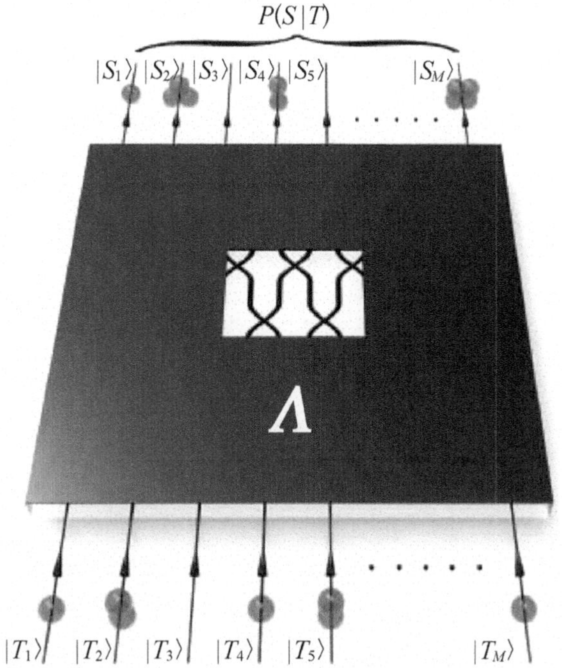

Fig. 1.64 Schematic diagram of boson sampling quantum device [79]

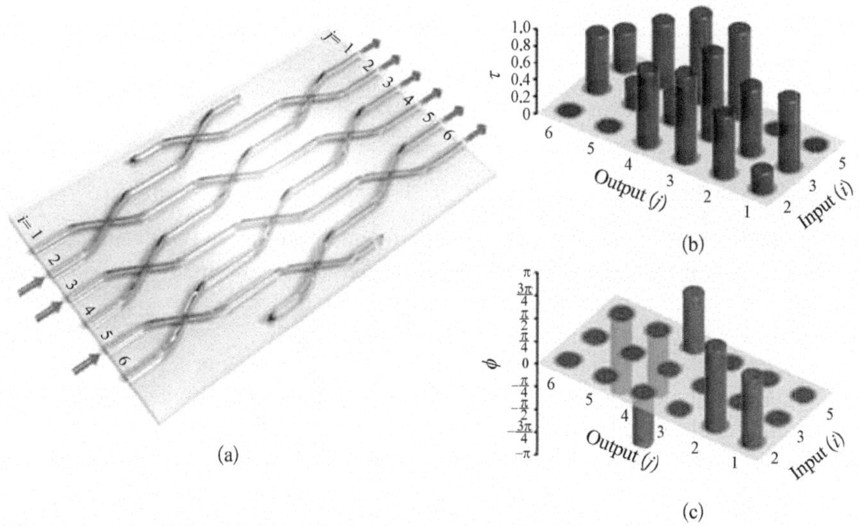

Fig. 1.65 Structure diagram of optical quantum interference network [79]

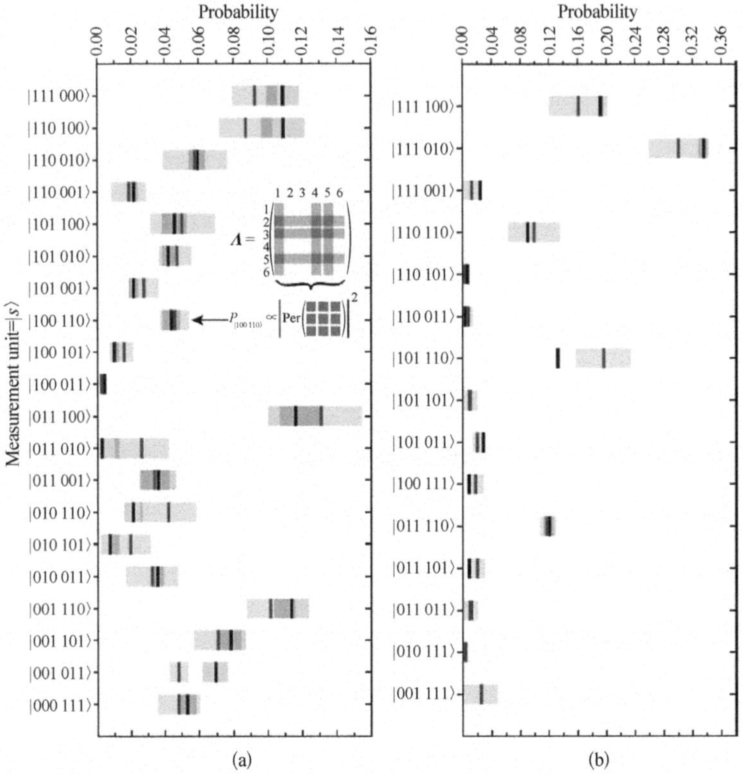

Fig. 1.66 Plot of results of bosonic sampling experiment [79]

difference is quantified using $d^{(N)}\left(p^{\exp}, p^{\text{th}}\right) = \frac{1}{2}\sum_{s}\left|p_s^{\exp} - p_s^{\text{th}}\right|$: when $d^{(\text{N})} = 0$, it means that theoretical calculations and experimental tests are in perfect agreement; when $d^{(\text{N})} = 1$, it indicates a complete lack of agreement. In the three-photon and four-photon cases, the calculation results are $d^{(3)} = 0.094 \pm 0.014$ and $d^{(4)} = 0.097 \pm 0.004$.

The imperfection of input photons also affects the performance of the quantizer. In terms of parametric down-conversion (PDC) entanglement sources, the form of output state can be expressed as $|00\rangle + \lambda|11\rangle + \lambda^2|22\rangle$. Although $\lambda \leq 1$, for a three-particle input state, it is entirely possible to produce imperfect input states such as $|T\rangle = |0121010\rangle$ and $|T\rangle = |012020\rangle$, thus resulting in errors. In addition, if the input photons are not completely indistinguishable, the interference behavior of the photons can also be affected to cause errors.

1.5.1.4 Photonic Boson Sampling in Tunable Lines

In an experiment by Andrew White's team in Australia [80], they carried out the boson sampling of two- and three-photons in a 6×6 tunable line. In the experiment, the probability distributions of the output states of n input photons entering the evolution matrix U were calculated and compared by classical computation and boson sampling experiments respectively. Assuming that Alice uses the classical computer and Bob boson sampling, the final comparison to be made is: (i) the speed with which each of the two methods is completed for the case where n is relatively large; (ii) the accuracy of the computation (sampling). For (i), since Alice's computational complexity increases exponentially as the number of n photons increases, Bob will win the race. For (ii), there is no effective way to verify this accuracy so far, but we can consider the case: select a large n, let Bob finish the boson sampling and Alice compute and verify Bob's result.

In the experiment, a 6×6 matrix composed of two orthogonally polarized 3×3 fused optical fiber-type beam splitters was built, and the computational basis vector was $\{1, \ldots, 6\} = \{|H_1\rangle, |V_1\rangle, |H_2\rangle, |V_2\rangle, |H_3\rangle, |V_3\rangle\}$, which allows the use of a polarization controller to prepare the input states.

In Fig. 1.67, (a) is a diagram of the 6×6 unitary matrix U, (b) is its equivalent line, and (c) is a sketch map of the experimental facilities. Two pairs of entangled photons are generated through PDC, and photon 4 is directly measured as a trigger signal. Photons 1, 2 and 3 are fed into the interference network, and $a_l \ldots b_3$ unitary operations can be implemented by polarization control; $U^{(3)}$ is realized with a 3×3 fiber-optic beam splitter, and the six output ports are measured through an SPAD. From the experimental data of transmitting interference of single- and two-photons, Alice obtained the reconstructed unitary matrix U, and then computed the permanent of U to get the probability distribution of the output photons. Bob, on the other hand, performed boson sampling experiment directly, during which the differences in the quantum efficiency of the individual detectors could introduce errors into the experimental results; to remove the possible bias in the experimental results that this difference might introduce, they used a visibility $V_T = \frac{P_T^C - P_T^Q}{P_T^C}$ to replace the original method of output probability distribution; P_T^C and P_T^Q corresponded to the output probability distributions of fully distinguishable and fully indistinguishable particles, respectively (these two quantities could be measured experimentally by adjusting the delay of the optical path).

The experimental results of three-photon boson sampling are shown in Fig. 1.68. To compare the difference in the results obtained by Alice and Bob through their respective methods, they introduced $L^1 = \frac{1}{C(m,n)} \sum_T \left| V_T^A - V_T^B \right|$ to quantify this difference: for the two-photon input state, there was $\overline{L}^1 = 0.021 \pm 0.001$; for the three-photon input state, there was $\overline{L}^1 = 0.122 \pm 0.025$.

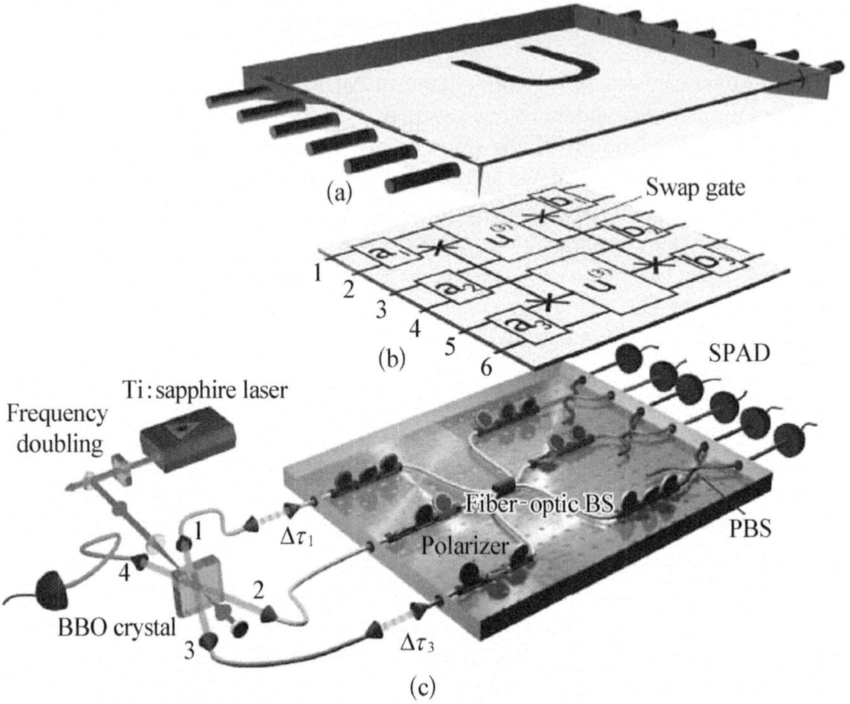

Fig. 1.67 Experimental scheme and apparatus for boson sampling

1.5.1.5 Boson Sampling on Arbitrarily Integrated Optical Quantum Circuits

Fabio Sciarrino's team in Italy similarly completed the boson sampling of three fully homogeneous photons on an integrated 5×5 optical chip in 2012 [81]. Figure 1.69 is the schematic representation of a 5×5 quantum interference network. In their experiment, 10 directional couplers with different transmittances were prepared on fused silica substrates by laser writing technique in order to achieve a stable quantum interference network.

In the experiment, in addition to independently controlling 10 directional couplers with different transmittances, 15 different phases must be precisely controlled, i.e., $\alpha_i, \beta_i \in [0, \pi]$. When processing the waveguide chip, the geometry of the waveguide had to be changed to obtain different transmittances, but at the same time, the optical path might also be changed, resulting in phase deviations. In specific experiment, the team modulated the transmittances and phases of the directional couplers by changing the spacing between the waveguides and the bending shapes of the waveguides. As shown in Fig. 1.70, different phases were obtained by controlling the shapes of the waveguides in the light gray region, and different transmittances were obtained by varying the spacing between the waveguides in the gray region.

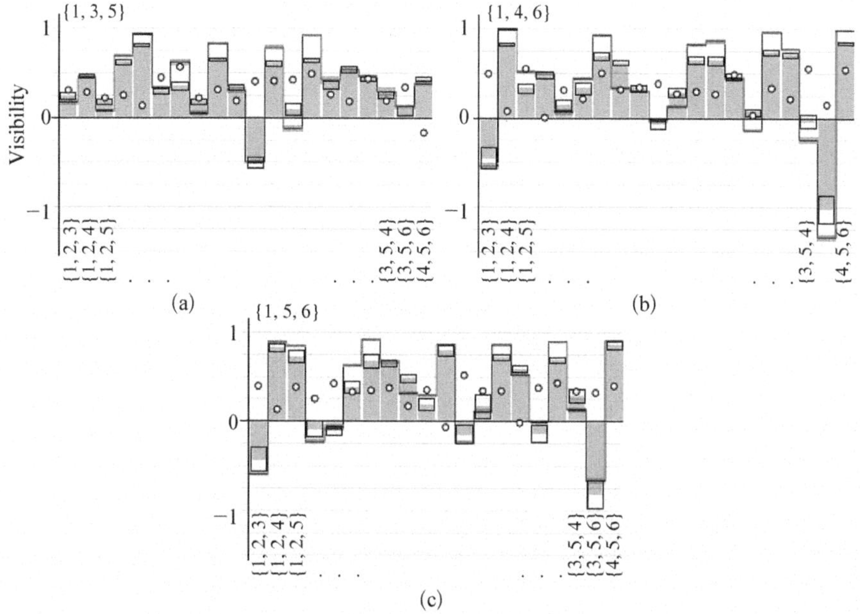

Fig. 1.68 Experimental results of three-photon boson sampling [80]

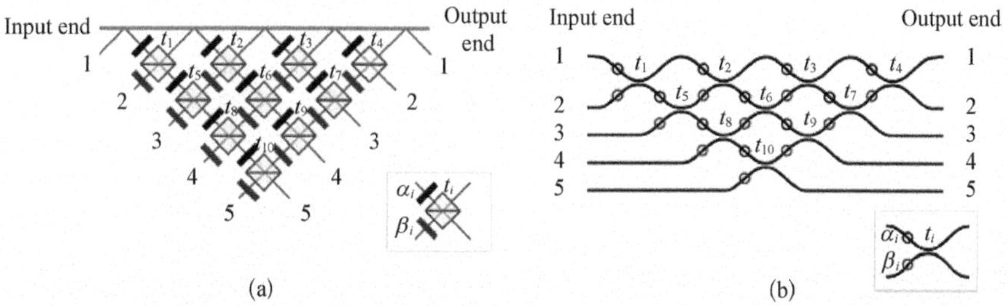

Fig. 1.69 Schematic of a 5 × 5 quantum interference network [81]

Once the chips are fabricated, the parameters of this quantum interference network need to be characterized theoretically and experimentally. As shown in Fig. 1.71, they quantified the parameters of this waveguide chip using a photon preparation and detection system consisting of detector (APD), filter (IF), polarization controller (PC), polarizing beam splitter (PBS), entanglement source (PDC), half-wave plate (HWP), delayer, etc. With respect to a single photon input, they used $\mathrm{P}^1{}_{\exp}(i,K)$ to describe the experimental probability of a single photon coming to port i and going out of port K; similarly, the results calculated from the sampling matrix theory were given by $P_t^1(i, K)$ (see Fig. 1.72a). The probability distributions of the experimentally and theoretically obtained single-photon output states were denoted by p and q, respectively, which could be used to represent the overlap $S^1_{\exp,t} = 0.946 \pm 0.005$

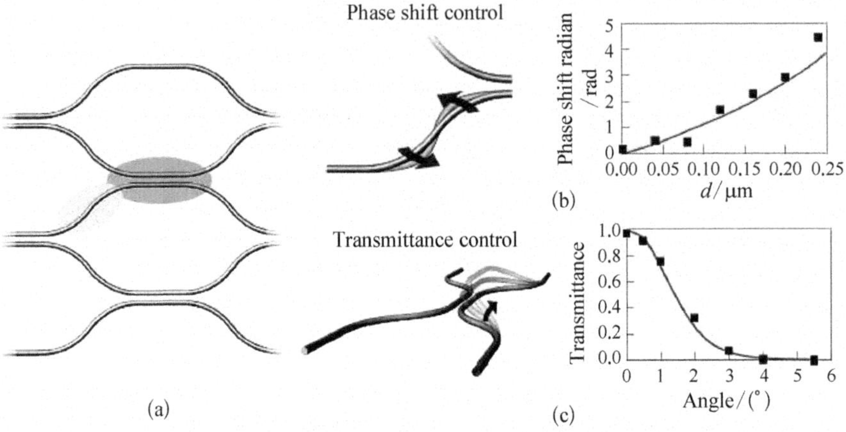

Fig. 1.70 Modulation of waveguide directional coupler transmittance and phase shifter

between the actually prepared matrix and the ideal matrix. For two-photon input, the situation was more complex, including a total of 100 different interferences. Similarly, $P^2_{\exp}(i, j, K, L)$ and $P^2_t(i, j, K, L)$ signify the experimental and theoretical probability distributions, respectively, and ultimately yield an overlap of $S^2_{\exp,t} = 0.901 \pm 0.025$, as drawn in Fig. 1.72b.

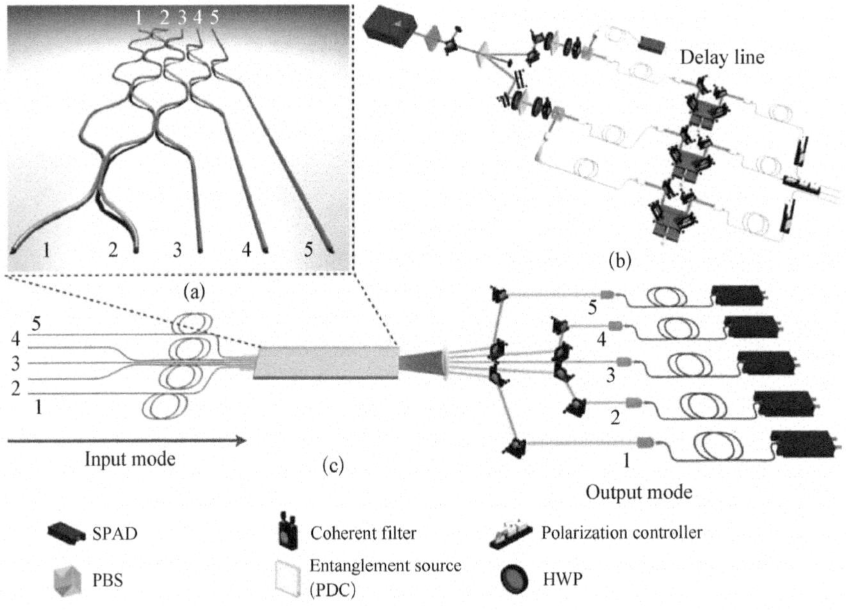

Fig. 1.71 Schematic diagram of experimental setup [81]

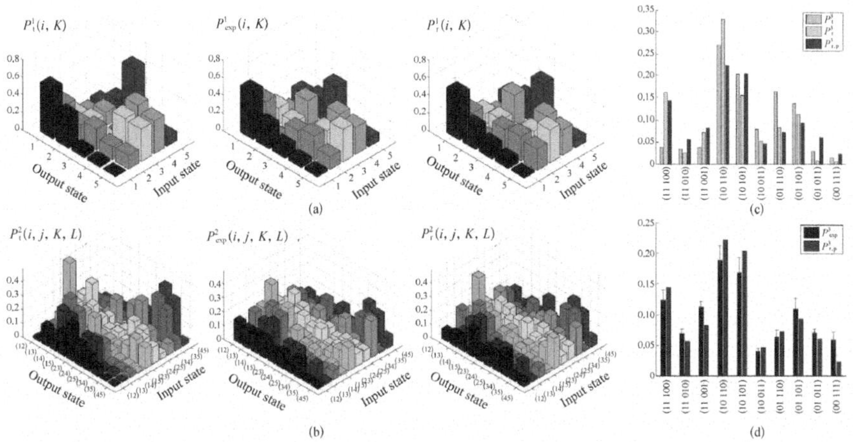

Fig. 1.72 Results of single-photon/two-photon/triple-photon boson sampling experiments [81]

It is perfectly feasible to reconstruct the evolution matrix U using experimental results of single- and two-photon input states, and the team also ended up comparing the degree of overlap between the ideal and reconstructed matrices, $S^1_{\mathrm{exp},r} = 0.990 \pm 0.005$ and $S^2_{\mathrm{exp},r} = 0.977 \pm 0.027$, respectively. What's more, the team performed the boson sampling of three-photon input states. Figure 1.72c indicates the probability distribution of the output state when the three-photon input states were theoretically computed to be $|10101\rangle$; Fig. 1.72d is a comparison map of the experimental output state P^3_{exp} and the reconstructed state $P^3_{r,p}$ when the input state is the three-photon state $|10101\rangle$. The experimentally obtained overlap is $S^3_{\mathrm{exp},rp} = 0.983 \pm 0.045$, wherein the subscript rp stands for $P^3_{r,p}$, which is the probability distribution deduced from the reconstructuring matrix based on the consideration that photons are still somewhat distinguishable in the actual experiment.

Although the above four experiments simply demonstrate that useful calculations can be achieved using quantum systems even without quantum logic operations, their impact on the research of light quantum computing is undoubtedly far-reaching. Scientists from all over the world are currently working on achieving boson sampling for larger photon numbers. Theoretical research show that, with the increase of N, when the number of photons for boson sampling experiments reaches 20 or so, boson sampling based on optical systems would likely be the first "quantum computer" to outperform classical computers in a real sense.

1.5.2 Topological Quantum Error Correction Experiments [21]

Quantum computers, due to its powerful parallel computing capacity beyond the limits of classical computers, have become the coveted goal of quantum physicists in the twenty-first century. However, the biggest problem recognized by the academic community as having long plagued its physical implementation—"decoherence effect" (due to the inevitable coupling of quantum computers with the environment, a variety of noise is produced to cause numerous errors in the computational process)—has not been solved yet. To overcome this difficulty, quantum error correction techniques need to be developed. Threshold theory or threshold effect, as the most important result in quantum error correction theories, elaborates that in a quantum computer, it is possible to realize arbitrarily long and precise quantum computations if P is less than some threshold value P_C under the circumstance that every gate would go wrong at probability P. However, most quantum fault-tolerant calculations with high fault-tolerance thresholds ($10^{-4} \sim 10^{-2}$) require long-range strong interparticle interactions and are therefore experimentally difficult to achieve, while fault-tolerant quantum computing schemes that require only local interactions typically correspond to low fault-tolerance thresholds. For traditional concatenated coding based on two-dimensional crystal lattice qubits and nearest-neighborhood logic gates, the highest known fault-tolerance threshold is 2.02×10^{-5}.

In such a two-dimensional crystal lattice, it is preponderant to perform topological quantum error correction based on the theoretical framework of cluster-state topological quantum computation. The scheme exploits the topological nature of three-dimensional (3D) cluster states, which themselves have a structure that is highly robust to errors, allowing us to implement quantum computation and simultaneous error correction through local measurements. By combining active error correction and topological methods, it is possible to obtain a fault-tolerance threshold of up to 0.7–1.1% [19, 20] and a loss-tolerance threshold of up to 24.9% [82]. The topological quantum error correction scheme has immensely lowered the requirement for operational accuracy and reached a level that is achievable with existing experimental techniques. To implement practical topological quantum computation, we need to build a topological cluster state with a larger number of quantum bits. The three-dimensional structure of the cluster state can be mapped onto two-dimensional space plus one-dimensional time while requiring only nearest-neighborhood interactions. In addition, a different topological quantum computation scheme has recently been developed: it employs non-Abelian anyons to perform quantum computation and achieves fault-tolerant properties through passive stabilization caused due to band gaps in the basic state energy bands. Although some simple quantum error-correcting codes (QECC) have been implemented in NMR, ion trap, and linear optical systems, the puzzles and challenges of how to experimentally realize topological quantum error-correcting codes are what we always confronted with.

Through years of hard work, our team has innovated the technique for the preparation of multiphoton entangled states, and successfully prepared an ultra-high brightness pulsed entangled light source using the Bell-State synthesizer mentioned above, and, with a uniquely designed new eight-photon interferometer that can suppress noise, experimentally prepared for the first time an eight-photon cluster state with the topological properties required by topological error correction. In 2012, based on this topological cluster state, we first experimentally implemented topological quantum error correction [21]: first, each photon was locally measured and then the measurement results were analyzed to construct the fault symptoms of corresponding errant particles, thus protecting corresponding topological quantum correlations. We proved that if only one qubit was in error, it could be identified and corrected; if all qubits were in error with an equal probability, the probability of valid errors in the protected quantum correlation would be significantly reduced. These results signify that we have successfully demonstrated the principle of topological quantum error correction and proved its practicality. In this section, we will describe the experiment in detail.

1. Topological quantum error correction theory

When the most essential interaction diagram is embedded in the 3D structure of so-called unit cell complex, we are able to exploit the 3D cluster states for quantum error correction and fault-tolerant quantum computation [83]. Each unit cell complex includes body, face, edge, and vertex. Quantum bits are encoded in the edges and faces of the unit cell compound form. In the graph state, each face is associated with its corresponding quantum bits at the surrounding boundaries by edges representing Ising interactions. As shown in Fig. 1.73a for a square unit cell (elementary lattice cell) structure, the dashed line refers to the edges of the unit cell complex and the solid line denotes the interactions between face qubits and edge qubits. Qubits (represented by small spheres) are encoded in the faces and edges of a square unit cell made up by a cube, 6 square planes, 12 edges, and 8 vertices. The interacting edges represented by solid lines connect all the face qubits and edge qubits to form an 18-qubit cluster state $|G_{18}\rangle$ with six independent face stabilizers $K_f(f = 1, 2, \ldots, 6)$. According to the definition of the stabilizer code introduced in Sect. 1.3.4, we find that if we multiply the six-face stabilizer codes together, we can reduce all the Z operators and obtain an average value $\langle X_1, X_2, \ldots X_6\rangle = 1$. This equation implies a very important outcome: although X-measurement of each face qubit would lead to a random result $\pm$ 1, multiplying the X-measurements of all the face qubits of the closing surface would obtain a deterministic value of 1. That is, for any closing surface F, there is a deterministic quantum topology association $C_F \equiv \left\langle \underset{f\in F}{\otimes} X_f \right\rangle = 1$, wherein f is the surface of F. Figure 1.73b shows a larger unit cell complex consisting of $5 \times 5 \times \mathrm{T}$ square unit cells. The solid point represents a local Z-measurement, which actually removes the measured quantum bits from the cluster state so as to create a "defect" that encodes a quantum correlation. A hollow dot indicates a Z error, and a hollow dot square unit cell means that it contains an incorrect association. This structure allows us to encode a logic bit in topological quantum computation and propagate it in the

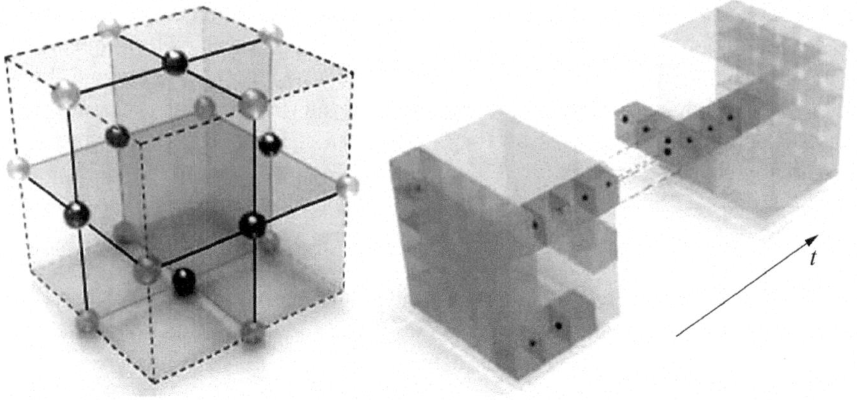

Fig. 1.73 Schematic representation of topological cluster states [21]

direction T that represents the dimension tensed by time t. First, a "defect" along the t-direction (see the solid point in Fig. 1.73b) is generated by a local Z-measurement. Then, it is used through the quantum topology association $C_{F_D} = 1$ on the closing surface F_D around the defect and the boundary of the unit cell complex to encode a logical bit. The evolution of the logical quantum state from time t_1 to t_2 is achieved by performing a local X-measurement of all other qubits between t_1 and t_2. In its true sense, quantum computation requires larger unit cell complexes and more "defects", of which the quantum algorithm is realized by appropriate braiding operations on the "defects".

Since the quantum topology association $C_{F_D} = 1$ holds for any closing surface containing defects, quantum computation based on this structure is possible. The error correction capability of topological error correction arises from the topological property Z_2 coherence of sufficiently large three-dimensional unit cell complexes. For a given defect-containing closing surface F_D, there are multiple equivalent homology closing surfaces that all have the same topological association $C_{F_D} = 1$ We protect the topological quantum correlation by redundant information. This is illustrated below using a simplified 2D model. In Fig. 1.74, the two-dimensional plane includes two point defects (●). The boundary of the defined plane is the sum of all closing surfaces around it. For example, the boundary of a face/(shaded area) is the sum of e_1 and e_2, $\partial f = e_1 + e_2$. Three closing surfaces e_1, e_2, and e_3 are each insufficient to form the boundary of a face due to the presence of point defects. Similarly, the boundary of a line is defined as the sum of its endpoints. As the three curves are closed, they have no boundaries, i.e. $\partial e_1 = \partial e_2 = \partial e_3 = 0$. We can see that the closing surface e_2 can switch to e_1 continuously and vice versa. In other words, the difference between e_1 and e_2 is only two different boundaries $e_2 = \partial f - e_1$ on one face, and we call such relation homology. On the contrary, because of the defect on the right side, e_3, e_1, and e_2 are completely different things.

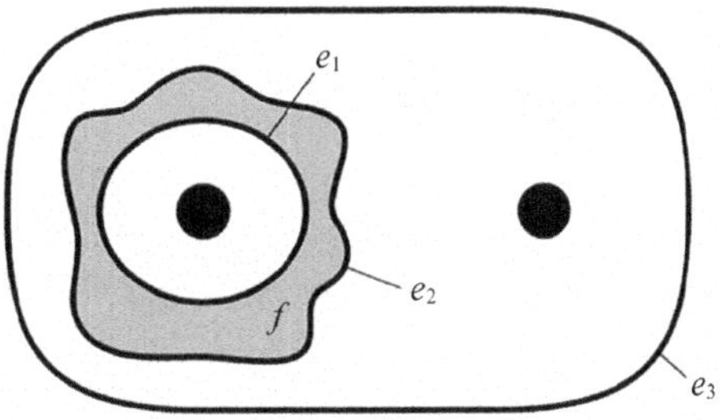

Fig. 1.74 Schematic representation of topological homology in a two-dimensional system [21]

The definition of homology in higher dimensional space is similar to this. In three dimensions, the boundaries of a stereoscopic surface are the sum of the surrounding faces. A closing surface, such as the surface of a ball, has no boundaries, i.e. $\partial f = 0$. If the difference between two surfaces F and F' is only $\partial V = F' - F$, then they are coherent.

In topological cluster state calculations, the error correction scheme involves only the local X-measurements of output $\lambda = \pm 1$. The result of the computation is the product $R(F) = \prod_{a \in F} \lambda_a$ of all outputs on a closing surface $F \to \partial F = 0$. In almost all coding schemes, the ability to correct errors is reflected in the redundancy of the coding. The computation result for a particular qubit can be obtained not only from a particular surface F, but also from the set of all surfaces that are homological with it. This is because two homological facets F and F' have the same association result $R(F) = R(F')$ when no errors occur. So, every individual V should have $R(\partial V) = 1$. If $R(\partial V) = -1$, it means an error has occurred in V, and thus $R(\partial V)$ can be regarded as an error feature. These error features also have a geometric interpretation and are related to homotopy: two homological curves e and e′ play the same role in topological error correcting quantum computation: Two homological curves e and e' play the same role in the topological error correcting quantum computation. Obviously, in topological error correction, we only need to correct Z errors, since any X error has no effect on the protected association, either because an X error is X-measured immediately after it occurs, or this X error is equivalent to the multiplication of multiple Z errors. Ultimately, as a measurement-based process, when we apply topological error correction to topological quantum computation, its error correction is performed for the classical outputs of X-measurements rather than remaining cluster states.

2. Simplified topological cluster state

The unit cell complex structure shown in Fig. 1.73b encodes a quantum logic bit propagation process via the topological association $C\,F_D = 1$ and is robust to local Z-errors, but it contains 25 square unit cells and 180 physical quantum bits, which is far beyond the reach of current experimental techniques. To demonstrate this process, we designed a simplified graph state $|G_8\rangle$ to simulate the unit cell complex displayed in Fig. 1.73b. The topology of $|G_8\rangle$ in Fig. 1.75b consists of four basic bodies $\{v, w, y, z\}$, six faces $\{f_1, f_2, f_3, f_4, f_5, f_6\}$, two edges $\{e_7, e_8\}$, and two vertices $\{s, t\}$. The

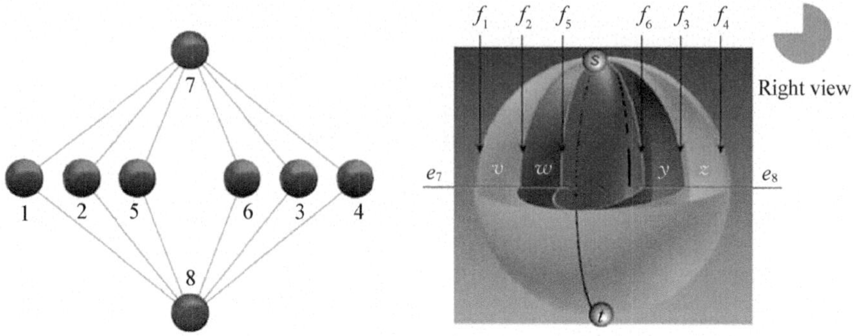

(a) (b)

Fig. 1.75 Cluster state $|G8\rangle$ and its topological structure [21]

six faces have a common edge $e_7 \cup e_8$. Any two of the six faces can form a closing surface F. The body in the middle is absent and the protected topological association used to demonstrate the defect shown in Fig. 1.75b is $C\,F_D = \langle X_5X_6\rangle = 1$.

In this simplified unit cell complex structure, the topological association $C_{FD} = 1$ has been redundantly encoded multiple times, which can be obtained by averaging any $\langle X_iX_j\rangle$, wherein $i \in \{1,2,5\}$ and $j \in \{3,4,6\}$. In addition, there are four other closing surfaces that do not contain "defects", and their topological associations are $\langle X_1X_2\rangle = \langle X_2X_5\rangle = \langle X_3X_6\rangle = \langle X_3X_4\rangle = 1$. This redundant information can be used as an error symptom for single-particle bit errors when the mean value of any one or more of these associations is -1. As shown in Table 1.4, any single qubit Z error can be located and corrected by these error symptoms. Thus, from the perspective of topological quantum error correction, the cluster state $|G_8\rangle$ resembles the structure of the unit cell complex described in Fig. 1.73b. Although the structure of the unit cell complex exhibited in Fig. 1.75b is so small that it cannot propagate a single logical bit, it still protects individual quantum correlations from Z errors.

3. Experimental preparation of eight-photon topological cluster states

In the experiment, we prepared eight-photon cluster states by means of linear optics. The first step of the experiment was the preparation of ultra-high brightness and

Table 1.4 Cluster state $|G8\rangle$ and its error symptoms $\langle X_iX_j\rangle$

Qubits with Z errors	$\langle X_1X_2\rangle$	$\langle X_2X_5\rangle$	$\langle X_3X_6\rangle$	$\langle X_3X_4\rangle$
1	-1	1	1	1
2	-1	-1	1	1
3	1	1	-1	-1
4	1	1	1	-1
5	1	-1	1	1
6	1	1	-1	1

fidelity pulsed entangled photon pairs. As shown in Fig. 1.76a, a UV femtosecond pulsed laser with a power of 915 mW was passed through a BBO crystal to obtain an ultra-high brightness entangled photon pair $|\phi\rangle = \frac{1}{\sqrt{2}}(|HH\rangle + |VV\rangle)$ with the aid of the Bell-State synthesizer described earlier. The four entangled photon pairs are 1–2, 3–4, 5–6, and 7–8 (see Fig. 1.76b). Using these four pairs of entangled sources, two four-photon picture states were prepared: the first one was a four-photon GHZ state, and the coherent superposition of photon 2 and photon 4 on PBS produced the resulting $\left(\left|H^{\otimes 4}\right\rangle_{1\sim 4} + \left|V^{\otimes 4}\right\rangle_{1\sim 4}\right)/\sqrt{2}$. At the same time, let photon 6 and photon 8 be coherently superimposed on the PDBS ($T_H = 1$, $T_V = 1/3$) and then let them pass through two other PDBS ($T_H = 1/3$, $T_V = 1$) for equilibrium amplitude. By the combination of these three PDBS, a controlled phase gate operation could be achieved with a success probability of 1/9, and the prepared four-photon cluster state was $[|HH\rangle_{56}(|HH\rangle_{78} + |VV\rangle_{56}) + |VV\rangle_{56}(|HH\rangle_{78} - |VV\rangle_{56})]/2$. After that, photon $4'$ and photon $6'$ were superimposed on another PBS. When eight photons were simultaneously detected at the output port, the following eight-photon topological cluster state was obtained:

$$|\psi\rangle = \frac{1}{2}\left[|H\rangle_{1\sim 6}^{\otimes 6}(|HH\rangle_{78} + |VV\rangle_{78}) + |V\rangle_{1\sim 6}^{\otimes 6}(|HH\rangle_{78} - |VV\rangle_{78})\right] \quad (1.48)$$

Doing the Hadamard operation on all the quantum bits of the cluster state represented by Eq. (1.48) yields the state indicated by Fig. 1.75a. Note that we used a star-shaped eight-photon interferometer and that all the photons interfering on the PBS and PDBS have the same bandwidth, which are key factors in reducing noise in the experiment. In order to ensure perfect overlap in time and space when interfering with photons, we first spectrally filtered all the photons: photons 1, 3, 5, and 7 passed through a 8 nm fwhm filter, and photons 2, 4, 6, and 8 through a 2.8 nm fwhm filter. A single-mode fiber was then used to collect the photons, which is also a spatial filtering process. Furthermore, the optical path delay was finely tuned so that all interfering photons could reach the PBS and PDBS simultaneously.

The measurement of photons was done with a polarization analyzer comprising a 1/4 wave plate, a half-wave plate, and a PBS crystal that could theoretically project any single-photon polarized quantum state onto the basis vector. We placed a single mode fiber coupler and a single photon detector at each output port of the PBS to collect the detected photons. The output signals from the single photon detector were sent into a homemade field-programmable gate array-based multichannel coincidence counter to simultaneously measure all 256 complete basis vectors of the eight photons. The experimentally obtained eight-body coincidence counting rate was 3.2 pieces per hour. The eight-body coincidence counting obtained from our measurements under the {|H > ,|V>} basis vector are shown in Fig. 1.77a, and the SNR ratio of the experimental results (defined as the ratio of the average value of the desired components divided by the average value of the unwanted components) is 200:1. This clearly shows that we have successfully fabricated eight-photon cluster states.

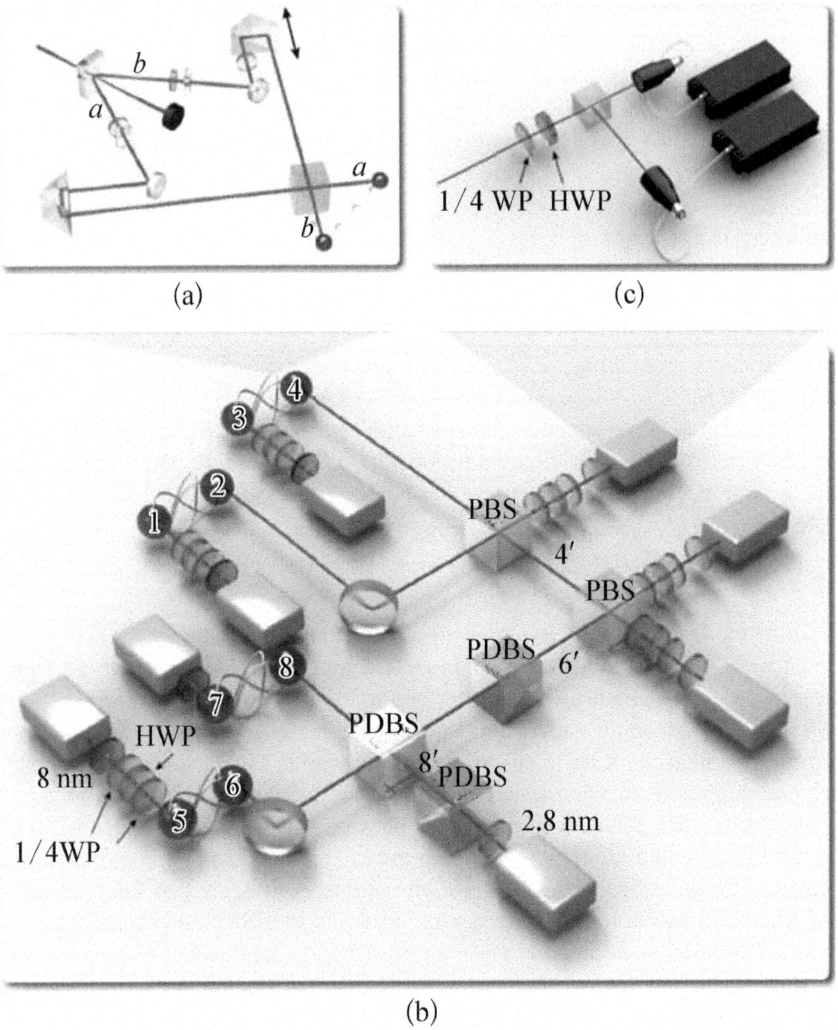

Fig. 1.76 Diagram of experimental setup for preparing eight-photon cluster states and implementing topological quantum error correction [21]

To fully characterize the prepared eight-qubit cluster states, we employed an entanglement eyewitness to verify whether the experimentally prepared states were true multiparticle entangled states. If there exists a sizable measurement W that satisfies the expectation that all two-body separable states are positive and the expectation for the prepared entangled state is negative, it is called entangled eyewitness operator. We devised a special entangled eyewitness operator $W = \frac{1}{2} - |\Psi\rangle\langle\Psi| + |\Psi'\rangle\langle\Psi'|$, wherein $|\Psi'\rangle$ is the orthogonal state of $|\Psi\rangle$, i.e., $\langle\Psi'|\Psi\rangle = 0$. The eyewitness operator was then decomposed onto a series of local von Neumann (or projection) measurements:

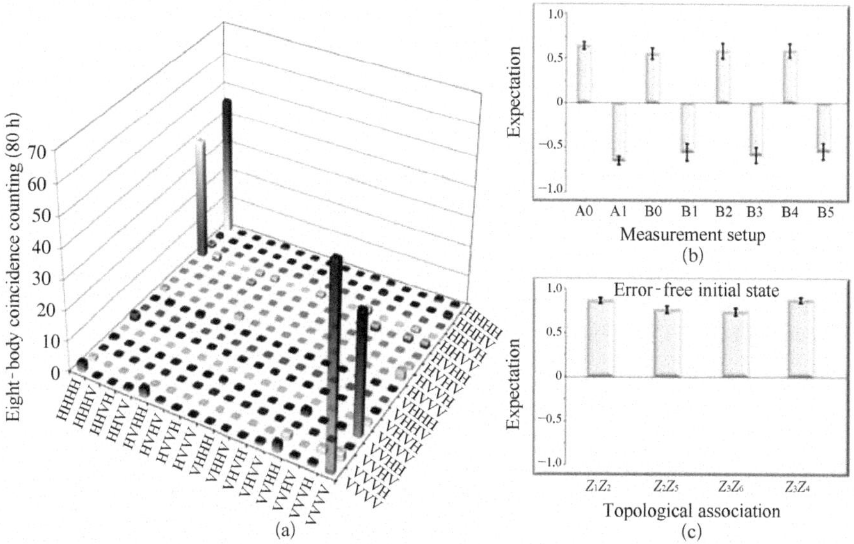

Fig. 1.77 Experimental results of eight-photon cluster states [21]

$$
\begin{aligned}
W &= \frac{1}{2} - |\Psi\rangle\langle\Psi| + |\Psi'\rangle\langle\Psi'| \\
&= \frac{1}{2} - \left[\frac{1}{4}(|H\rangle\langle H|^{\otimes 6} - |V\rangle\langle V|^{\otimes 6})_{1\sim 6} \otimes (X_7X_8 - Y_7Y_8)\right] \\
&\quad - \frac{1}{12}\left(\sum_{k=0}^{5}(-1)^k M_k^{\otimes 6}\right)_{1\sim 6} \otimes \left(|H\rangle\langle H|^{\otimes} - |V\rangle\langle V|^{\otimes 2}\right)_{78}
\end{aligned}
\tag{1.49}
$$

In the formula, $M_k = \cos\left(\frac{k\pi}{6}\right)X + \sin\left(\frac{k\pi}{6}\right)Y$. With this special entanglement eyewitness operator, we lessened the measurement operators to eight, which greatly reduced the difficulty of the experiment. As shown in Fig. 1.77, the mean value of the obtained entanglement eyewitnesses is $<W> = -0.105 \pm 0.023$, up to 4.5 standard deviations. Therefore, the fidelity of the eight-photon cluster states we prepared is $<F> 0.5- <W> = 0.605 \pm 0.023$, a result that reaffirms that the cluster states we prepared are true multiparticle entangled states.

4. Experimental realization of topological quantum error correction

Building around the prepared eight-photon cluster states, we were able to achieve topological error correction by means of local measurements of single qubits and classical error correction operations. In our experiment, we worked on states $|\psi\rangle$ given in Eq. (1.48) rather than on states $|G_8\rangle$, both differing from each other by a set of Hadamard operations $H^{\otimes 8}$. Therefore, the quantum correlation $<X_5X_6>$ be protected corresponded to the experimental association $<Z_5Z_6>$. Similarly, the redundant quantum correlations $<X_iX_j>$ experimentally all corresponded to $<Z_iZ_j>$ respectively. Besides, the X error was used in place of the Z error in the experiment.

In the experiment, the quantum noise channel of polarized quantum bits was simulated by placing a half-wave plate between two 1/4-wave plates. By randomly rotating the angle of the HWP on $\pm\,\theta$, it was possible to simulate a noise channel with an error probability of $P = \sin^2(2\theta)$ for the occurrence of bit flips. First, we studied the case in which only one photon out of six {1, …, 6} had an *X*-error. See Fig. 1.78 for experimentally measured error symptoms, each graph representing one qubit occurring with an *X*-error, and each set of data measured for 80 h. The error bars obtained by analyzing Poisson statistical errors on the original experimental data represent a standard deviation. For comparison, the results of quantum correlations were plotted in Fig. 1.77c when there were no errors. By comparison, it was possible to precisely locate the qubit where *X*-error occurred.

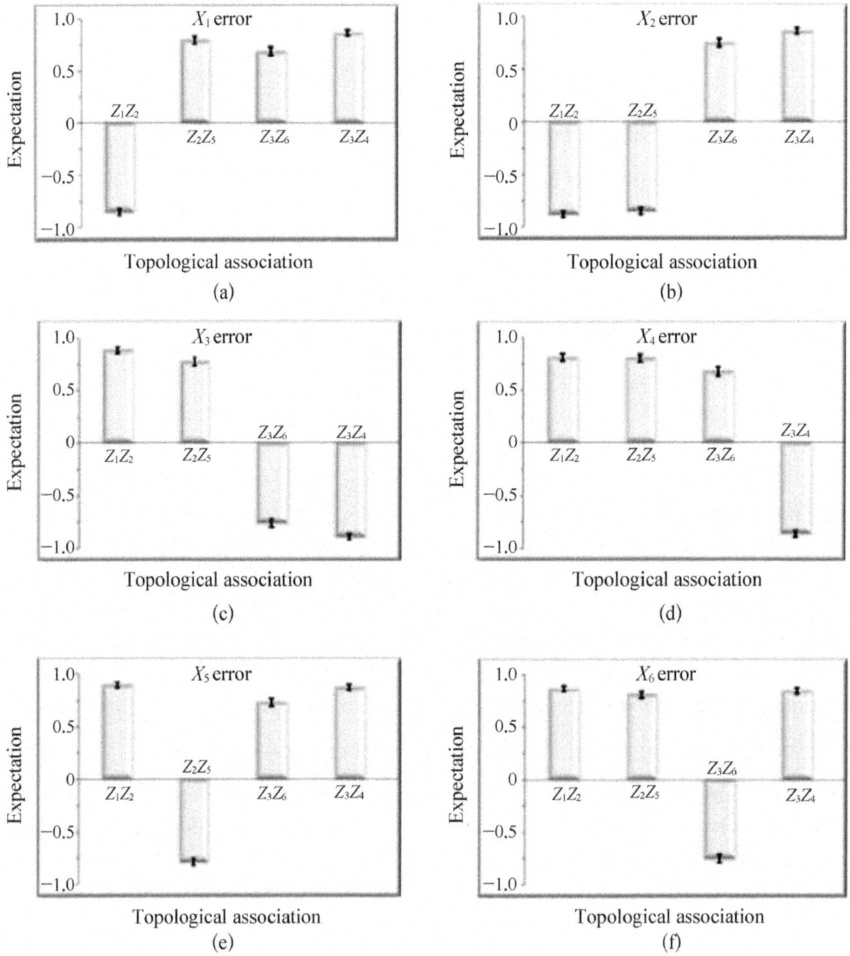

Fig. 1.78 Experimental results for error symptoms in topological error correction [21]

Next, we considered the case where six photons {1, …, 6} had X-error at the same time with the same probability p, and we also took the probability that the protected topological association would go wrong into account, i.e., probability $\langle Z_5Z_6\rangle = -1$. In the absence of error correction methods, the probability of an error in the association $\langle Z_5Z_6\rangle$ is $P = 1 - p^2 - (1-p)^2$, and the probability of an error in the association $\langle Z_5Z_6\rangle$ with error correction is as follows:

$$P = 1 - \left[p^6 + (1-p)^6\right] - \left[6p(1-p)^5 + 6p^5(1-p)\right] - \left[9p^2\left(1-p^4\right) + 9p^4(1-p)^2\right] \tag{1.50}$$

When p is small, the probability of making an error in the protected association would be greatly reduced. In the experiment, the entire measurement process was carried out over 80 days, requiring extremely stable experimental facilities due to the low counting rate of the eight photons. The flaws in the experiment mainly stemmed from the errors caused due to multiple pair effects when preparing the entanglement source. Moreover, the imperfect coherent superposition of photons on the PBS and PDBS would also degrade the quality of the experimental results. Nevertheless, our experimental results have successfully demonstrated the feasibility of topological quantum error correction, as shown in Fig. 1.79, and are in good agreement with the theoretical predictions, allowing the protected association $\langle Z_5Z_6\rangle$ being significantly to be more resistant to noise.

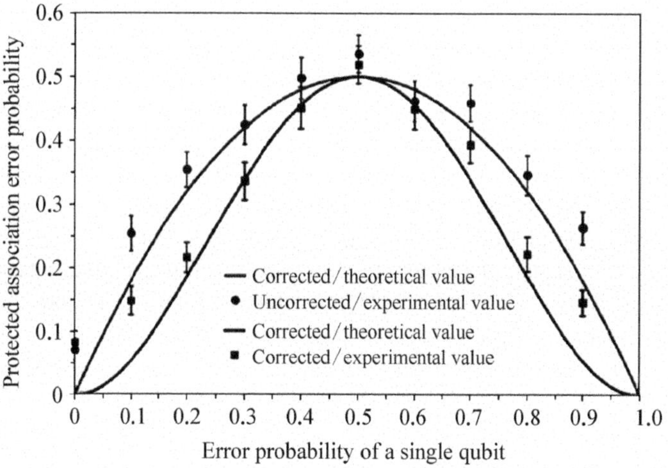

Fig. 1.79 Experimental results of topological quantum error correction [21]

1.5.3 Experimental Realization of Blind Quantum Computation (BQC)

Although so far, with rapidly developing technologies, scientists have implemented quantum computers with a small number of quantum bits in ion traps, linear optics, superconducting quantum circuits, and solid-state systems, the challenges to successfully implement a truly useful quantum computer are still enormous. Thus, initially, it is only possible to build very few quantum computers that need to work under special conditions, i.e., quantum servers. So, how do we make use of these quantum servers to ensure that they fulfill their computational tasks without compromising the data of remote users? It would be natural to wonder if we could use the cryptography that has become more established in classical systems. The answer is no, because the security of a classical cipher depends only on the mathematical computational complexity associated with that cipher protocol and the limits of eavesdroppers' capabilities. In other words, if the mathematical complexity of the cryptographic protocols is low or the eavesdropper's ability is high, there is a risk of information leakage, and so the absolute security of user data cannot be guaranteed in principle using classical cryptography. To crack this puzzle, scientists have introduced a new approach to quantum computing—blind quantum computation [84]. The BQC protocol cleverly ties quantum computers with quantum cryptography so that a user who does not have a strong quantum computing power can not only fulfill a computing task through the quantum server but also ensure absolute data security.

We know that cluster states are the core resource for one-way quantum computation, and that deterministic universal quantum computation can be realized as long as single qubit measurements are performed on the cluster states during the computation process. Different measurement schemes correspond to different quantum algorithms. A natural idea arises that the whole quantum computation process would be safe if the measurement process can be guaranteed to be safe. In order to achieve this purpose, we need to find a special entanglement state that will not reveal computational information and can be used for one-way quantum computation. This is called a blind cluster state. Its preparation process is as follows: all qubits are initialized into state $\left|\theta_j\right\rangle = \frac{1}{\sqrt{2}}\left(|0\rangle + e^{i\theta_j}|1\rangle\right)$, in which $|0\rangle$ and $|1\rangle$ denote the computational basis vectors of physical qubits, and θ_j is a phase chosen randomly from $\left\{0, \frac{\pi}{4}, \ldots, \frac{7\pi}{4}\right\}$ at an equal probability; then, a C-phase gate is acted between each particle and its nearest neighbor. To implement blind quantum computation, we must act a series of successive algorithm-dependent single qubit measurements on blind cluster states. For example, measure the first qubit in state $|\theta_1\rangle$ under the basis vector $|\pm\delta_1\rangle = \frac{1}{\sqrt{2}}\left(|0\rangle \pm \mathrm{e}^{\mathrm{i}\delta_1}|1\rangle\right)$. This is equivalent to performing a rotation $R_z(-\delta_1 + \theta_1)$ on the encoded single-bit input state $|+\rangle$, followed by an H transformation. Here, $R_z(\Phi) = \mathrm{e}^{-\frac{-i\phi\sigma_z}{2}}$, $H = \frac{\sigma_x+\sigma_z}{\sqrt{2}}$, $\sigma_x, \sigma_y, \sigma_z$ are Pauli operators. As long as θ_I is unknown, the entire rotation is absolutely confidential.

This feature of blind cluster states allows users to "delegate" their tasks to a quantum server to do the computation, while ensuring that the user data is completely private throughout computation. Users only need the ability to prepare the initial states of single qubits—to prepare each particle j into the state $|\theta_j\rangle$ and transmit them to a quantum server—a process that in principle does not require quantum storage or the ability to perform quantum gate operations. After the initial state preparation and transfer process, they only need to issue measurement instructions to the quantum server via a classical channel. The quantum server first implements controlled phase gate operations on the quantum state received from users, and then performs a single-qubit measurement according to users' measurement instructions. During each interaction, the server would make a projection measurement of $|\pm\delta_j\rangle$ for single qubits according to the instructions. The measurement basis vector is determined by $\delta_j = \Phi_j + \theta_j + \pi r_j$, wherein Φ_j is a rotation angle desired to be performed on the target and r_j is a random number selected from {0,1} for the purpose of hiding the last measurements from eavesdropping. These classical measurement instructions compensate for the initial random rotating angle θ_j as well as for the unwanted "byproducts" of the previous measurement.

Next, we will present the blind quantum computing experiment done in 2012 by Anton Zeilinger's team in Austria [85]. The experiment is based on a linear optical experimental platform for a simple reason: users need to transmit quantum states to a quantum server, and quite naturally, photons are the best supporter to accomplish this task. In the experiment, they implemented the concept of blind quantum computation using prepared four-photon blind cluster states, including single-qubit/two-qubit gates, Deutsch's algorithm, and Grover's algorithm.

A resource-optimized blind quantum computation scheme was adopted in the experiment. The theory shows that it is not necessary to keep all the quantum bits secret from the servers during the preparation of blind cluster states; in other words, it is not necessary to prepare all the initial qubits as $|\theta_j\rangle$. For the four-photon blind cluster state, the users only need to prepare one or two qubits to any $|\theta_j\rangle$ to implement fully confidential single- and two-qubit circuits and simple quantum algorithms. They fixed θ_1 and θ_4 to 0 and varied the values of θ_2 and θ_3 to obtain a four-qubit linear blind cluster state:

$$\begin{aligned}\left|\Phi^{\hat{\theta}}\right\rangle = \frac{1}{2}[&|+00+\rangle_{1234} + \mathrm{e}^{i\theta_3}|+01-\rangle_{1234}\\ &+\mathrm{e}^{i\theta_2}|-10+\rangle_{1234} - \mathrm{e}^{i(\theta_2+\theta_3)}|-11-\rangle_{1234}]\end{aligned} \tag{1.51}$$

In the above formula, $\hat{\theta} = (n_2, n_3)$, $(\theta_2, \theta_3) = \left(\frac{n_2\pi}{4}, \frac{n_3\pi}{4}\right)$. The experimental setup is shown in Fig. 1.80, where they first generate a four-photon entanglement source by pumping the same BBO crystal back and forth, and then post-select through the polarization beam splitter to obtain a four-photon entangled state, which differs from $\left|\Phi^{\hat{\theta}}\right\rangle$ only the local unitary transform $H \otimes I \otimes I \otimes H$.

$$\begin{aligned}\left|\Phi_L^{\hat{\theta}}\right\rangle = \frac{1}{2}[&|0000\rangle_{1234} + e^{i\theta_3}|0011\rangle_{1234}\\ &+e^{i\theta_2}|1100\rangle_{1234} - e^{i(\theta_2+\theta_3)}|1111\rangle_{1234}]\end{aligned} \tag{1.52}$$

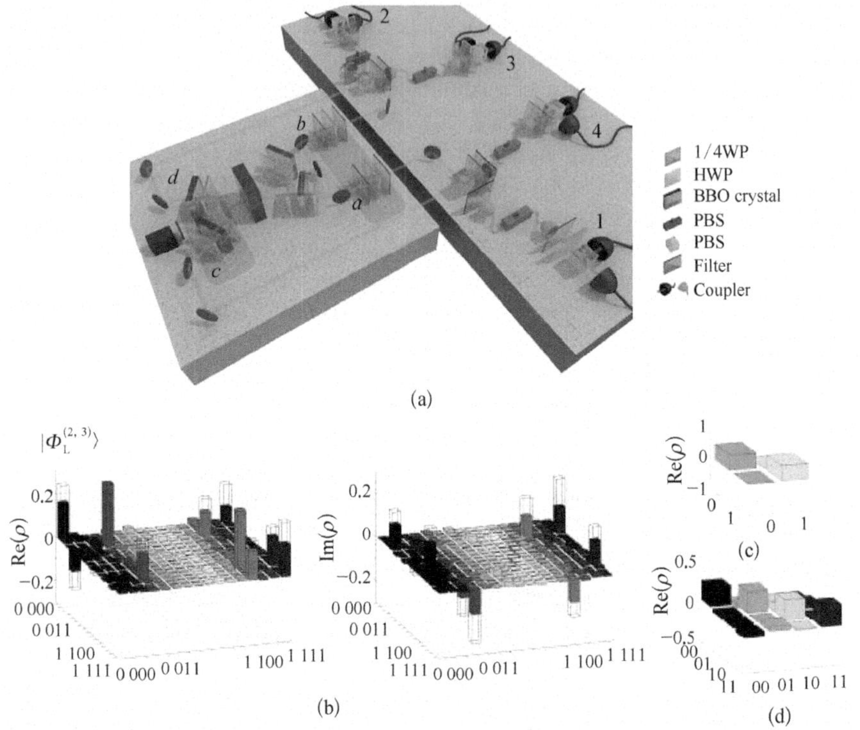

Fig. 1.80 Experimental setup for blind quantum computation and preparation results of blind cluster states

Users need to prepare the states according to θ_j. In this experiment, they prepared a variety of four-photon blind cluster states in the form of $\left|\Phi_L^{(2,n)}\right\rangle, n = 0 \sim 7$ and $\left|\Phi_L^{(6,m)}\right\rangle, m = 0 \sim 4$. In addition, they selected 1962 different measurement bases from a number of measurement basis vectors and obtained 31,392 measurements, thus enabling all possible branches of computation (i.e., r_j traversed all cases). To validate the blind cluster states prepared, they used the ultra-complete quantum state tomography technique to reconstruct the density matrix of the four-photon blind cluster states, as plotted in Fig. 1.80. The maximum fidelity of the experimentally obtained blind cluster states was 0.679 ± 0.004.

The four-photon linear cluster state $\left|\Phi^{\hat{\theta}}\right\rangle_{\rightarrow}$ could be used to achieve an arbitrary single-qubit unitary transformation (see Fig. 1.81). Take X rotation as an example: by measuring the first qubit under the basis vector $\sigma_x, \sigma_y, \sigma_z$, the second qubit could be prepared on the states $|0\rangle$, $|+_i\rangle$ and $|+\rangle$ respectively, in which $|+_i\rangle = 1/\sqrt{2}(|0\rangle + i|1\rangle)$; next, by measuring the second and third qubits, they obtained $R_x(-\phi_3)R_z(-\phi_2)|\Psi_{\text{in}}\rangle$, wherein $R_x(\alpha) = \exp(-\frac{i\alpha\sigma_x}{2})$. Fixing θ_2 and varying θ_3 enabled them to implement blind single-qubit X rotation.

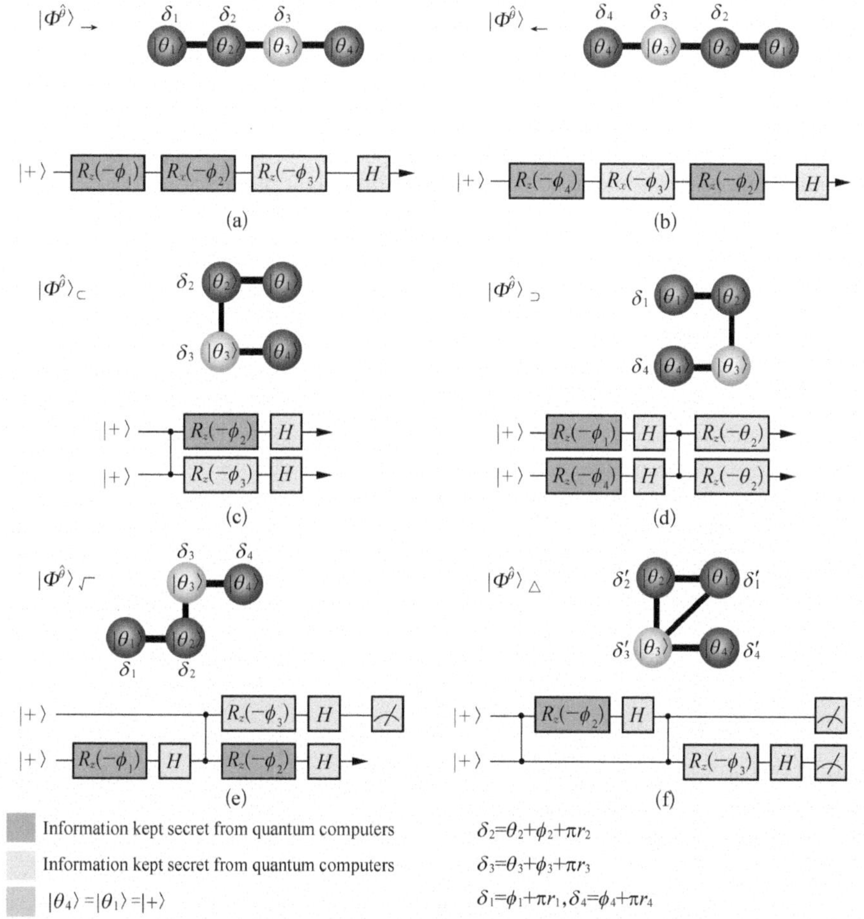

Fig. 1.81 Blind quantum computation lines and corresponding measurement schemes [85]

Further, they experimentally verified the security of the blind quantum computation by using eight blind cluster states prepared $\left|\Phi_{\mathrm{L}}^{(2,n)}\right\rangle$. When the computation was totally confidential, the Holevo information was 0; when the computation was totally unconfidential, the Holevo information was 3. The experimentally obtained Holevo information was 0.169 ± 0.074, well verifying the confidentiality of the scheme. As shown in Fig. 1.81, they also implemented various types of quantum computations such as two-qubit gate operations, Grover's algorithm, and Deutsch-Josza algorithm using different $\left|\Phi^{\hat{\theta}}\right\rangle$ and measurement schemes in the experiment. Interested readers are suggested to read the original literature further. We are not going to dwell on it here.

Although the experiment has theoretically validated the feasibility of the blind quantum computation scheme, many questions still need to be addressed for real application: how to select θ_j completely randomly? How to prepare the initial state without noise? How do we solve photon loss? Nevertheless, the implementation of blind quantum computation will have a profound impact on future information security, cloud quantum computing, etc. As long as users can communicate properly with quantum servers, they will be able to enjoy the powerful computing power of quantum computers while maximally safeguarding their privacy.

1.5.4 Optical Implementation of Shor's Algorithm

As we have described in Sect. 1.2.1, Shor's quantum factoring algorithm is by far the most important quantum algorithm that enables the quantum computer to exhibit its unparalleled advantages. Researchers have implemented a proof-of-principle verification $15 = 3 \times 5$ of the Shor algorithm in linear optics. In this section, we will introduce two experiments with the optical Shor's algorithm, one platform-based experiment done by our team and the other waveguide-based experiment done by O'Bnen's team at the University of Bristol, UK.

1.5.4.1 Platform-Based Experiment with Shor's Algorithm [14]

1. The principle of Shor's algorithm [24]

Suppose that our task is: to decompose $\mathrm{N} = \mathrm{p} \times \mathrm{g}$, where N is an l-bit number whereas p and g are both prime numbers.

Algorithmic strategy:

Step 1: Pick a random number a (0 < a < N), with no factor in common with N.

Step 2: calculate the period r of function $f(x) = a^x \mathrm{mod} N$, that is, r is the smallest integer to meet the condition $a^r \mathrm{mod} N = 1$. This function $f(x)$ is known as modular exponential function (MEF).

Step 3: Calculate the greatest common divisor between N and $a^{r/2} \pm 1$ using a simple Euclid algorithm. This common divisor has a larger than 50% probability of being the prime factor to be found. This step can be done quickly with just a classical computer.

The real need for the quantum computer to play an accelerating role in Shor's algorithm lies in the second step: how to find the period quickly. Below we will describe the second step of the quantum logic circuit. As listed in Fig. 1.82a, we need two qubit registers: the first register has $n = 2[\log_2 N]$ qubits, and the second one has $m = [\log_2 N]$. The qubits of the first register are first all initialized in the $|0\rangle$ state, and then a quantum Hadamard gate is applied to them to transform the state into the coherent superposition states $2^{-n/2} \sum_{x=0}^{2^n-1} |x\rangle$ of all 2 numbers from 0 to $2^n - 1$.

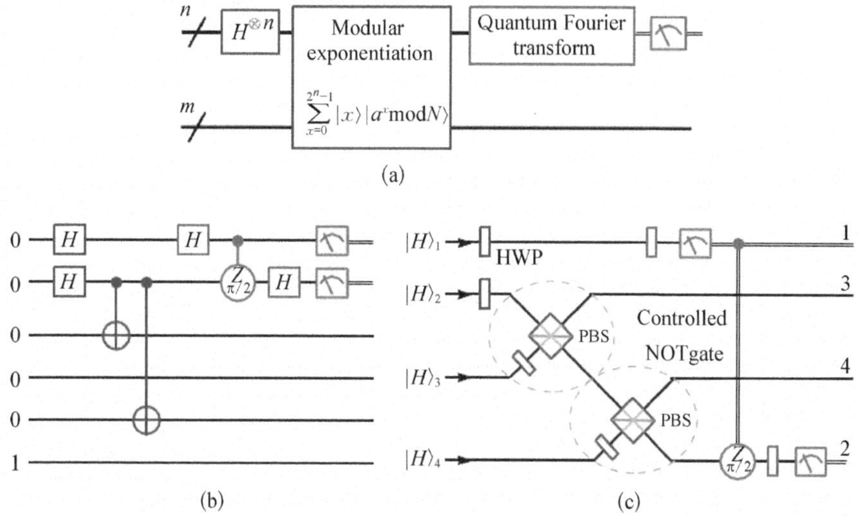

Fig. 1.82 Quantum circuit for Shor's algorithm [14]

The modular exponential function $a^x \mathrm{mod} N$ is applied for each $|x\rangle$ to get a joint state.

$$2^{-n/2} \sum_{0}^{2^n-1} |x\rangle \left| a^x \text{ mod } N \right\rangle \tag{1.53}$$

It is worth noting that we only run the MEF once to get a coherent superposition state of all the values from 0 to $2^n - 1$ with the direct products of its exponential model, a highly entangled 2^n-bit quantum state. This is what "quantum parallelism" is all about. The next step is to do a Fourier transform on n qubits of the first register to get the following:

$$2^{-n} \sum_{y=0}^{2^n-1} \sum_{x=0}^{2^n-1} e^{2\pi ixy/2^n} |y\rangle \left| a^x \text{ mod } N \right\rangle \tag{1.54}$$

In the above formula, state $|y\rangle$ of $y = c2^n/r$ will be interference-enhanced, while the other $|y\rangle$ will be interference-diminished. Therefore, we are able to deduce the required period r with a high probability. This quantum Fourier transform can quickly complete the transformation of $O(2^n)$ elements using only n^2 quantum logic gates, which is the main reason for quantum acceleration in Shor's algorithm.

2. Experimental demonstration

Figure 1.82b gives a simplified quantum circuit. In this simplified circuit diagram, we only need four quantum bits, two quantum Hadamard gates, and two controlled Not gates to achieve $15 = 3 \times 5$. In the experiment, we used single photons as qubits and coded 0 and 1 by the horizontal and vertical polarization of the photons. We noted that since the target qubits of the CNOT in the line were always fixed at H, it could be implemented in an easier and more efficient way. As indicated in Fig. 1.82c, this CNOT gate can be implemented with only a PBS and a HWP. We also noted that the quantum Fourier transform could as well be implemented in a more efficient way. The theory shows that the quantum Fourier transform can be replaced by semiclassical line when followed immediately by a measurement. These simplifications have greatly lowered the experimental needs for entangled quantum gates. The final quantum circuit diagram is plotted in Fig. 1.82c.

The experimental setup is illustrated in Fig. 1.83, where a focused UV laser pulse passes through two barium borate (BBO) crystals to generate two entangled photon pairs by the SPDC process. We disentangled the photons with a polarizer and prepared them on a pure direct product state H. The photons after passing through the HWP were coherently superimposed on the PBS. Then, we implemented necessary single- and two-qubit gates. To ensure good spatial and temporal overlaps, the photons were filtered via a narrow band filter ($FWHW = 3.2$ nm) to extend their coherence time and collected into a single mode fiber to ensure good mode matching.

In the experiment, we had a 50% probability of success in factorizing $15 = 3 \times 5$ with a fidelity of up to $99 \pm 1\%$. The successful implementation of the Shor's algorithm represents an important step towards practical quantum computing. Nevertheless, there are daunting challenges still, including how to coherently manipulate more qubits, how to design more complex quantum computation circuits, and how to implement practical quantum error-correcting techniques.

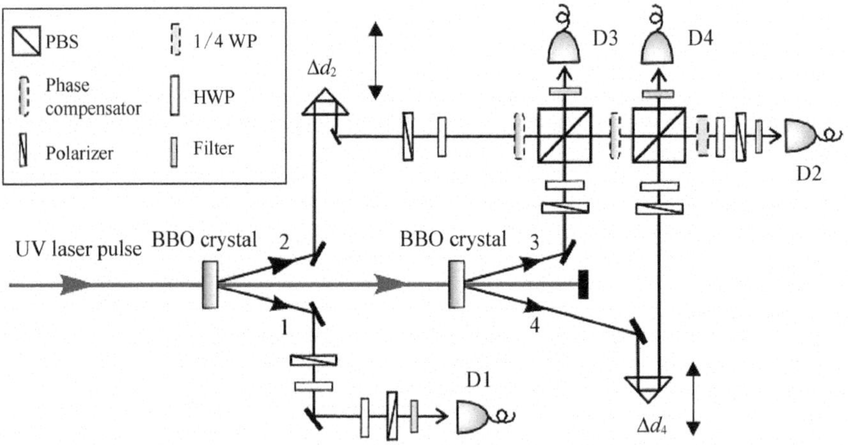

Fig. 1.83 Diagram of the experimental setup for implementing Shor's algorithm [14]

1.5.4.2 Shor's Algorithm Experiment Based on Photonic Chips [86]

Although the Shor's algorithm has been implemented in liquid NMR systems and platform-based linear optical systems, these systems are difficult to scale to large numbers of qubits due to purity, size, and stability limitations. In 2009, the O'Brien team implemented Shor's algorithm for the first time on a waveguide chip, which is a milestone on the way to miniaturized and integrated research on optical quantum computing.

Figure 1.84 displays the experiment facilities. The waveguide chip includes two non-deterministic *CZ* gates (with a success probability of 1/9) and six single-qubit Hadamard gates. The computational procedure is as follows: First, send four photons into "0" or "1" waveguide to complete initial state preparation $|\Psi_{\text{in}}\rangle = |0\rangle_{x_1}|0\rangle_{x_2}|0\rangle_{f_1}|1\rangle_{f_2}$, and realize the Hadamard gate by a directional coupler with reflectance of 2, which causes each qubit to be in the quantum superposition state of 0 and 1. The whole state is thus on a coherent superposition of all possible four-qubit state output states, which also reflects quantum parallelism. The core steps of the Shor's algorithm are then performed by two independent *CZ* gates, each consisting of three 1/3 directional couplers, resulting in a highly entangled output state. Finally, the computational results are obtained by performing x_1 and x_2 measurements and classical processing of the output state qubits.

They prepared four photons with a center wavelength of 790 nm by the SPDC process, coupled them into a waveguide chip via a fiber-optic docking array, and then

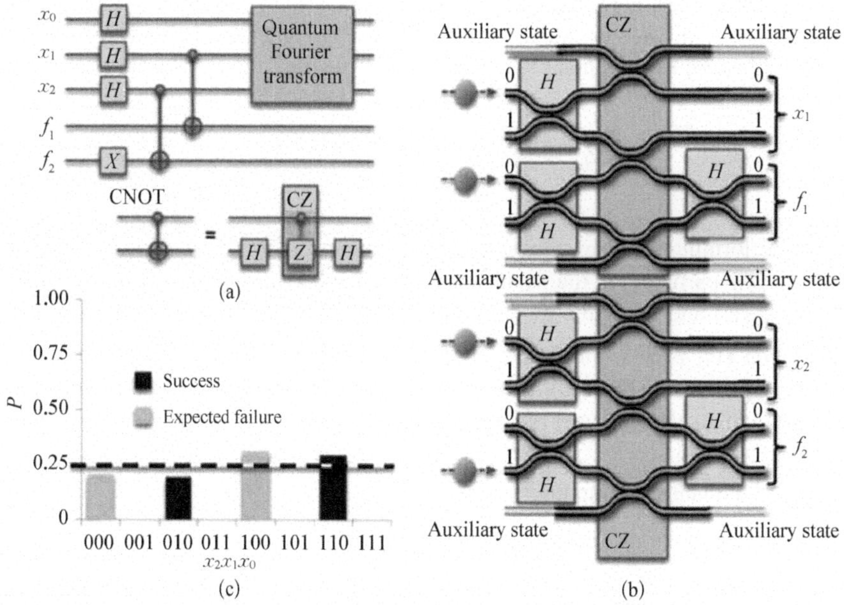

Fig. 1.84 Experimental setup for Shor's algorithm based on photonic chips

measured them with a single photon detector. The experiment yielded coincidence rate is 100 Hz per measurement (integration time 30 s). They obtained the results when the input state was $|\Psi_{in}\rangle$ and the measured output qubits were x_1 and x_2, as shown in Fig. 1.84c. Four binary outputs after including x_0 are 000, 010, 100, and 110. Output results 010 and 110 indicate that the Shor's algorithm correctly decomposed $15 = 3 \times 5$; output result 100 signifies a mediocre solution $15 = 3 \times 5$; 000 is an inherently error output in a Shor's algorithm. The fidelity of the above results is $(99 \pm 1)\%$ relative to the ideal case. Similar to the platform experiments mentioned earlier, the probability of success for this quantum circuit is 0.5, and repeating the calculation n times will raise the probability of success to $1 - (1/2)^{n}$.

This experiment demonstrates that one can implement a pre-compiled Shor's algorithm on a waveguide chip, proving the feasibility of integrated waveguides for quantum computing. Building on existing devices and with more efficient single photon sources and single photon detectors, it is hopeful to achieve larger scale integrated quantum circuits and quantum computation.

1.5.5 One-Way Quantum Computation Experiment Based on Cluster States

1.5.5.1 Theoretical Schemes [15]

Cluster states are a core resource for one-way quantum computation and can be generated by Ising nearest-neighbor interactions. The interaction can be expressed by the following Hamiltonian:

$$H_{\text{int}} = g(t) \sum_{(a,a')} \frac{1 + \sigma_z^{(a)}}{2} \frac{1 - \sigma_z^{(a')}}{2} \tag{1.55}$$

In the formula, $g(t)$ can characterize the strength of the time-dependent interactions. All particles of the system are initialized to the quantum state $|+\rangle = \frac{1}{\sqrt{2}}(|0\rangle + |1\rangle)$ and the initial direct product state is $|+\rangle_1 \otimes |+\rangle_2 \otimes \cdots \otimes |+\rangle_n$. The interaction H_{int} is then turned on, and after a period of time T, $\int_0^T g(t)dt = \pi$, which is equivalent to performing a unitary operation S on the initial state. At this point, the cluster state $|\Psi\rangle_C$ is successfully prepared. For $|\Psi\rangle_C$, we have $\sigma_x^{(a)} \otimes_{a' \in \text{ngbh}(a)} \sigma_z^{(a')} |\Psi\rangle_C = \pm|\Psi\rangle_C$. Next, it is described how to use $|\Psi\rangle_C$ to perform generalized quantum computations.

1. Transmission of information in quantum circuits

The information of a single qubit in a cluster state can be invisibly transferred from one grid point to another. Let's illustrate it by the following example. Assuming n is an odd number, the initial state of the system is $|\Psi_{in}\rangle_1 \otimes |+\rangle_2 \otimes \cdots \otimes |+\rangle_n$, and the entanglement operation S acts on the initial state to form a cluster state of n particles. Next, we measure σ_x for particles j = 1, …, n-1, and the quantum state of the remaining unmeasured particle n becomes a state similar to the initial coded state $|\Psi_{\text{in}}\rangle$, with only one unitary transformation between them. The exact form of the unitary transformation $U_\Sigma \in \{1, \sigma_x, \sigma_z, \sigma_x\sigma_z\}$ is determined by the results of the measurement. It is clear from this that the measurement of the preceding physical qubits enables the transfer of quantum information.

2. Arbitrary rotation of quantum bits

Based on a similar principle, a chained cluster state consisting of five physical bits can achieve an arbitrary rotation $U_R \in SU(2)$, which is analyzed as follows.

We consider this rotation in the Eulerian representation $U_R(\xi, \eta, \zeta) = U_x(\zeta)U_z(\eta)U_x(\xi)$, wherein $U_z(\alpha) = \exp\left(-i\alpha\frac{\sigma_z}{2}\right)$, $U_x(\alpha) = \exp\left(-i\alpha\frac{\sigma_x}{2}\right)$, and the initial state of the system is $|\Psi\rangle_{1,\ldots,5} = |\Psi_{\text{in}}\rangle_1 \otimes |+\rangle_2 \otimes \cdots \otimes |+\rangle_5$. After the action of S, the whole state becomes the following:

$$\begin{aligned} S|\Psi\rangle_{1,\ldots,5} = {} & \frac{1}{2}|\Psi_{\text{in}}\rangle_1|0\rangle_2|-\rangle_3|0\rangle_4|-\rangle_5 \\ & - \frac{1}{2}|\Psi_{\text{in}}\rangle_1|0\rangle_2|+\rangle_3|1\rangle_4|+\rangle_5 \\ & - \frac{1}{2}\left|\Psi_{\text{in}}^*\right\rangle_1|1\rangle_2|+\rangle_3|0\rangle_4|-\rangle_5 \\ & + \frac{1}{2}\left|\Psi_{\text{in}}^*\right\rangle_1|1\rangle_2|-\rangle_3|1\rangle_4|+\rangle_5 \end{aligned} \tag{1.56}$$

In the formula, $\left|\Psi_{\text{in}}^*\right\rangle = \sigma_z|\Psi_{\text{in}}\rangle$, by measuring particles 1 ~ 4, we can rotate the state and transfer it invisibly to particle 5. For particles 1 ~ 4, we choose measurement basis vector $B_j\left(a_j\right) = \left\{\frac{|0\rangle_j + e^{ia_j}|1\rangle_j}{2}, \frac{|0\rangle_j - e^{ia_j}|1\rangle_j}{2}\right\}$ and set the measurement result as $s_j, s_j \in \{0, 1\}$. Then, after setting the result of measurement as $s_j, s_j \in \{0, 1\}$ and we suppose that $s_j = 0$ is the result of measuring the first state in $B_j(a_j)$. The state after the measurement becomes $|s_1\rangle_{a_1,1} \otimes |s_2\rangle_{a_2,2} \otimes |s_3\rangle_{a_3,3} \otimes |s_4\rangle_{a_4,4} \otimes |\Psi_{\text{out}}\rangle_5$, wherein $|\Psi_{\text{out}}\rangle = U|\Psi_{\text{in}}\rangle$. Let $a_1 = 0$ and the form of U is $U = \sigma_x^{s_2+s_4}\sigma_z^{s_1+s_3}U_R\left[(-1)^{s_1+1}\alpha_2, (-1)^{s_2+1}\alpha_3\right]$. It can be seen here that an arbitrary Euler transformation namely arbitrary rotation of the input bits can be achieved by reasonably selecting the measurement basis vector (feedforward) according to the measurement results. For example, in terms of $U_R(\xi, \eta, \zeta)$, we can do the following steps: (i) measure quantum bit 1 under $B_1(0)$, (ii) measure quantum bit 2 under $B_2\left((-1)^{s_1+1}\xi\right)$, (iii) measure quantum bit 3 under $B_3\left((-1)^{s_2+1}\xi\right)$, and (iv) measure quantum bit 4 under $B_4\left((-1)^{s_3+1}\xi\right)$. The final rotational operation is: $U'_R(\xi, \eta, \zeta) = \sigma_x^{s_2+s_4}\sigma_z^{s_1+s_3}U_R(\xi, \eta, \zeta)$. The extra rotation operation can be compensated off at the end of the calculation. It can also be seen here that since it is possible to perform an arbitrary rotation operation on a quantum bit, it means that the preparation of the input state can be achieved by a rotation operation (a series of single-particle

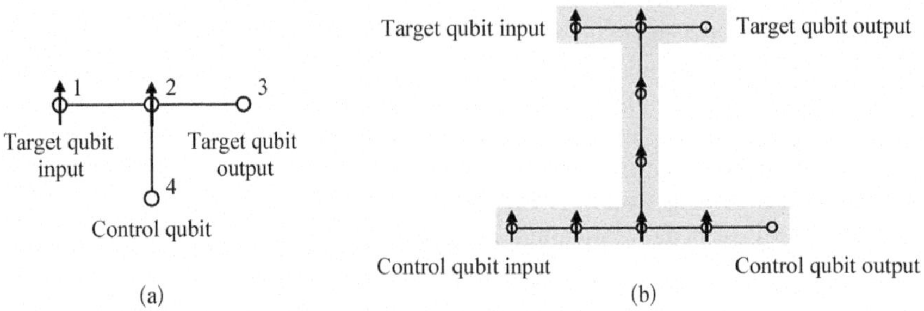

Fig. 1.85 Realization of C-NOT gate operation via single qubit measurement [15]

measurements) without the need to prepare the corresponding input state at the very beginning.

3. Realization of C-NOT

We need a four-qubit cluster state as shown in Fig. 1.85a to implement the C-NOT operation, in which c is the control bit and t is the target bit.

$$\mathrm{CNOT}(c, t_{\mathrm{in}} \rightarrow t_{\mathrm{out}}) = |0\rangle_{cc}\langle 0| \otimes 1^{(t_{\mathrm{in}} \rightarrow t_{\mathrm{out}})} + |1\rangle_{cc}\langle 1| \otimes \sigma_x^{(t_{\mathrm{in}} \rightarrow t_{\mathrm{out}})} \tag{1.57}$$

Suppose that particle 4 is a control qubit and the system's initial state is $|i_1\rangle_{z,1} \otimes |i_4\rangle_{z,4} \otimes |+\rangle_2 \otimes |+\rangle_3$. When the entanglement operator S acts on this state, the entire state would be tangled up. Then, particles 1 and 2 are put under σ_x measurement, the result is $s_j \in \{0, 1\}$. After the measurement, the remaining state is $|s_1\rangle_{x,1} \otimes |s_2\rangle_{x,2} \otimes U_\Sigma^{34} |i_4\rangle_{z,4} \otimes |i_1 + i_4 \bmod 2\rangle_{z,3}$, wherein $U_\Sigma^{34} = \sigma_z^{(3)^{s_1+1}} \sigma_x^{(3)^{s_2}} \sigma_z^{(4)^{s_1}}$. According to the measurements s_1 and s_2, the operation on target particles can be determined. It can be seen that in the whole process it has realized a typical CNOT gate, definitely including an additional rotation operation that changes with the measurement results.

1.5.5.2 Experimental Realization [16]

In 2005, Zeilinger's team in Austria successfully prepared for the first time a cluster state of four photons and successfully demonstrated one-way quantum computation based on this cluster state.

1. Preparation of four-photon cluster states

The experimental setup for preparing the four-photon cluster state is shown in Fig. 1.86.

A pulsed laser beam pumps the BBO crystal back and forth, producing two pairs of entangled photons simultaneously. After realizing single-photon rotation and two-photon interference operations by the HWP, PBS, and other optical elements, they successfully prepared a four-photon cluster state. In order to verify that the four-photon cluster state was indeed successfully prepared, they reconstructed the density

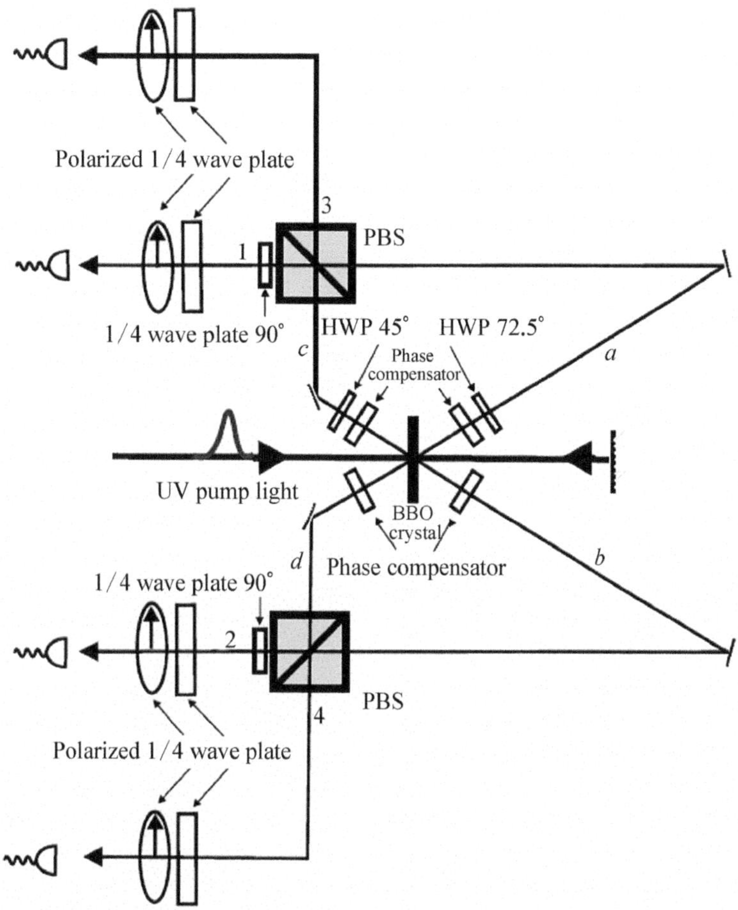

Fig. 1.86 Experimental setup for preparing a four-photon cluster state [16]

matrix of the prepared state by quantum state tomography and obtained its fidelity as $F = 0.63 \pm 0.02$. After doing a local $H_1 \otimes I_2 \otimes I_3 \otimes H_4$ operation on $|\Phi_{\text{cluster}}\rangle$, they were able to obtain the four-photon cluster state indicated in Fig. 1.87a and the horseshoe cluster state displayed in Fig. 1.87c. The linear cluster state as shown in (a) and the horseshoe cluster state as shown in Fig. 1.87c. If a local $H_1 \otimes H_2 \otimes H_3 \otimes H_4$ operation was done on $|\Phi_{\text{cluster}}\rangle$ and the numbers of photons 2 and 3 were swapped at the same time, the square cluster state as shown in Fig. 1.87e could be obtained.

2. Experimental realization of single-qubit rotation

As we can see in Fig. 1.87b, the input qubits are coded as $|\Psi_{\text{in}}\rangle = |+\rangle_{1E}$. By projection measurements of particles 1, 2, and 3, the measurement basis vector angles are selected as α, β, and γ, respectively, and the output state is $|\Psi_{\text{out}}\rangle =$

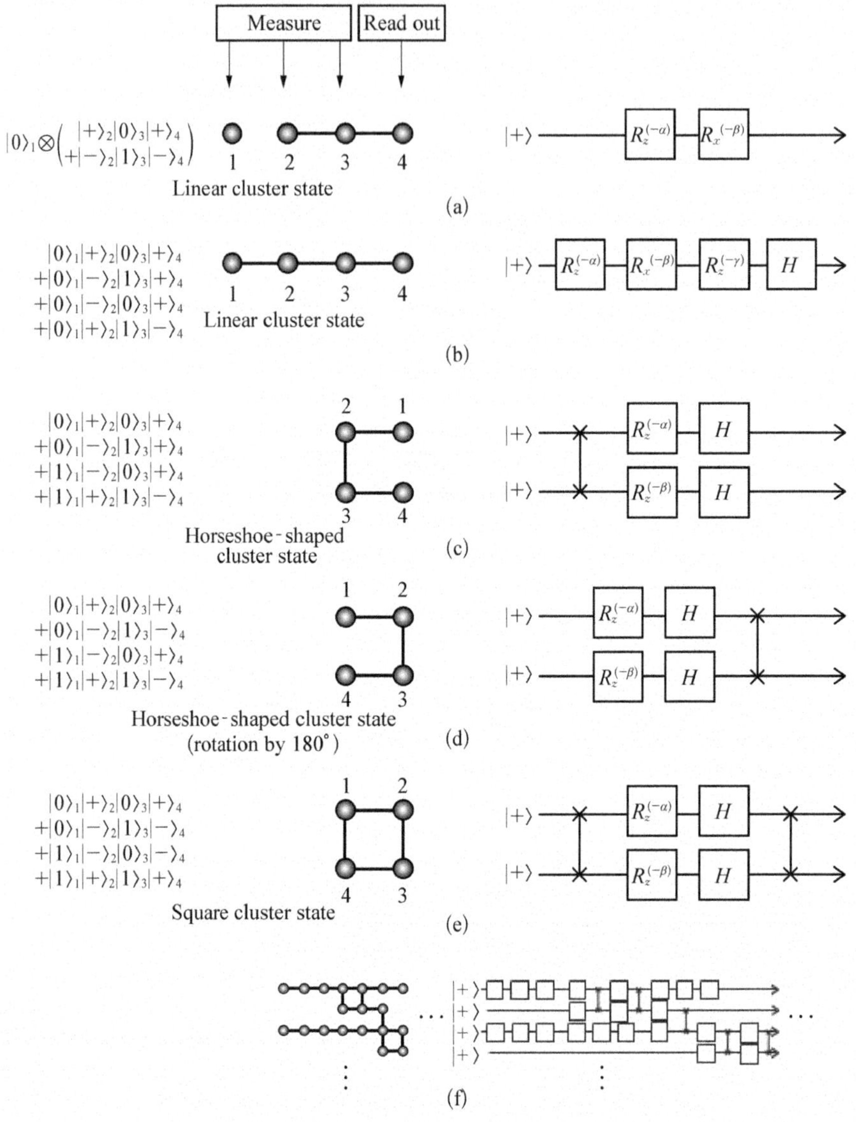

Fig. 1.87 Cluster-state quantum computation circuit [16]

$H R_z(-\gamma) R_x(-\beta) R_z(-\alpha)|\Psi_{\text{in}}\rangle$. In the experiment, they measured six different rotation angles and gave the fidelity of the output states. The experimental measurements are listed in Table 1.5.

Table 1.5 Results of single qubit rotation experiment [16]

α	B	γ	Output state (experimental basis vector)	Fidelity (single qubit state tomography)	Fidelity (four-qubit state tomography)
0	0	0	$\|H\rangle$	0.97 ± 0.03	0.81 ± 0.03
π	0	$-\frac{\pi}{2}$	$\|V\rangle$	0.93 ± 0.01	0.91 ± 0.02
0	$-\frac{\pi}{2}$	$-\frac{\pi}{2}$	$\frac{1}{\sqrt{2}}(\|H\rangle + \|V\rangle) = \|+\rangle$	0.58 ± 0.08	0.63 ± 0.03
$\frac{\pi}{2}$	0	$-\frac{\pi}{2}$	$\frac{1}{\sqrt{2}}(\|H\rangle - \|V\rangle) = \|-\rangle$	0.87 ± 0.04	0.84 ± 0.03
0	$-\frac{\pi}{2}$	-0	$\frac{1}{\sqrt{2}}(\|H\rangle - i\|V\rangle) = \|R\rangle$	$0.99^{+0.01}_{-0.06}$	0.78 ± 0.03
$\frac{\pi}{2}$	0	0	$\frac{1}{\sqrt{2}}(\|H\rangle + i\|V\rangle) = \|L\rangle$	$0.99^{+0.01}_{-0.02}$	0.94 ± 0.03

3. Two-qubit gate

In quantum computation, an important function of two-qubit logic gates is to change the entanglement between quantum bits, such as C-NOT gates and C-phase gates. In cluster-state quantum computation, two-qubit gates are capable of being implemented by two-dimensional cluster states, such as the horseshoe and square cluster states in Fig. 1.87c–e. Their quantum circuits contain single qubit rotations and two-qubit C-phase operations, which can, under certain conditions, allow the formation of entanglement between two independent coded quantum bits. Of course, the ability to create an entanglement depends on particular quantum circuits and initial states. In the square form, for example, as shown in Fig. 1.87e, the input quantum state is $|\Psi_{\text{in}}\rangle = |+\rangle_{1E}|+\rangle_{2E}$ and the output state is $|\Psi_{\text{out}}\rangle = C-\text{phase}(H_1 \otimes H_2)\left[R_z(-\alpha) \otimes R_z(-\beta)\right]C\text{−phase}\,|\Psi_{\text{in}}\rangle$, wherein α and β depend on the choice of the measurement basis vectors for photons 1 and 4. Under the experimental basis vector, if photons 1 and 4 are measured as $|V\rangle_1$ and $|H\rangle_4$, it corresponds to the result that the measurement basis vectors are $B_1(\pi)$ and $B_4(0)$ and the measurement output is zero. For a quantum circuit of square form, the whole operation is such that $|+\rangle_{1E}|+\rangle_{2E} \rightarrow |+\rangle_{1E}|-\rangle_{2E}$, and the output direct product state consists of photons 2 and 3. Under the experimental basis vector, photons 2 and 3 will be ended up in the direct product state $|V\rangle_2|H\rangle_3$. The reason for the absence of entanglement is that during the measurement of particles 1 and 4, there is not only a C-phase operation but also a series of rotational operations. The experimental results are shown in Fig. 1.88a, and the fidelity obtained is 0.93 ± 0.01.

In cluster-state quantum computation however, we need two-qubit operations that can lead to entanglement. Take horseshoe-shaped cluster states for example (Fig. 1.87c). The input state is still $|\Psi_{\text{in}}\rangle = |+\rangle_{1E}|+\rangle_{2E}$ while the output state becomes $|\Psi_{\text{out}}\rangle = (H_1 \otimes H_2)\left[R_z(-\alpha) \otimes R_z(-\beta)\right]C\text{-phase}\,|\Psi_{\text{in}}\rangle$. When the

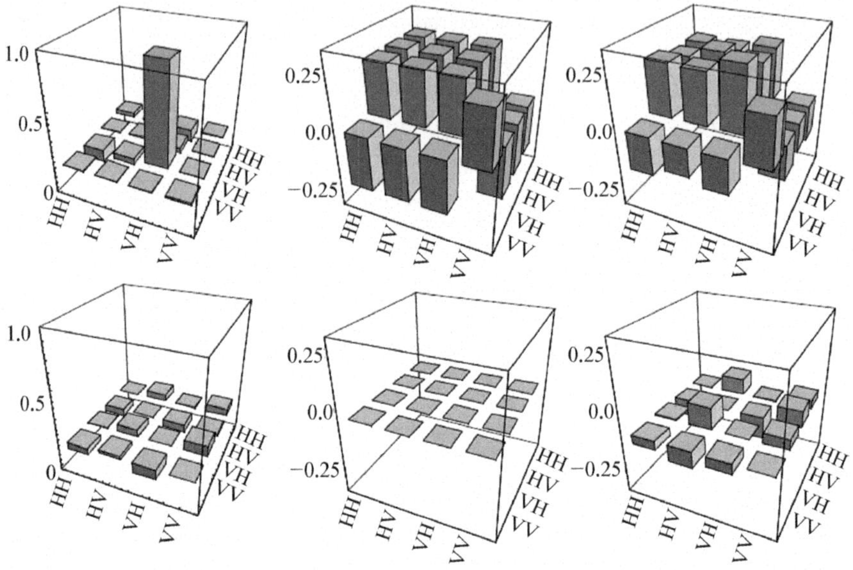

Fig. 1.88 Results of a two-qubit manipulation of the square cluster state [16]

measurement basis vector is chosen to be $B_2(0)$ and $B_3(0)$, and the measurement result is 0, the whole transformation process is like this:

$$|+\rangle_{1E}|+\rangle_{2E} \rightarrow \frac{1}{\sqrt{2}}(|0\rangle_{1E}|+\rangle_{2E} + |1\rangle_{1E}|-\rangle_{2E}) \tag{1.58}$$

It can be seen in the formula 1.58 that the output state encoded on photons 1 and 4 is a maximally entangled state. They used quantum state tomography to characterize the density matrix of the output state, as shown in Fig. 1.88; the left column is the density matrix of the output state obtained theoretically and the right column is the density matrix of the output state obtained experimentally with a fidelity of 0.84 ± 0.03.

By implementing single-qubit rotation and a two-qubit gate operation, the experiment has verified in principle the feasibility of cluster-state one-way quantum computation. Finally, based on the prepared cluster states, they also implemented a simplified version of the Grover search algorithm, which will not be repeated here.

1.5.6 One-Way Quantum Computation Based on "Non-Cluster States"

Measurement-based quantum computing, also known as one-way quantum computation, has attracted immediate and extensive attention from researchers in the field of quantum information since its debut in 2001. There are two necessary conditions for this quantum computing model: first, it is possible to prepare universal quantum states that are independent of the actual algorithm; second, perform feed-forward selection of single-qubit measurements based on measured effects. As we described previously, for measurement-based quantum computation, a universal highly-entangled quantum state namely cluster state was initially used as a core resource for computation, and it has long been thought to be the only kind of resource that can be used for universal quantum computing. Recent research, however, has indicated that one-way quantum computation in the correlation space allows numerous new entangled states to become universal resources for quantum computation so that we are not limited to the cluster state model alone. In this new model of computation, quantum information is stored in a virtual system (correlation space) that is mapped by projection measurements to the Hilbert space where the physical bits reside.

This discovery has greatly enriched the kinds of general-purpose resources that can be used for one-way quantum computing. These novel quantum states has largely improved the feasibility of quantum computation in various physical systems, such as linear optical systems, ultracold atomic systems, polarized molecular systems in optical lattices, or strongly-correlated many-body physics systems. Among them, a major advantage of the linear optical system is that it can effectively prepare a determined type of universal resource quantum states by the CNOT gate. In the strongly-correlated many-body systems, unlike cluster states, the universal resource for quantum computation in the correlation space can be a Hamiltonian ground state that contains two-body interactions, such as the well-known AKLT state. Thus, the proposal of one-way quantum computing in correlated space will stimulate widespread interest in studying universal resources for a variety of physical systems. The study of these new universal resources can bring new directions to our study of fundamental physical problems and open a new chapter in the research of one-way quantum computation.

In 2011, we successfully completed a one-way quantum computation experiment based on "non-cluster states" [87]. Our team first fabricated a four-photon polarized entanglement state for a single-qubit rotation and demonstrated a method to counteract the measurement randomness, which is a unique feature of "non-cluster state" quantum computing. Next, we implemented the operation of two-qubit entanglement gates using the hyper-entanglement state of a four-photon six-qubit polarization path and demonstrated a simple Deutsch algorithm based on it. Finally, by measuring the two-body correlation functions of the prepared entangled state and the entropy of the monomer, we proved that its entanglement properties were completely different from those of the cluster state, but at the same time it still satisfied the universality requirement for performing one-way quantum computation.

1. Theoretical scheme [88, 89]

The universal one-dimensional computational scheme is expressed by the matrix product state (MPS):

$$|\Psi\rangle = \sum_{s_i} \langle r|A[|s_n\rangle]\cdots A[|s_1\rangle]|l\rangle|s_1,\ldots,s_n\rangle \qquad (1.59)$$

In the formula above, $A[|s_i\rangle]$ is a single-qubit operator and $|l\rangle$ and $|r\rangle$ represent the left and right boundary vectors in the two-dimensional vector correlation space. A projective measurement acting on particle i with a local result of $|\phi_i\rangle$ will introduce an operator $A[|\phi_i\rangle] = \sum_{s_i} \langle\phi|s_i\rangle A[|s_i\rangle]$ in the correlation space. This operator operation can be used to perform tasks such as initial state preparation, single-qubit rotation, and readout of logic bits. These one-dimensional quantum circuits can be coupled to form two-dimensional resource states, enabling two-qubit entangled logic gates.

To implement the single-qubit logic gate operation, we take a four-qubit matrix product state as an example (see Fig. 1.89a). The tensor matrices corresponding to the black quantum bits are $\mathbf{A}[|H\rangle] = \hat{H}\cos\theta$ and $\mathbf{A}[|V\rangle] = \hat{H}\hat{Z}\cos\theta$, while the corresponding tensor matrice for the gray quantum bits are $\mathbf{B}[|H\rangle] = \hat{H}$ and $\mathbf{B}[|V\rangle] = \hat{H}\hat{Z}$. We use $l\rangle = |+\rangle$ and $|r\rangle = |0\rangle$ to represent the left and right boundary vectors of the matrix product state, in which $|+\rangle = \frac{1}{\sqrt{2}}(|0\rangle + |1\rangle)$. Here, $|\mathrm{H}\rangle$ and $|\mathrm{V}\rangle$ stand for the horizontal and vertical polarization states of the photon and are used to encode $|0\rangle$ and $|1\rangle$, while $\hat{H}$ refers to the Hadamard gate, and $\hat{\mathrm{Z}}$ indicates the Pauli matrix σ_z. Substituting these parameters into Eq. (1.59), we can have the following:

$$\begin{aligned}|\Psi_4\rangle = {} & c|H\rangle_1(c|H\rangle + s|V\rangle)_2(c|H\rangle|P\rangle + s|V\rangle|M\rangle)_{34} \\ & + s|V\rangle_1(c|H\rangle - s|V\rangle)_2(c|H\rangle|M\rangle + s|V\rangle|P\rangle)_{34}\end{aligned} \qquad (1.60)$$

In the above formula, $|P\rangle = \frac{1}{\sqrt{2}}(|H\rangle + |V\rangle)$, $|M\rangle = \frac{1}{\sqrt{2}}(|H\rangle - |V\rangle)$. For the sake of simplicity, we assume that $c = \cos\theta$ and $s = \sin\theta$. For any $\cos\theta \neq 0$ and $\sin\theta \neq 0$

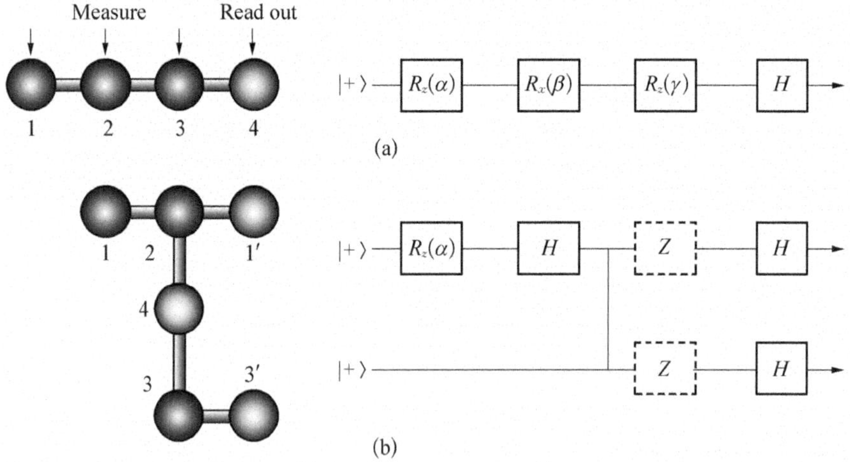

Fig. 1.89 Non-cluster states and their quantum circuits [87]

circumstance, such a matrix product state can be used to realize the quantum circuit of single-qubit logic gates. In the experiment, if we select $\theta = \frac{\pi}{6}$, then $c = \frac{\sqrt{3}}{2}$, $s = \frac{1}{2}$. In this condition, if the tensor matrix A acts on the qubit, it is able to perform the quantum information processing task. The last particle not only plays a reading out role but also maps the information carried in the correlation space onto the physical bits, i.e., $|0\rangle_c \rightarrow |P\rangle_p$ and $|1\rangle_c \rightarrow |M\rangle_p$.

To realize two-qubit entanglement gates, two-dimensional "non-cluster" states need to be prepared, as shown in Fig. 1.89b. The two-dimensional "non-cluster" states are prepared by coupling two one-dimensional matrix direct product states (1–2–1′ and 3–3′) at grid point 4. In the coupling process, particle 4 is first prepared onto |P⟩ and then two *CZ* gates are applied to quantum bits 2–4 and 3–4. The quantum bits 1′ and 3′ are denoted by the photon's path degrees of freedom $|u\rangle$ and $|d\rangle$. The corresponding tensor matrices are $\mathbf{B}[|u\rangle_{1'}] = \mathbf{B}\left[|P\prime\rangle_{3\prime}\right] = \hat{H}$ and $\mathbf{B}[|d\rangle_{1'}] = \mathbf{B}\left[|M\prime\rangle_{3\prime}\right] = \hat{H}\hat{Z}$, respectively, wherein $\left|P'\right\rangle = \frac{1}{\sqrt{2}}(|u\rangle + |d\rangle)$, $\left|M'\right\rangle = \frac{1}{\sqrt{2}}(|u\rangle - |d\rangle)$. Then, this "non-cluster state" can be written as:

$$|\Psi_6\rangle = |H\rangle_4|\mu\rangle_{121'}|\nu\rangle_{33'} + |V\rangle_4\hat{Z}_2|\mu\rangle_{121'}\hat{Z}_3|\nu\rangle_{33'} \tag{1.61}$$

where

$$\begin{aligned} |\mu\rangle &= c^2|HHu\rangle + cs|VHu\rangle + cs|VHd\rangle - s^2|VVd\rangle \\ |\nu\rangle &= \frac{1}{2}(c|Hu\rangle + s|Vd\rangle) \end{aligned} \tag{1.62}$$

$\hat{Z}_i$ means the Pauli matrix σ_z that acts on qubit *i*.

2. 'Non-cluster state' test preparation

In our experiment, we prepared entangled photons using a type II PDC technique. The femtosecond pulsed laser (~150 fs, 76 MHz, 788 nm) was converted into UV pulses by doubling the frequency crystal Ll B3O5 to pass through two nonlinear crystals (BBOs) to produce two pairs of entangled photons in paths 1–2 and 3–4.

First, we will tell how to prepare the four-qubit "non-cluster state" $|\Psi_4\rangle$ given in Fig. 1.90. We prepared the two-photon entangled states as $\frac{1}{\sqrt{2}}(|H\rangle_1|P\rangle_2 + |V\rangle_1|M\rangle_2)$ and $\frac{1}{\sqrt{2}}(|H\rangle_3|P\rangle_4 + |V\rangle_3|M\rangle_4)$ through HWP. Then, we placed a polarizing beam cube (PBCa) in optical path 2 and optical path 3; for PBCa, $T_v = \left(\frac{s}{c}\right)^2$ and $T_h = 1$, wherein T_v and T_h denote the permeability of vertical and horizontal polarization. So, the two-qubit entangled state can be written as follows:

$$\begin{aligned} &\frac{1}{\sqrt{2}}[|H\rangle_1(c|H\rangle + s|V\rangle)_2 + |V\rangle_1(c|H\rangle - s|V\rangle)_2] \\ &\frac{1}{\sqrt{2}}(c|H\rangle_3|P\rangle_4 + s|V\rangle_3|M\rangle_4) \end{aligned} \tag{1.63}$$

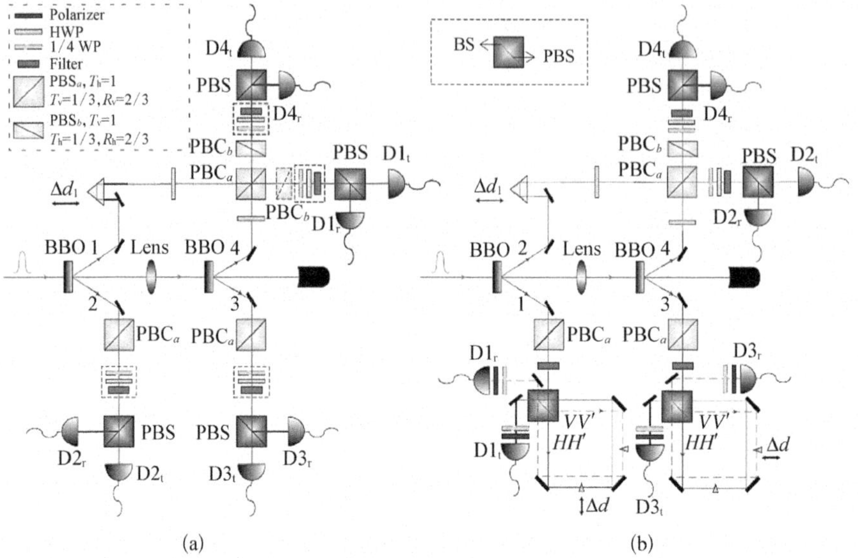

Fig. 1.90 Experimental setup for preparing "non-cluster state" $|\psi 4\rangle$ and "non-cluster state" $|\psi 6\rangle$ [87]

Using these two pairs of entangled states, a controlled phase gate is acted on photons 1 and 4 by a set of PBC_s. Specifically, this consists of a piece of coherent $PBC_a\left(T_v = \frac{1}{3}, T_h = 1\right)$, and two $PBC_b\left(T_v = 1, T_h = \frac{1}{3}\right)$ used to balance the amplitudes, one each for paths 1 and 4. We noted that the role of path-1 PBC_b was simply to adjust the amplitudes of $|H\rangle_1$ and $|V\rangle_1$. By removing the PBC_b from optical path 1, the $|\psi 4\rangle$ state could be accurately prepared. Next, we describe how to prepare six-qubit "non-cluster state" $|\psi 6\rangle$: based on the prepared state $|\psi 4\rangle$, we swapped the labels of paths 1 and 2; then we let photons 1 and 3 enter the PBS, because its function was to reflect V polarization through H polarization. After a post-selection process, the final state could be accurately prepared into the state $|\psi 6\rangle$.

To obtain the fidelity of the prepared four-photon "non-cluster state", we obtained its density matrix by ultra-perfect quantum state tomography to describe this state. Quantum state tomography requires each photon be projected onto $\{|H\rangle, |V\rangle, |P\rangle, |M\rangle, |R\rangle, |L\rangle\}$ for measurement, and there is a total of 1296 different measurement vectors for the four-photons. With these data, we reconstructed the density matrix of the state using maximum likelihood. Figure 1.91a and c show the real and imaginary parts of the experimentally obtained density matrix; (b) and (d) display the real and imaginary parts of the theoretical density matrix. The final fidelity obtained is $F = \langle \Psi_4|\rho|\Psi_4\rangle = 0.74 \pm 0.1$. For the six-qubit "non-cluster state", if we adopt quantum state tomography, it would be an impossible task to measure too many basis vectors and take too much time to complete it. So, we chose to decompose the "non-cluster state" directly into 36 observable quantities and finally got the fidelity $F = \langle \Psi_4|\rho|\Psi_4\rangle = 0.73 \pm 0.1$ by measuring the 36 data points.

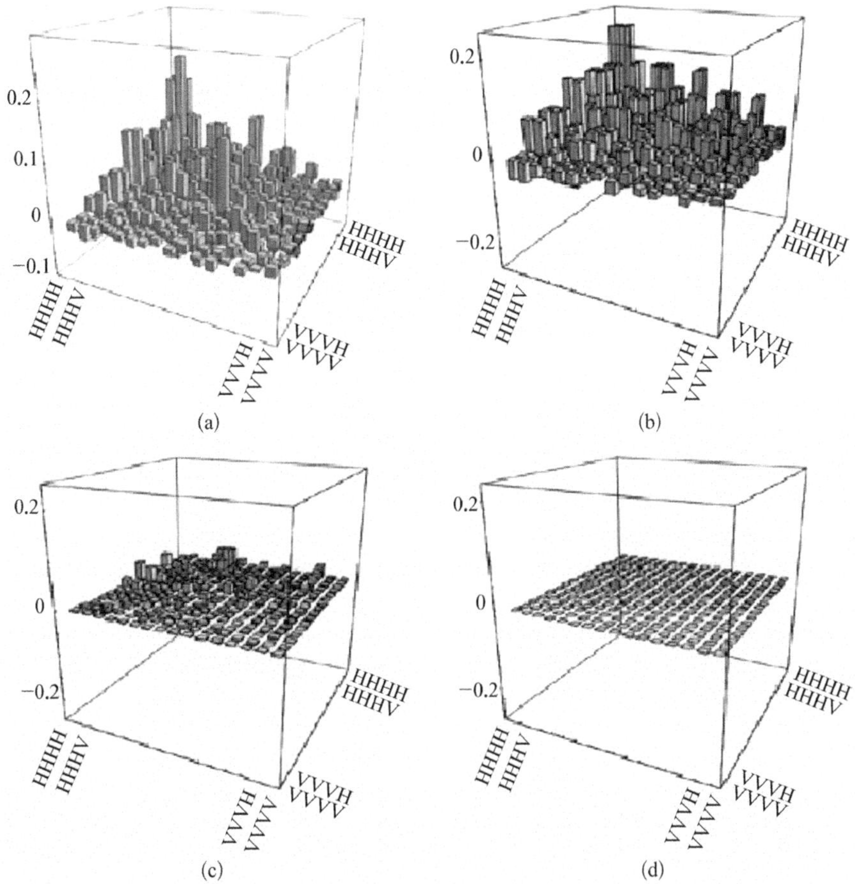

Fig. 1.91 Density matrix of the four-photon "non-cluster state" $|\psi 4\rangle$ [87]

Finally, to prove that the prepared "non-cluster state" has different entanglement properties from the cluster state, we also measured the two-body correlation function and local entanglement of the "non-cluster state". The two-body correlation function is defined as $Q^{ij}_{ab}(|\Psi\rangle) = \langle\Psi|\hat{a}_i \otimes \hat{b}_j|\Psi\rangle - \langle\Psi|\hat{a}_i|\Psi\rangle\langle\Psi|\hat{b}_j|\Psi\rangle$, wherein $\hat{a}_i$ and $\hat{b}_j$ are the Pauli operators acting on qubits i and j, respectively. The local entanglement $E(\boldsymbol{\rho}_i) = 2(1 - \mathrm{Tr}\boldsymbol{\rho}_i^2)$ between qubit i and another qubit is defined by the local entropy, wherein p_i is the reduced density matrix of qubit i. The experimentally obtained two-body correlation functions of the "non-clustered state" are both non-zero and both the local entropies are not 1, clearly indicating that the entanglement properties of the "non-cluster state" are different from those of the cluster state.

3. Single qubit rotation

To achieve an arbitrary SU(2) single-qubit rotation, we measured $|\psi_4\rangle$ under the basis vector $\mathbf{B}(\zeta) = \{|\zeta_0\rangle, |\zeta_1\rangle\}$, wherein $|\zeta_0\rangle = s|H\rangle + \mathrm{i}c\tan\zeta/2|V\rangle$, $|\zeta_1\rangle =$

$c|H\rangle$ − is $\cot\zeta/2|V\rangle$. For simplicity, we defined the value of the output r_j=0 if the measurement result is $|\zeta_0\rangle$ and r_j=1 if the measurement result is $|\zeta_1\rangle$. First, assume that all measurements are zero. Although it will reduce the success probability of a gate in each measurement step, but this does not prevent us from demonstrating how a single qubit gate works. A single-qubit rotation $R_z(\zeta) = \exp(-i\zeta\sigma_z/2)$ can be achieved by applying a Hadamard transform in each measurement step. Based on the prepared four-qubit state, we measured physical bits 1, 2 and 3 consecutively, i.e., $B_1(\alpha)$, $B_2(\beta)$, $B_3(\gamma)$. By these operations, the input state $|\Psi_{\text{in}}\rangle_c = |+\rangle$ of the correlation system was transformed into $|\Psi_{\text{out}}\rangle_c = \hat{H}R_z(\gamma)R_x(\beta)R_y(\alpha)|+\rangle$. The output physical state became $|\Psi_{\text{out}}\rangle_p = R_z(\gamma)R_x(\beta)R_y(\alpha)|+\rangle\big|_{0\to H,1\to V}$ under the mapping of the tensor matrix B. Six different rotational operations were performed to obtain six different output states, and their fidelities were measured. The final average value of the fidelity obtained was 0.86 ± 0.01, much larger than the theoretical threshold 2/3, proving that we have perfectly implemented single-qubit rotation.

In cluster-state one-way quantum computation, the erroneous measurement results are introduced only by an additional Pauli operator, so that they can be compensated for by feedforward control of the subsequent measurement basis vector. The additional operator arising from "non-cluster state" one-way quantum computation is different from the Pauli operator, and so the simple feedforward technique is no longer applicable. Fortunately, the theory has proved it is effective in correcting errors in finite steps by trying again and again until a successful strategy is used. We have considered a simple case: perform rotation operation $\hat{H}R_z(\alpha)|+\rangle$ on two-qubit state $|\lambda_{34}\rangle = c|H\rangle_3|P\rangle_4 + s|V\rangle_3|M\rangle_4$ and four-qubit state $|\Psi_4\rangle$ and compare their success probabilities. For the two-qubit state, qubit 1 is measured under the basis vector *B*(α). When the output $r_1 = 0$, the success probability of obtaining its $\hat{H}R_z(\alpha)|+\rangle$ is $P_s(\alpha)$; however, when the output $r_1 = 1$, a rotation $\hat{H}R_z(\alpha')|+\rangle$ that contains an error angle is obtained. For the four-qubit state, we first measured qubit 1 under the $B(\alpha)$ basis vector. When $r_1 = 0$, we measured qubits 2 and 3 under the *Z* basis vector and the output state was $\hat{X}^{r_3}\hat{Z}^{r_2}\hat{H}R_z(\alpha)|+\rangle$. When $r_1 = 1$, we measured qubit 2 under the *Z* basis vector and the output result was r_2, then qubit 3 under the basis vector $\mathbf{B}[(-1)^{r_2}(\alpha - \alpha\prime)]$. When $r_3 = 0$, we had probability of $P_s(\alpha - \alpha')$ to get $Z^{r_2}\hat{H}R_z(\alpha)|+\rangle$. The final result was equivalent to the rotation operation we envisioned. Thus, by the two extra qubits, we could increase the probability of success from $P_s(\alpha)$ to $P_s(\alpha) + [1 - P_s(\alpha)] \cdot P_s(\alpha - \alpha')$.

We have also experimentally verified this process using the prepared four-photon "non-cluster state", and the results are shown in Fig. 1.92a. It can be seen that the theoretical and experimental results agree perfectly, and the probability of success is indeed greatly increased by the compensation strategy.

4. Two-qubit entanglement gate

Like single-qubit rotations, two-qubit entanglement gates are also fundamental modules of universal quantum computation, and we experimentally implemented two-qubit C-Phase gates based on six-qubit "non-cluster state" $|\Psi_6\rangle$. First, we measured qubit 1 under the basis vector *B*(α). When r1 = 0, the measurement

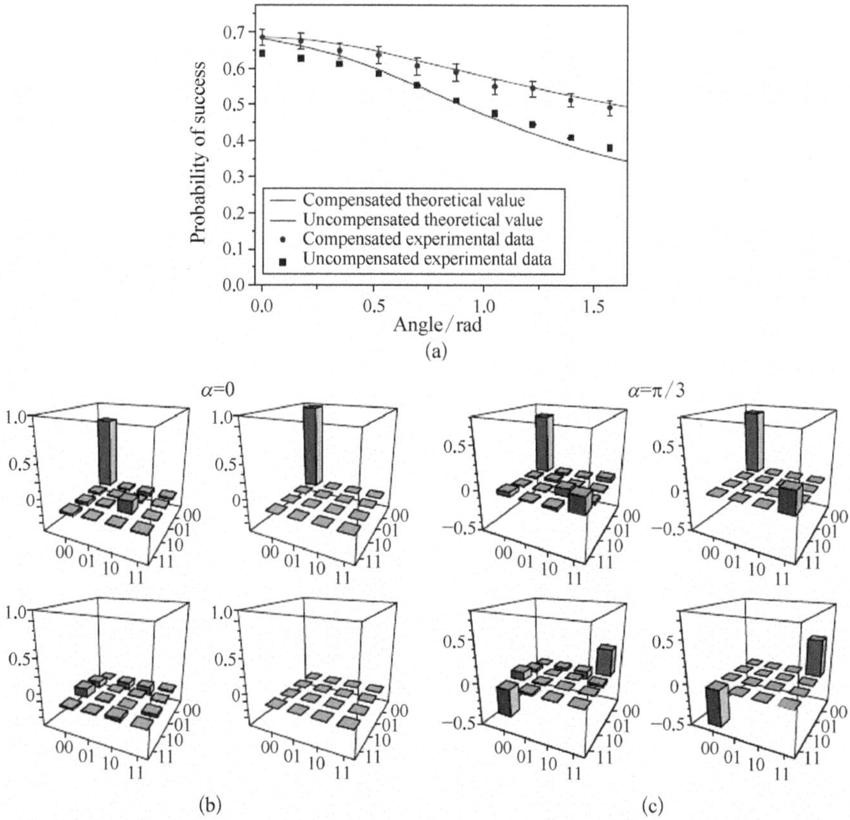

Fig. 1.92 Results of one-qubit rotation operation and two-qubit gate experiment results [87]

initializes the input state of correlation space to $\hat{H}R_z(\alpha)|+\rangle_t \otimes |+\rangle_c$, wherein t and c represent the target qubit and control qubit (1–2–1′ and 3–3′) of the two computational schemes. Next, we measured qubits 2 and 3 under the basis vector $\mathbf{B}(\frac{\pi}{2}) = \{s|H\rangle + \mathrm{i}c|V\rangle, c|H\rangle - \mathrm{i}s|V\rangle\}$. When the output result $r_2 = r_3 = 0$, we measured qubit 4 under the Y basis vector $\{|H\rangle + \mathrm{i}|V\rangle, \cdot|H\rangle - \mathrm{i}|V\rangle\}$. After these measurements, the logical state carried by the correlation space would change from $|\Psi_{\text{in}}\rangle = HR_z(\alpha)|+\rangle_t \otimes |+\rangle_c$ to $|\Psi_{\text{out}}\rangle = (\hat{H} \otimes \hat{H}) \cdot (\hat{Z} \otimes \hat{Z})^{r_4} \cdot CZ|\Psi_{\text{in}}\rangle$, wherein $CZ = I \otimes |0\rangle\langle 0| + Z \otimes |1\rangle\langle 1|$ was a controlled phase gate. If $r_2 \neq 0$ or $r_3 \neq 0$, we would measure qubit 4 under the Z basis vector and remove the coupling between two one-dimensional lines. After the measurement, there would be a local rotation error on each qubit. As mentioned in the single-qubit revolving gate, if we had more qubits, the error could be effectively corrected using the strategy of trying until success.

In our experiment, we only considered the case where r2 = r3 = 0. When $\alpha = 0$, $r_4 = 0$ and $\alpha = \pi/3$, $r_4 = 0$, quantum state tomography was employed to obtain the density matrices of the output states. The density matrices of the theoretical and experimental physical output states are given in Fig. 1.92b and c. These states are very close to the theoretical expectations, with a fidelity of 0.88 ± 0.02 and 0.84 ± 0.03.

5. Deutsch algorithm

The Deutsch algorithm, also known as Deutsch-Jozsa algorithm, is a highly significant decisional quantum algorithm. Experimentally, we demonstrated the Deutsch algorithm using the six-qubit "non-cluster state" $|\psi_6\rangle$. A practical computational task was taken as an example: how to distinguish between two different functions $f(x)$ in a black box? The function $f(x)$ is one of the following two functions: When the input is an N-bit binary number X, the output is a constant namely a constant function for all Xs, or half 0 and half 1, that is, a balanced function. Using the classical method, in some cases, we need to try $2^{N-1} + 1$ times to get what we want. However, with Deutsch algorithm, only one attempt is needed to get the answer. In the two-qubit case, the algorithm is expressed as $|x\rangle|y\rangle \rightarrow |x\rangle|y \oplus f(x)\rangle$, wherein $|x\rangle$ is the input state of query and $|y\rangle$ is an auxiliary input state. By preparing $|x\rangle|y\rangle$ into $|+\rangle|-\rangle$, we obtained the output state $\left[(-1)^{f(0)}|0\rangle + (-1)^{f(1)}|1\rangle\right]|-\rangle$, and applied the Hadamard operator on both output states simultaneously. If $f(x)$ was a constant function, the final result would be $|\mathrm{H}\rangle|\mathrm{V}\rangle$; if it was a balanced function, the final result would be $|\mathrm{V}\rangle|\mathrm{V}\rangle$. So, we determined the type of $f(\mathrm{x})$ by identifying the output states.

In the experiment, the constant function and the balanced function were $f_c = I \otimes I$ and $f_b = g_{\mathrm{CNOT}}$. We used 3–3′ to encode query logic bits and 1–2–1′ to encode auxiliary logic bits. Furthermore, we only considered measurements $r_1 = r_2 = r_3 = r_4 = 0$. Measuring bit 1 under basis vector $B(\pi)$ and bits 2, 3, and 4 under basis vector $B(0)$ means carrying out the transformation $H_q \otimes R_z(\pi)_a$ of them. In combination with the mapping of logic bits from correlation space to physical bits, the transformation of input states can be written as $\left(\hat{H}_q \otimes \hat{H}_a\right)\left(I_q \otimes R_z(\pi)_a\right)$, which has realized the preparation of auxiliary input states. See Fig. 1.93a for the readouts of constant functions and measurement results. Similarly, the measurement of bit 1 under basis vector $B(\pi)$ and bits 2, 3, and 4 under basis vector $B\left(\frac{\pi}{2}\right)$ is same as gate operation on input states $\left(\hat{H}_q \otimes \hat{H}_a\right)g_{\mathrm{CNOT}}\left(I_q \otimes R_z(\pi)_a\right)$, which has realized balanced functions (Fig. 1.93b). In the experiment, the probability of successfully judging the constant function and balanced function was 99 ± 1 and $75 \pm 2\%$ respectively.

1.5.7 Experiment on Solving a System of Linear Equations

The system of linear equations has been applied in almost every field of science and engineering, including numerical computation, signal processing, economics, and computer science. For example, weather forecasting, which is closely related to our daily life, requires the creation and solution of linear equation groups containing millions of variables to simulate and predict various physical parameters (temperature, pressure, humidity, etc.) in the atmosphere. Highly accurate weather forecasting, on the other hand, requires solving equation groups with huge amounts of

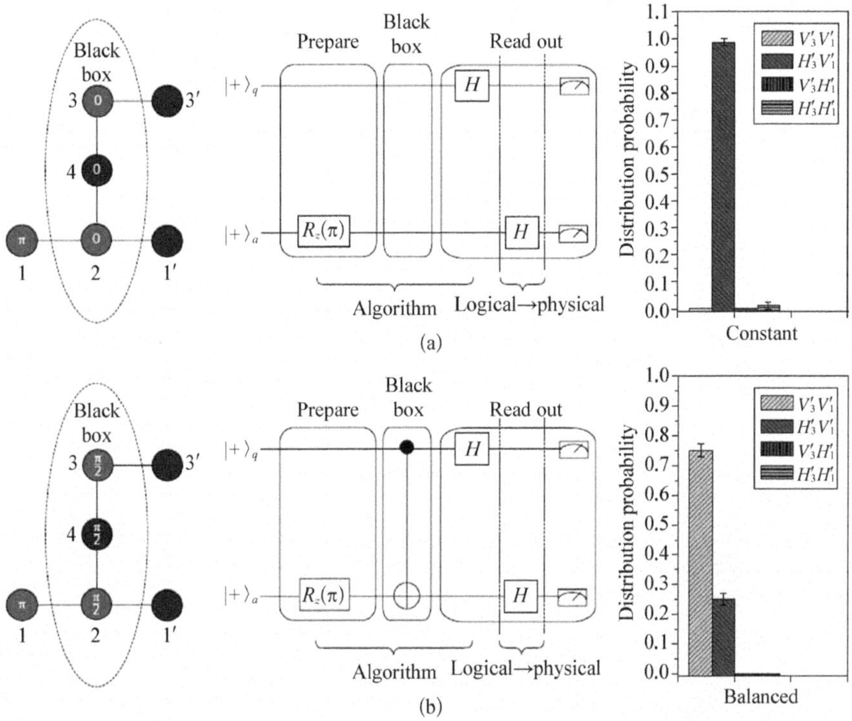

Fig. 1.93 Quantum circuits and experimental results of the Deutsch algorithm [87]

data, which would take at least several hundred years, even with the world's fastest supercomputers, if it were an equation group with 10^{24} variables.

In 2009, Seth Lloyd, a professor at MIT, et al. proposed a quantum algorithm for solving the system of linear equations [90], arguing that the parallelism of quantum computation would bring exponential acceleration, far exceeding the speed of existing computers. According to the theoretical proof, if one is only concerned with the expected value of the physical or engineering parameters under the solution of a system of linear equations instead of the complete solution to the equation group, the time complexity of using a quantum computer would be reduced to O(lgN), with N as the number of independent variables of the equation group. For example, it only takes a quantum computer with a GHZ clock frequency 10 s to solve a linear system of equations with 10^{24} variables.

In 2013, our team successfully operated a quantum circuit to solve a 2×2 system of linear equations using a multi-photon entanglement manipulation technique, and demonstrated the feasibility of this algorithm in principle for the first time. In the following sections, we will introduce this experiment [91].

1. Theoretical schemes

For the equation group $Ax = b$, given an $N \times N$ Hermitian matrix A and a vector b of N-order, one can solve x. For a quantum system, the vector b can be regarded as a quantum state $|b\rangle = \sum_i b_i|i\rangle$ and the corresponding solution of x in the quantum system can be written as

$$|x\rangle = c\mathbf{A}^{-1}|b\rangle, c^{-1} = \left\|\mathbf{A}^{-1}|b\rangle\right\| \tag{1.64}$$

The entire quantum algorithm includes three subsystems: an auxiliary system consisting of a single qubit with an initial state of $|0\rangle$; a qubit register comprising n qubits and initialized to $|0\rangle^{\otimes n}$; and an input state $|\mathrm{b}\rangle$. Expanding the input state $|\mathrm{b}\rangle$ by the eigenvector $|u_j\rangle$ of A can obtain $|b\rangle = \sum_{j=1}^{N} \beta_j|u_j\rangle$, wherein $\beta_j = \langle u_j|b\rangle$.

The specific quantum algorithm is given in Fig. 1.94, made up by the following three steps: phase estimation, controlled rotation, and reverse phase estimation. Phase estimation can be used to determine the eigenvalue λ_j of matrix A, which is a controlled unitary operation changing with a basis vector and able to map the eigenvalue to the memory. Phase estimation is an operation on the input state $|\mathrm{b}\rangle$. Using the memory bits as control, we get $\sum_{j=1}^{N} \beta_j|u_j\rangle|\lambda_j\rangle$, wherein the λ_j in $|\lambda_j\rangle$ is expressed by the binary system. At this point, the eigenvalue information is stored in the register qubits.

In the second step, we extract the eigenvalue λ_j^{-1} of matrix A^{-1} by a controlled rotation operation. The key to this step is an auxiliary qubit with an initial state $|0\rangle$. This auxiliary bit is controlled by register qubits to perform a rotational operation to obtain the following qubit state $\sum_{j=1}^{N} \beta_j|u_j\rangle|\lambda_j\rangle\left(\sqrt{1-\frac{c^2}{\lambda_j^2}}|0\rangle + \frac{c}{\lambda_j}|1\rangle\right)$. The last step is to perform a reverse phase estimation. By reversing the gate operation of the first step, it is possible to disentangle the register qubits, i.e. the register qubits are

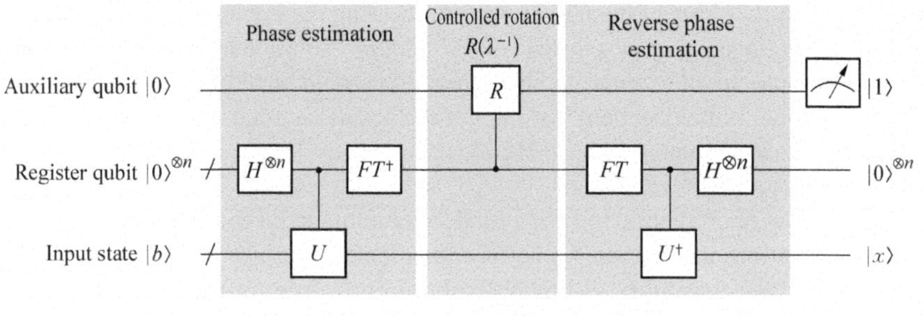

Fig. 1.94 Quantum circuit for solving a linear system of equations [91]

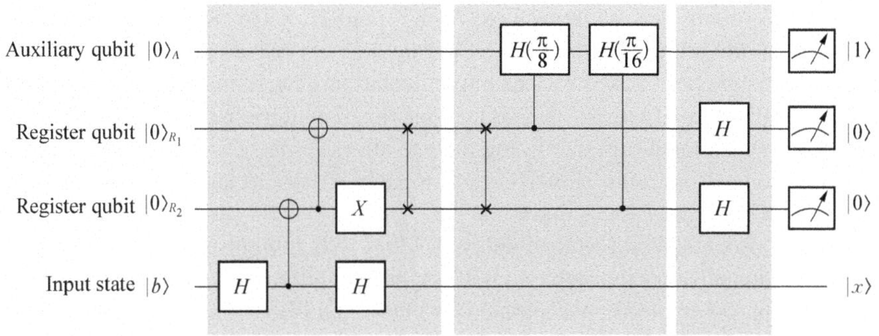

Fig. 1.95 Quantum circuit for solving a system of 2 × 2 linear equations [91]

all reset to the initial state $|0\rangle^{\otimes n}$. After this step, the state of the system becomes $\sum_{j=1}^{N} \beta_j |u_j\rangle \left(\sqrt{1-\frac{c^2}{\lambda_j^2}}|0\rangle + \frac{c}{\lambda_j}|1\rangle\right)$; post-selective measurement is performed on the auxiliary qubits. When $|1\rangle$ is detected, the outcome is $\sum_{j=1}^{N} C(\beta_j/\lambda_j)|u_j\rangle$, which is proportional to our desired moral computation $|x\rangle = cA^{-1}|b\rangle$.

2. Solutions to the system of 2 × 2 linear equations

To solve the system of 2 × 2 linear equations, we first designed a simplified quantum circuit as drawn in Fig. 1.95, chose matrix $A \begin{pmatrix} 1.5 & 0.5 \\ 0.5 & 1.5 \end{pmatrix}$, and took input states $|b_1\rangle = \frac{1}{\sqrt{2}}\begin{pmatrix} 1 \\ 1 \end{pmatrix}$, $|b_2\rangle = \frac{1}{\sqrt{2}}\begin{pmatrix} 1 \\ -1 \end{pmatrix}$, $|b_3\rangle = \begin{pmatrix} 1 \\ 0 \end{pmatrix}$. The eigenvalues of matrix A were 1 and 2, and the corresponding eigenvectors were $\frac{1}{\sqrt{2}}\begin{pmatrix} 1 \\ -1 \end{pmatrix}$, $\frac{1}{\sqrt{2}}\begin{pmatrix} 1 \\ 1 \end{pmatrix}$. This matrix is chosen to facilitate the use of two register qubits for characterization.

Phase estimation can be divided into two C-NOT gates, a swap gate, and three single-particle rotation gates. The controlled rotation $R(\lambda^{-1})$ includes two steps: first, find out the state $|1/\lambda_j\rangle$ based on $|\lambda_j\rangle$, which can be achieved experimentally by the swap gates; second, implement the controlled unitary gate $H(\theta)$ operation, wherein $\mathbf{H}(\theta) = \begin{vmatrix} \cos 2\theta & \sin 2\theta \\ \sin 2\theta & -\cos 2\theta \end{vmatrix}$; and finally, use the semi-classical approach of a single-bit rotation controlled by the measurements to achieve the reverse phase estimation.

Experimentally, we adopted the horizontal polarization ($|\mathrm{H}\rangle$) and vertical polarization ($|\mathrm{V}\rangle$) of photons to encode the logics $|0\rangle$ and $|1\rangle$. We noted that the target bits of the C-NOT gate were fixed during phase estimation, and so it was possible to implement a simplified C-NOT operation using the PBS and the HWP. By this operation, any input state $\alpha|H\rangle + \beta|V\rangle$ and the direct product state of the target bit $|H\rangle$ evolved into an entangled state $\alpha|H\rangle|H\rangle + \beta|V\rangle|V\rangle$. The $\mathrm{R}(\lambda^{-1})$ process includes two controlled unitary gates: $H(\pi/8)$ and $H(\pi/16)$. Rather than decomposing them into multiple difficult C-NOT gates, we experimentally adopted a more efficient entanglement-based way of implementation: interfering the register qubits and the auxiliary qubit via PBS 5, and then feeding the interfered auxiliary qubit to a polarization-correlated self-stabilizing Sagnac interferometer to achieve the desired controlled unitary operation. Finally, the computation was declared successful when the auxiliary qubit was measured to be $|1\rangle$.

The experimental facilities are displayed in Fig. 1.96. UV pulsed light with a wavelength of 394 nm passes through two BBO crystals to produce two pairs of entangled photons. The four photons are separated on PBS 1 and PBS 2, and prepared onto the corresponding initial states by the HWPs. Among them, photon 1 is initialized to the input state $|b\rangle$ through the HWP; photons 2 and 3 are both initialized to $|0\rangle$ as two register qubits; and photon 4 is also initialized to $|0\rangle$ as an auxiliary qubit.

Before performing quantum computation, we calibrated the quantum circuit. The initial state of the whole system is $|H\rangle_A \otimes |H\rangle_{R_1} \otimes |H\rangle_{R_2} \otimes |H\rangle_b$. Theoretically, the four qubits would change to four-photon GHZ state after passing through the phase estimation and $R(\lambda^{-1})$ process. We observed the Hong–Ou–Mandel interference of the four photons after passing through the polarizing beam splitters 3, 4, and 5, respectively. The experimentally obtained fidelity of the four-photon GHZ state was 0.65, thus proving that the computational quantum line is successful. After correction, we performed quantum computations for different input states $|b\rangle$, corresponding to three input states $|b_1\rangle = \frac{1}{\sqrt{2}}(|H\rangle + |V\rangle)$, $|b_2\rangle = \frac{1}{\sqrt{2}}(|H\rangle - |V\rangle)$, $|b_3\rangle = |H\rangle$, and the fidelity of the computations was 0.993, 0.825, and 0.836, respectively.

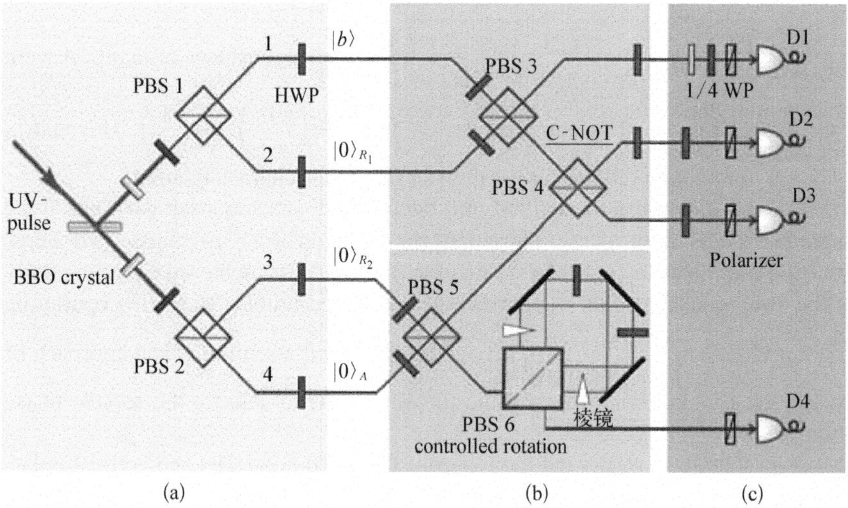

Fig. 1.96 Diagram of experimental facilities [91]. **a** Qubit initialization; **b** Phase estimation; **c** Measurement

In summary, this experiment solved the system of 2×2 linear equations on a small-scale optical quantum computer and perfectly implemented each of the steps required for this quantum algorithm. There is reason to believe that as technology advances, based on this algorithm, we will be able to find out the solution of more complex systems of linear equations using more efficient single photon sources and single photon detectors.

References

1. Bouwmeester D, Pan J W, Daniell M, et al.Observation of Three-Photon Greenberger-Horne-Zeilinger Entanglement, Phys. Rev. Lett., 1999, 82: 1345.
2. Pan J W, Daniell M, Gasparoni S, et al. Experimental Demonstration of Four-Photon Entanglement and High-Fidelity Teleportation, Phys. Rev. Lett., 2001, 86: 4435.
3. Zhao Z, Chen Y A, Zhang A N, et al. Experimental demonstration of five-photon entanglement and open-destination teleportation, Nature, 2004, 430: 54.
4. Zhang Q, Goebel A, Wagenknecht C, et al. Experimental quantum teleportation of a two-qubit composite system, Nature Physics, 2006, 2: 678.
5. Lu C Y, Zhou X Q, Gühne O, et al. Experimental entanglement of six photons in graph states, Nature Physics, 2007, 3: 91.
6. Yang T, Zhang Q, Zhang J, et al. All-Versus-Nothing Violation of Local Realism by Two-Photon, Four-Dimensional Entanglement, Phys. Rev. Lett., 2005, 95, 240406.
7. Chen K, Li C M, Zhang Q, et al. Experimental Realization of One-Way Quantum Computing with Two-Photon Four-Qubit Cluster States, Phys. Rev. Lett., 2007, 99: 120503.
8. Vallone G, Pomarico E, Martini F D, et al. Active One-Way Quantum Computation with Two-Photon Four-Qubit Cluster States, Phys. Rev. Lett., 2008, 100: 160502.
9. Ceccarelli R, Vallone G, Martini F D, et al. Experimental Entanglement and Nonlocality of a Two-Photon Six-Qubit Cluster State, Phys. Rev. Lett., 2009, 103: 160401.
10. Gao W B, Xu P, Yao X C, et al. Experimental Realization of a Controlled-NOT Gate with Four-Photon Six-Qubit Cluster States, Phys. Rev. Lett., 2010, 104: 020501.
11. Yao X C, Wang T X, Xu P, et al. Observation of eight-photon entanglement, Nature Photonics, 2012, 6: 225–228.
12. O'Brien J L, Pryde G J, White A G, et al. Demonstration of an all-optical quantum controlled-NOT gate, Nature, 2003, 426: 264-267.
13. Kiesel N, Schmid C, Weber U, et al. Linear Optics Controlled-Phase Gate Made Simple, Phys. Rev. Lett., 2005, 95: 210505.
14. Lu C Y, Browne D E, Yang T, et al. Demonstration of a Compiled Version of Shor's Quantum Factoring Algorithm Using Photonic Qubits, Phys. Rev. Lett., 2007, 99: 250504.
15. Raussendorf R, Briegel H J. A One-Way Quantum Computer, Phys. Rev. Lett., 2001, 86: 5188.
16. Walther P, Resch K J, Rudolph T, et al. Experimental one-way quantum computing, Nature, 2005, 434: 169-176.
17. Dennis E, Kitaev A, Landahl A, et al. Topological quantum memory, J. Math. Phys., 2002, 43: 4452.
18. Freedman M H, Kitaev A, Larsen M J, et al. Topological quantum computation, Bull. Amer. Math. Soc., 2003, 40: 31-38.
19. Raussendorf R, Harrington J. Fault-Tolerant Quantum Computation with High Threshold in Two Dimensions, Phys. Rev. Lett., 2007, 98: 190504.
20. Bombin H, Martin-Delgado M A. Topological Computation without Braiding, Phys. Rev. Lett., 2007, 98: 160502.
21. Yao X C, Wang T X, Chen H Z, et al. Experimental demonstration of topological error correction, Nature, 2012, 482: 489–494.

22. Aguado M, Brennen G K, Verstraete F, et al. Creation, Manipulation, and Detection of Abelian and Non-Abelian Anyons in Optical Lattices, Phys. Rev. Lett., 2008, 101: 260501.
23. Lu C Y, Gao W B, Gühne O, et al. Demonstrating Anyonic Fractional Statistics with a Six-Qubit Quantum Simulator, Phys. Rev. Lett., 2009, 102: 030502.
24. Shor P W. Polynomial-Time Algorithms for Prime Factorization and Discrete Logarithms on a Quantum Computer, SIAM Journal on Scientific and Statistical Computing, 1997, 26(5), 1484–1509.
25. Ekert A, Jozsa R. Quantum computation and Shor's factoring algorithm, Rev. Mod. Phys., 1996, 68: 733–753.
26. Gerjuoy E. Shor's factoring algorithm and modern cryptography. An illustration of the capabilities inherent in quantum computers, American Journal of Physics, 2005, 73: 521.
27. Vandersypen L M K, Steffen M, Breyta G, et al. Experimental realization of Shor's quantum factoring algorithm using nuclear magnetic resonance, Nature, 2001, 414: 883-887.
28. Bennett C H, Brassard G, Crépeau C, et al. Teleporting an unknown quantum state via dual classical and Einstein-Podolsky-Rosen channels, Phys. Rev. Lett., 1993, 70: 1895–1899.
29. Grover L K. Quantum Mechanics Helps in Searching for a Needle in a Haystack, Phys. Rev. Lett., 1997, 79: 325–328.
30. Knill E, Laflamme R, Milburn G J. A scheme for efficient quantum computation with linear optics, Nature, 2001, 409: 46.
31. Hong C K, Ou Z Y, Mandel L. Measurement of subpicosecond time intervals between two photons by interference, Phys. Rev. Lett., 1987, 59: 2044.
32. Lloyd S. Almost any quantum logic gate is universal, Phys. Rev. Lett., 1995, 75: 346.
33. Scheel S, Lütkenhaus N. Upper bounds on success probabilities in linear optics, New J. Phys., 2004, 6: 51.
34. Eisert J. Optimizing linear optics quantum gates, Phys. Rev. Lett., 2005, 95: 040502.
35. Ralph T C, White A G, Munro W J, et al. Simple scheme for efficient linear optics quantum gates, Phys. Rev. A, 2002, 65: 012314.
36. Zou X B, Pahlke K, Mathis W. Teleportation implementation of nondeterministic quantum logic operations by using linear optical elements, Phys. Rev. A, 2002, 65: 064305.
37. Scheel S, Pachos J, Hinds E A, et al. Quantum gates and decoherence, e-print quant-ph, 2004: 0403152.
38. Rudolph T, Pan J W. A simple gate for linear optics quantum computing, e-print quant-ph, 2001: 0108056.
39. Ralph T C,Langford N K, Bell T B, et al. Linear optical controlled-NOT gate in the coincidence basis, Phys. Rev. A, 2002, 65: 062324.
40. Hofmann H F, Takeuchi S. Quantum phase gate for photonic qubits using only beam splitters and postselection, Phys. Rev. A, 2002, 66: 024308.
41. Ralph T C. Scaling of multiple postselected quantum gates in optics, Phys. Rev. A, 2004, 70: 012312.
42. Pittmann T B, Jacobs B C, Franson J D. Probabilistic quantum logic operations using polarizing beam splitters, Phys. Rev. A, 2001, 64: 062311.
43. Gottesman D, Chuang I L. Demonstrating the viability of universal quantum computation using teleportation and single-qubit operations, Nature, 1999, 402: 390.
44. Yoran N, Reznik B. Deterministic Linear Optics Quantum Computation with Single Photon Qubits, Phys. Rev. Lett., 2003, 91: 037903.
45. Popescu S. An optical method for teleportation, e-print quant-ph, 1995: 9501020.
46. Nielsen M A, Optical Quantum Computation Using Cluster States, Phys. Rev. Lett., 2004, 93: 040503.
47. Browne D E, Rudolph T. Resource-Efficient Linear Optical Quantum Computation, Phys. Rev. Lett., 2005, 95: 010501.
48. Weinfurter H, Experimental Bell-State Analysis, Europhys. Lett., 1994, 25: 559.
49. Scully M O, Lamb W E. Quantum Theory of an Optical Maser. I. General Theory, Phys. Rev., 1969, 179: 368.

50. Kok P, Braunstein S L. Postselected versus nonpostselected quantum teleportation using parametric down-conversion, Phys. Rev. A, 2000b, 61: 042304.
51. Brown R H, Twiss R Q. Correlation between Photons in two Coherent Beams of Light, Nature, 1956, 177: 27.
52. Giacomini S, Sciarrino F, Lombardi E, et al. Active teleportation of a quantum bit, Phys. Rev. A, 2002, 66: 030302(R).
53. Pittman T B, Jacobs B C, Franson J D. Demonstration of feed-forward control for linear optics quantum computation, Phys. Rev. A, 2002a, 66: 052305.
54. Gisin N, Ribordy G, Tittel W, et al. Quantum cryptography, Rev. Mod. Phys., 2002: 74.
55. Scully M O, Zubauiry M S, QUANTUM OPTICS, Cambridge Universigy Press, 1997.
56. Liu Y, Chen T Y, Wang J, et al. Decoy-state quantum key distribution with polarized photons over 200 km, OPTICS EXPRESS, 2010, 18: 8.
57. Kwiat P G, Mattle K, Zeilinger A, et al. New High-Intensity Source of Polarization-Entangled Photon Pairs, Phys. Rev. Lett., 1995, 75: 4337–4341.
58. Langford N K, Ramelow S, Zeilinger A, et al. Efficient quantum computing using coherent photon conversion, Nature, 2005, 478: 360–363.
59. Pan J W, Chen Z B, Lu C Y, et al. Multiphoton entanglement and interferometry, Rev. Mod. Phys., 2012, 84: 777–838.
60. Fedrizzi A, Herbst T, Zeilinger A, et al. A wavelength-tunable fiber-coupled source of narrowband entangled photons, Opt. Express, 2007, 15: 15377-15386.
61. Liao Y B. Polarization Optics, Tsinghua University Press, 2003.
62. Hong C K, Ou Z Y, Mandel L. Measurement of sub-picosecond time intervals between two photons by interference, Phys. Rev. Lett., 1987, 59: 2044.
63. Kim Y H, Kulik S P, Chekhova M V, et al. Experimental entanglement concentration and universal Bell-state synthesizer, Phys. Rev. A, 2003, 67: 010301.
64. Gao W B, Lu C Y, Yao X C, et al. Experimental demonstration of a hyper-entangled ten-qubit Schrödinger cat state, Nature Physics, 2010, 6: 331 – 335.
65. Matthews J C F, Politi A, O'Brien J L, et al. Manipulation of multiphoton entanglement in waveguide quantum circuits, Nature Photonics, 2009, 3: 346.
66. Politi A, Cryan M J, O'Brien J L, et al. Silica-on-silicon waveguide quantum circuits, Science, 2008, 320 (5876): 646-649.
67. Marshall G D, Politi A, O'Brien J L, et al. Laser written waveguide photonic quantum circuits, Optics Express, 2009, 17, 15: 12546-12554.
68. Zhang D Y. Principles of Quantum Information Physics, Science Press, 2008.
69. Hadfield R H. Single-photon detectors for optical quantum information applications, Nature Photonics, 2009, 3: 696-705.
70. Morton G A. Photomultipliers for scintillation counting, RCA Rev., 1949, 10: 525-553.
71. Cova S, Longoni A, Andreoni A. Towards picoseconds resolution with single-photon avalanche diodes, Rev. Sci. Inst., 1981, 52: 408-412.
72. Fukuda D, Fujii G, Numata T, et al. Photon number resolving detection with high speed and high quantum efficiency, Metrologia, 2009, 46: S288-S292.
73. Rosfjord K M, Joel K W Yang, Dauler E A, et al. Nano-wire single-photon detector with an integrated optical cavity and anti-reflection coating, Opt. Express, 2006, 14: 527–534.
74. Kardynal B E, Hees S S, Shields A J, et al. Photon number resolving detector based on a quantum dot field effect transistor, Appl. Phys. Lett., 2007, 90: 181114.
75. Blakesley J C, See P, Shields A J, et al. Efficient single photon detection by quantum dot resonant tunneling diodes, Phys. Rev. Lett., 2005, 94: 067401.
76. García-Patrón R, Fiurášek J, Grangier P, et al. Proposal for a Loophole-Free Bell Test Using Homodyne Detection, Phys. Rev. Lett., 2004, 93: 130409.
77. Aaronson S, Arkhipov A. The computational complexity of linear optics, Proceedings of the ACM Symposium on Theory of Computing, 2011: 333–342.
78. Tillmann M, Dakić B, Heilmann R, et al. Experimental boson sampling, Nature Photonics, 2013, 7: 540–544.

79. Spring J B, Metcalf B J, Humphreys P C, et al. Boson Sampling on a Photonic Chip, Science, 2013, 339, 6121: 798-801.
80. Broome M A, Fedrizzi A, Rahimi-Keshari S, et al. Photonic Boson Sampling in a Tunable Circuit, Science, 2013, 339, 6121: 794-798.
81. Crespi A, Osellame R, Ramponi R, et al. Integrated multimode interferometers with arbitrary designs for photonic boson sampling, Nature Photonics, 2013, 7: 545–549.
82. Barrett S D, Stace T M. Fault tolerant quantum computation with very high threshold for loss errors, Phys. Rev. Lett., 2010, 105: 200502.
83. Raussendorf R, Harrington J, Goyal K. A fault-tolerant one-way quantum computer, Ann. Phys., 2006, 321: 2242.
84. Broadbent A, Fitzsimons J, Kashefi E. Universal Blind Quantum Computation, Foundations of Computer Science (IEEE Computer Society, Los Alamitos, CA), 2009: 517–526.
85. Barz S, Kashefi E, Broadbent A, et al. Demonstration of Blind Quantum Computing, Science, 2012, 335, 6066: 303-308.
86. Politi A, Jonathan C F Matthews, J L O'Brien. Shor's Quantum Factoring Algorithm on a Photonic Chip, Science, 2009, 325, 5945: 1221.
87. Gao W B, Yao X C, Cai J M, et al. Experimental measurement-based quantum computing beyond the cluster-state model, Nature Photonics, 2011, 5: 117-123.
88. Gross D, Eisert J. Novel schemes for measurement-based quantum computation, Phys. Rev. Lett., 2007, 98: 220503.
89. Gross D, Eisert J. Quantum computational webs, Phys. Rev. A, 2010, 82: 040303(R).
90. Harrow A W, Hassidim A, Lloyd S. Quantum Algorithm for Linear Systems of Equations, Phys. Rev. Lett., 2009, 103: 150502.
91. Cai X D, Weedbrook C, Su Z E, et al. Experimental Quantum Computing to Solve Systems of Linear Equations, Phys. Rev. Lett., 2013, 110: 230501.

Chapter 2
New Progress in Quantum Optics and Atom Optics

Weiping Zhang

2.1 Introduction

The world today is developing into a new era of quantum technology founded on quantum physics thanks to the increasing scientific cognition level and technical control ability on the microscopic quantum systems. The invention of laser has opened a window to reveal and manipulate the internal structure of matter and also triggered explorations and contemplations about the nature of light. In the process of deepening the understanding of what is light and the light-matter interaction, two important branches of modern optics, quantum optics and atom optics, came into being. Arguably, the two disciplines have not only brought us new scientific mind-sets but also led to a number of cutting-edge technologies such as atomic clock and quantum information that has reshaped and will continue to reshape human society. If quantum physics has opened a door to understanding the objective world, emerging disciplines are guiding us on how to use the knowledge of quantum physics to manipulate and transform the world around us. With the development of quantum optics and atom optics, it is now possible to manipulate and construct the quantum states of photons and even atoms, molecules and matter at the quantum level, and further explore new phenomena or technologies brought about by the interactions between photons and the quantum states of matter. This kind of research has been given a concise but far-reaching name—quantum manipulation. These studies are currently at the forefront of the international community of atomic, molecular and optical physics (AMO).

Scientific research and technological development have always advanced side by side in an alternating and synergistic manner. As quantum manipulation research continues to deepen, new scientific and technological questions begin to emerge.

W. Zhang (✉)
School of Physics and Astronomy, Shanghai Jiao Tong University, 800 Dongchuan Road, Shanghai 200240, China
e-mail: wpz@sjtu.edu.cn

W. Zhang and Z.-B. Chen (eds.), *Special Topics in Quantum Optics*,
https://doi.org/10.1007/978-981-99-8454-1_2

For instance, intensive research on the quantum nature of light has contributed to the gradual mastery of techniques to manipulate the quantum properties of photons. The advancements of these technologies extend to the fields of optical communication and information science, and the new sub-discipline of quantum information science has thus been born. Meanwhile the development and technical demand of quantum information in return drive further integration of quantum optics and atom optics to a new height. Over these years, major breakthroughs one after another on this development have continued to inject fresh vitality into the international AMO community. Ever since laser cooling and trapping of atoms won the Nobel Prize in Physics in 1997, every few years the Nobel Prize in Physics would again be awarded to this field, including atomic Bose–Einstein condensation in 2001, coherence of light in 2005, and manipulation of single quantum systems in 2012. Such achievement has rarely been seen in other research fields. On the other hand, with research further progressing, new pursuit and desire for more precise manipulation methods and technical tools increasingly grow. As a result, new developments like quantum metrology and quantum optomechanics have emerged.

As of present, a number of monographs have provided systematic introductions to quantum optics and atom optics, covering the contents from the early stages of these two different but interconnected branches of optics from different research perspectives and in different styles. However, since both disciplines are always advancing rapidly, cutting-edge projects and research interests require tracking and summarizing the latest updates. The motive of writing about this subject in the "Advances in Optical Physics" series is our earnest endeavor to present some of the newest research developments in quantum optics and atomic optics in conjunction with our research progress and experience, with the hope that it will serve as a catalyst for the exchange and development of these disciplines.

Participants in the writing and collation of this subject are young scholars who have worked along my side for many years in scientific research in the corresponding fields, and they have accumulated extensive academic knowledge and research experience in their respective areas of expertise. Our presentation of this subject is divided into eight topics, each of which covers the background knowledge and research advances of a new direction, including our own research findings. These topics reveal some of the most pioneering research trends and developments in quantum and atomic optics in recent years. The first three topics deal with the quantum manipulation of matter (atoms or materials) by light and basically represents new advances in atom optics. In Sect. 2.2, the fledgling research topic of cavity optomechanics is introduced, and this section is authored by Ke-Ye Zhang. The study of ultracold atoms in highly excited states (also known as ultracold Rydberg atom), because of their special properties, is currently a popular research topic. In Sect. 2.3, the research topic and recent advances about the interaction between ultracold Rydberg atoms and light are systematically summarized by Jing Qian. Simulating the motion of electrons in electromagnetic fields with neutral ultracold atoms, and thus enabling quantum simulation of complex electronic behaviors in the physical system of condensed states, is a new direction of research in the quantum manipulation of ultracold atoms. Section 2.4 is written by Lu Zhou and describes at length progress in the research on how to generate artificial canonical potentials by light-atom interactions to induce spin–orbit coupling of ultracold atoms to achieve spin–orbit coupled quantum gases.

The last four sections focus on photon and quantum manipulation based on an atomic ensemble, illustrating a new trend of expanding and developing quantum optics into quantum metrology. This tendency is the inevitable result of the demand for high precision manipulation techniques in quantum information research, and reflects the natural integration of quantum manipulation theory and precision measurement technology of physical quantities. This development not only endows precision measurement science a new scientific significance because of its role in quantum manipulation, but also revolutionizes some key technologies for major national strategic needs, such as interferometers and gyroscopes in inertial guidance and navigation, atomic clocks used in GPS, and weak signal detection technology in submarine tracking. In fact, precision measurement has always been an important element in the research and development of physics. The pursuit of measurement accuracy of physical quantities has driven the growth of several disciplines, as exemplified by the likes of quantum optics and atom optics. Measurement is a physical process, and measurement accuracy is governed by the laws of physics. In this sense, the measurement accuracy of physical quantities is ultimately limited by the laws of quantum mechanics, i.e., a limit determined by Heisenberg's uncertainty principle. As early as the early 1980s, the measurement accuracy limits of devices for detecting weak signals such as the laser interferometer in the laser interferometer gravitational-wave observatory (LIGO) were theoretically analyzed and researched. In the following decades, many efforts have been devoted to improving phase measurement accuracy and sensitivity of interferometers. Restrictions over measurement accuracy of physical quantities by the laws of quantum mechanics are largely reflected in two aspects: The intrinsic uncertainty invested by the Heisenberg's uncertainty principle, which is a feature of physical laws; and the quantum limit restricting each measuring instrument because each is also a quantum system. The phase sensitivity of an optical interferometer based on coherent light, for example, is limited by shot noise—often called standard quantum limit (SQL)—which is the result of the quantum fluctuations in the number of photons determined by the Poisson quantum statistics of the coherent light source. Theoretical studies have demonstrated that the SQL of phase measurement can be surpassed by changing the quantum statistical properties of the light source. While the experimental implementation of such a quantum interferometer faces many technical drawbacks and challenges, efforts in experimenting with these interferometers and related quantum interference techniques have never ceased. Substantial progress has been achieved, which should be attributed to the continual improvements in photonic and atomic quantum manipulation techniques and the rapid development of quantum information technology in recent years. Therefore, the developments of quantum manipulation techniques based on quantum optics and atom optics are bound to be the driving force behind the rise of quantum metrology. Research in this area currently involves phase measurement and valuation methods, quantum limits on quantum noise and measurement, quantum light sources, quantum correlation and quantum interference, etc. Research advancements in this field will be outlined below through the work we have been involved with in recent years.

Section 2.5 introduces optical phase measurement and quantum noise. The part addressing the quantum limit of phase measurement accuracy, the relationship between it and the measured signal-to-noise ratio (SNR), how to break the SQL of measurement using quantum processes or quantum enhancement effects, and latest advances is written by Chun-Hua Yuan. Section 2.6, which is written by Jie-Fei Chen, describes the principle and recent progress of realizing narrow-band, controllable heralded single-photon sources via a cold atomic ensemble generated by magneto-optical trap (MOT) technology. In the context of atomic ensembles, the quantum manipulation of photons achieved via Raman scattering and four-wave mixing process to generate photon-atom and photon-photon quantum correlations is currently an international focus of research, motivated by laying a foundation for the development of quantum information, quantum metrology, and other related technology. Sections 2.7 and 2.8 include the experimental investigations we have made in this area over the years, with contributions from Li-Qing Chen, Jie-Tai Jing, Florian Hudelist, Ze-Yu Ou, and some students. This book is arranged by Weiping Zhang who also writes the introduction, and compiled and proofread by Chun-Hua Yuan.

Due to time constraints and our limited knowledge, there is still room for improving this book. Suggestions and criticisms will be gratefully received. Finally, I would like to thank the graduate students in my group for their efforts in doing research associated with this subject.

2.2 Cavity Optomechanics

2.2.1 Mechanical Effects of Light

Light is an electromagnetic wave, yet few people realize that when light strikes an object, there is a direct pressure effect in addition to the Lorentz force exerted on the internal charges of the object by the electromagnetic field it carries. This pressure is caused by the momentum of photons, and thus it is called radiation pressure. Take sunlight as an example, its power intensity is about 1.5 kW/m^2 after reaching the earth, and even if it is completely converted into radiation pressure, it is only 10^{-5} Pa, far less than a standard atmospheric pressure namely 10^5 Pa. So, its existence is not even noticeable in daily life.

In a vacuum, however, the effect of radiation pressure is obvious in places close to the sun, such as in the outer space. Four hundred years ago, Johannes Kepler (1571–1630) thought that the slender tail of a comet was blown by the "solar wind" and proposed in his book *Dissertatio cum Nuncio Sidereo* (Venice, 1610) the idea that a flying machine could sail between the stars through the "solar wind", just like a sailboat sails with the sea breeze. But there is no atmosphere in the universe, there is of course no "wind". Scientific research has now proved that this "solar wind" is actually the radiation pressure effect of sunlight, and Kepler's vision has been realized on a small pilotless spacecraft by NASA and some other research institutions. They designed a "solar sail" to collect radiation pressure to propel a flight vehicle. In the astrospace near the Earth, the "solar sail" with an area of 1 km^2 can make a 1 kg object obtain an acceleration of about 9 m/s^2, which is enough to resist the gravitational force of the earth. As the vehicle approaches the sun, the effect of radiation pressure will become more and more apparent, eventually achieving five times the speed of an ordinary rocket. More importantly, the "solar sail" will enable the flying machine to be powered by any star to navigate among galaxies, which is now impossible for humankind to achieve using conventional power.

The earliest theoretical prediction that light exerts a force on the object it hits was made independently by James Clerk Maxwell (1831–1879), the builder of electromagnetic theory, and the Italian scientist Adolfo Bartoli (1851–1896) in 1871 and 1876, so radiation pressure is also known as Maxwell-Bartoli pressure. Its pressure intensity is proportional to light intensity and inversely proportional to the speed of light, i.e., $P = I/c$. Due to the low engineering level at the time, experimental tests were not performed until 20 years later by the Russian scientist Pyotr Lebedev and two American scientists Ernest Fox Nichols and Gordon Ferrie Hull, respectively.

Because the radiation pressure effect of natural light is too weak, it is impossible to carefully study and use the optical force until the invention of laser. In the 1970s, scientists successfully manipulated the motion of tiny particles using the radiation pressure difference between different regions of a focused laser beam, which became the source of laser cooling technology, and the basic theoretical framework for using lasers to trap and cool atoms was established in the mid-1970s. By the 1980s, it was experimentally possible to cool atomic gases to the μK scale, which greatly improved the measurement accuracy of atomic spectra and facilitated research on atomic clocks. For this reason, the 1997 Nobel Prize in Physics was given to Steven Chu, Claudel Cohen-Tannoudli and William D. Phillips for their contributions in this field.

With further development of laser cooling technology, atomic gases were eventually cooled to the μK scale, verifying the theoretical prediction of Bose–Einstein condensation, a result that earned Eric A. Cornell, Wolfgang Ketterle and Carl E. Wieman the 2001 Nobel Prize in Physics. Today, laser beams have become the standard tool for capturing and manipulating quantum systems composed of atoms and molecules to the level of single quanta. The ability to prepare and observe exotic quantum states with this precision has led to a major step forward in the testing and application of quantum mechanics, for which Serge Haroche and David J. Wineland shared the 2012 Nobel Prize in Physics.

At the same time, the research of using optical force to precisely manipulate mechanical components is developing almost based on the same principle. Laser manipulation of mechanical components in the solid state is of greater interest to the field of engineering than manipulation of atoms. However, mechanical components are many orders of magnitude heavier than atomic gases, so it is difficult to demonstrate significant radiation pressure effects. In order to maximize the light intensity and thus the radiation pressure, an optical cavity or optical resonator becomes the preferred aid. For example, the incident laser power can be amplified by a factor of 10^4 using a Fabry-Peort optical cavity with a mirror reflectivity of 0.9999. Therefore, optical cavities or other equivalent structures have become one of the key elements in this research direction, and cavity structure-based optomechanics has been developed accordingly.

2.2.2 Optomechanics Based on Cavity Structure

Cavity structure-based optomechanics, or cavity optomechanics for short, is, as the name implies, a branch of science that combines optical cavities and mechanics. In 1980, a research group led by H. Walther at the Max Planck Institute for Quantum Optics in Germany observed the effect of radiation pressure on a mechanical component[1]. They use an optical cavity to enhance the laser beam and thus increase the radiation pressure, while a cavity mirror on one side of the cavity is suspended vertically to form a mechanical pendulum that is forced to oscillate under the action of the radiation pressure. This device is the core of the cavity optomechanical system studied in this book—suspended cavity mirror, as a mechanical component whose mechanical motion is associated with the optical properties of the cavity such as reflectivity and transmittance through the radiation pressure. The resonant frequency of the optical cavity depends on the position of the cavity mirror or other equivalent mechanical component. Once the radiation pressure generated by the optical field inside the cavity moves the component, the cavity becomes detuned, and the intensity of the optical field inside the cavity changes resulting in a change in the radiation pressure. This feedback mechanism allows the cavity optomechanics system to exhibit rich optomechanical effects, such as "optical spring" and "optical damping" [1].

Following is the brief analysis of its principle. As shown in Fig. 2.1, if we assume that the suspended cavity mirror has a mass of M and oscillates harmonically like a mechanical pendulum with resonant frequency Ω_{m} and mechanical damping Γ_{m}, when the cavity mirror is driven by the cavity field's radiation pressure force F_{RP}, its mechanical equation of motion can be expressed as

$$\frac{\mathrm{d}^2 x}{\mathrm{d}t^2} + \Gamma_{\mathrm{m}}\frac{\mathrm{d}x}{\mathrm{d}t} + \Omega_{\mathrm{m}}^2 x = \frac{F_{\mathrm{RP}}}{M} \tag{2.1}$$

where the radiation pressure force $F_{\mathrm{RP}} = SI/c$, depending on the steady-state cavity field intensity I (S is the cross-sectional area of the optical beam) whereas the optical field intensity in the optical cavity is determined by the cavity length, i.e., cavity mirror displacement x. Here, we will not write down the equation for cavity field intensity; detailed discussions can be found in literature [2]. It should be emphasized that the cavity mirror and the cavity field in the system each take different time to stabilize, which can be estimated by the reciprocal of the cavity mirror damping 1/

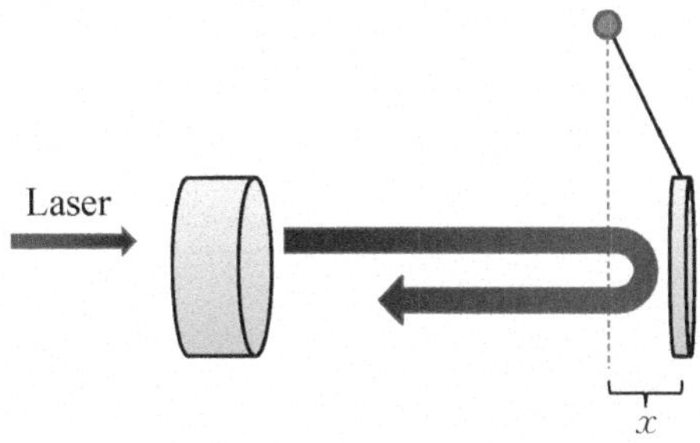

Fig. 2.1 Feedback principle of the cavity optomechanics system: the incident laser beam accumulates in the optical cavity, and its enhanced radiation pressure is sufficient to push the thin suspended cavity mirror and change the cavity length; in turn, the change in cavity length readjusts the intensity of the optical field in the cavity and the equilibrium position of the cavity mirror affected by the radiation pressure

Γ_m and the reciprocal of the optical cavity linewidth l/*k*, respectively. The oscillation frequency $\Omega_m/2\pi$ of the cavity mirror or other equivalent mechanical component ranges from 10 to 10^9 Hz, while the range taken for the mechanical quality factor Q_m is from 10^3 to 10^7. Therefore, the typical mechanical damping is about $\Gamma_m = \Omega_m/Q_m \approx 2\pi \times 10^{-2} \sim 10^2$ Hz, much smaller than the linewidth of the optical cavity, meaning that there is a delay in the response of the cavity mirror to the radiation pressure when the radiation pressure in the cavity field is changed. We introduce an effective time delay τ for this purpose, and the steady radiation pressure felt by the cavity mirror is actually $F_{RP}[x(t-\tau)]$. Conversely, the change in the cavity field after the cavity mirror moves is followed almost instantaneously. It indicates that Eq. (2.1) is a delay differential equation. If the delay τ is small, we can make the Taylor series of the right side of the Eq. (2.1)

$$\frac{d^2x}{dt^2} + \Gamma_m \frac{dx}{dt} + \Omega_m^2 x \approx \frac{F_{RP}(x(t))}{M} - \frac{\tau}{M}\frac{dx}{dt}\left.\frac{\partial F_{RP}(x)}{\partial x}\right|_{x=x(t)} + \quad (2.2)$$

The second term on the right side of Eq. (2.2) is proportional to the first-order derivative of displacement with respect to time, which can be perceived as the cavity damping being corrected by the radiation pressure, i.e., "optical damping" or "radiation pressure damping". The corrected cavity damping is

$$\Gamma_{RP} = \Gamma_m + \frac{\tau}{M}\left.\frac{\partial F_{RP}(x)}{\partial x}\right|_{x=x(t)} \quad (2.3)$$

Unlike the mechanical damping Γ_m of the cavity mirror, optical damping is not a constant but a function of mirror displacement, and its value can vary between positive and negative. Positive optical damping will help stabilize the cavity mirror while negative one will make the dynamics of the cavity mirror unstable, leading to self-induced oscillations [3] and even chaotic phenomena [4]. If the displacement x

of the cavity mirror is also small, we can expand the Taylor series of the first term on the right side of Eq. (2.3) as well, and its first-order term can be regarded as the correction of the harmonic oscillator's stiffness factor by the radiation pressure.

$$k_{RP} = M\Omega_{m}^{2} - \left.\frac{\partial F_{RP}(x)}{\partial x}\right|_{x=0} \tag{2.4}$$

It is "optical spring" effect. With the stiffness factor of the cavity position, multiple equilibrium positions may appear during the motion of the cavity and optical bistability [5] or multistability [2, 6], similar to that in nonlinear optics, thus arises from cavity field intensity. The difference is that in nonlinear optics it is the Kerr nonlinear medium that corrects the refractive index and effective optical path length, but here it is the cavity length that is altered.

In fact, the Soviet scientist V. Braginsky performed similar experiments in the microwave band more than a decade ago and later in his work proposed the idea of using this correction of the radiation pressure on the mechanical properties of the oscillator to implement weak force measurements [7]. Under his influence, R. Drever et al. suggested attaching such a device to a giant Michelson interferometer to detect gravitational waves [8]; because of the interferometer's super-long optical cavity length (~4 km), the light intensity accumulated in the cavity brings about massive radiation pressure to stabilize and cool the interferometric cavity mirror. The scheme is still in use today by two most famous gravitational wave detection projects, i.e., Laser Interferometer Gravitational-Wave Observatory (LIGO) in the United States, and "Vigra", a joint project of France and Italy.

2.2.3 Cavity Optomechanics and Weak Force Measurement

The creation of cavity optomechanics is to measure the weak displacement of cavity mirrors caused by gravitational waves. First, let us analyze the principle whereby the cavity optomechanical system is used for measuring weak force. Assuming that the suspended cavity mirror is sensitive to the weak signals to be detected, such as magnetic fields and gravitational fields, their effect is manifested as a weak force F_w exerted on the cavity mirror. The cavity mirror changes the cavity length by moving under its drive, which leads to phase shift and intensity change of the output cavity field. The cavity mirror varies in cavity length by moving under its drive, thereby leading to the phase shift and intensity change of output cavity field. Although the magnitude of the weak force is known from measurements of both light intensity change and phase shift, we only consider phase measurement in this book as phase shift is more sensitive to the displacement of the cavity mirror. Ideally, we have a one-to-one correspondence among phase shift Δ_{φ}, cavity mirror displacement x, and the weak force F_w to be measured.

$$\delta\varphi \propto x \propto F_w \tag{2.5}$$

So, by measuring the phase shift of the laser field we can obtain the magnitude of the weak force. In actual measurement however, the shot noise of the laser field will affect the measurement accuracy of its phase. If we use ξ_{sh} to represent the influence of shot noise in the dimension of displacement, we have

$$\delta\varphi \propto x + \zeta_{\text{sh}} \tag{2.6}$$

Apart from the weak force, the cavity mirror displacement is also affected by the laser field's radiation pressure and its own thermal motion. If we represent their effects by F_{RP} and η_{th} respectively, then

$$x \propto F_{\text{w}} + F_{\text{RP}} + \eta_{\text{th}} \tag{2.7}$$

These three types of noise will limit the actual accuracy of the cavity optomechanics measuring device. In most measurement experiments, the signal to be measured often has a specific frequency range, so spectral analysis of noise effects is of more practical interest. As in Sect. 2.2, we still assume that the motion of the cavity mirror is a single-mode harmonic oscillator driven by weak force as well as other noise forces, and its equation of motion is

$$\frac{\mathrm{d}^2}{\mathrm{d}t^2}x(t) + \Gamma_{\text{m}}\frac{\mathrm{d}}{\mathrm{d}t}x(t) + \Omega_{\text{m}}^2 x(t) = \frac{F_{\text{w}}(t) + F_{\text{RP}}(t) + \eta_{\text{th}}(t)}{M} \tag{2.8}$$

According to the Fourier analysis of spectral space, the equation is

$$x(\omega) = \chi(\omega)[F_{\text{w}}(\omega) + F_{\text{RP}}(\omega) + \eta_{\text{th}}(\omega)] \tag{2.9}$$

where the mechanical polarizability of the cavity oscillator is defined by Eq. (2.10)

$$\chi(\omega) = \frac{1}{M\left(\Omega_{\text{m}}^2 - \omega^2 - i\Gamma_{\text{m}}\omega\right)} \tag{2.10}$$

According to the relations (2.6) and (2.7) derived from previous analysis, the relationship among the measurement dI(ω) corresponding to the phase shift in the spectrum space, the weak force to be measured, and the noise can be obtained

$$\delta\phi(\omega) \propto \underbrace{\chi(\omega)F_{\text{w}}(\omega)}_{\text{Signal}} + \underbrace{\chi(\omega)\left[F_{\text{RP}}(\omega) + \eta_{\text{th}}(\omega)\right] + \zeta_{\text{sh}}(\omega)}_{\text{Measurement Noises}} \tag{2.11}$$

Equation (2.11) clearly shows the effect of three types of noise on the measurement. Thermal noise is much greater than the other two under usual conditions, so cooling the cavity mirror is often the primary measure to improve the accuracy of the cavity optomechanical measuring device. After the thermal noise is suppressed at very low temperatures, the effect of shot noise and radiation pressure noise stands out. Both of them have the quantum nature of light by nature, so they are quantum noise, which exists even at absolute zero. However, it is important to note that the noise F_{RP} arising from radiation pressure is proportional to the light intensity while the shot noise ξ_{sh} of the optical field is inversely proportional to the light intensity. This means that if the two kinds of noise are independent of each other, the sum of their effects will be minimized at a particular value of light intensity, which is the standard quantum limit (SQL) of the cavity optomechanical measuring device.

To estimate the value of this measurement limit, we assume that the thermal noise has been removed at low temperatures, the power intensity spectrum of the noise can be expressed as

$$S_n(\omega) = |\chi(\omega)|^2 S_{\mathrm{F}}(\omega) + S_{\zeta}(\omega) \tag{2.12}$$

It must be emphasized here that the above equation has already assumed that there is no quantum correlation between radiation pressure noise and shot noise, depending on the quantum nature of the incoming laser. Due to the positive definiteness of the spectral density, the relation can be derived from the fundamental inequality

$$S_n(\omega) \geq 2|\chi(\omega)|\sqrt{S_{\mathrm{F}}(\omega) S_{\zeta}(\omega)} \tag{2.13}$$

However, quantum mechanics sets limits to the measurement accuracy at the microscopic scale. Based on the uncertainty relation of quantum mechanics, one can derive [9]

$$S_{\mathrm{F}}(\omega) S_{\zeta}(\omega) \geq \frac{\hbar^2}{4} \tag{2.14}$$

Hence, the SQL of the additional noise is

$$S_n(\omega) \geq \hbar|\chi(\omega)| \tag{2.15}$$

In the resonance case in particular, we can find

$$S_n^{\mathrm{SQL}}(\omega_{\mathrm{m}}) = \frac{\hbar}{m\gamma\omega_{\mathrm{m}}} \tag{2.16}$$

The SQL corresponding to the uncertainty of displacement measurement is [10]

$$\Delta x_{\mathrm{SQL}} = \sqrt{\hbar/2m\omega_{\mathrm{m}}} \tag{2.17}$$

It is exactly the same as the zero-point fluctuation of the quantum harmonic oscillator. What needs to be emphasized is that this SQL is not a fundamental limit and can be breached by using the squeezed optical field to establish a correlation between the two noises or by applying a measurement scheme that is immune to quantum noise. It will be further discussed in Sect. 2.5.

2.2.4 *Cavity Optomechanics in Quantum Region*

Quantum mechanics is a century old. It is probably the most successful theory constructed by humankind so far, yet its explanations and roots are still criticized by many, and the predictions it gives are often difficult to test experimentally. This is because the scale of the phenomena described in quantum mechanics is far from the scale of daily life; quantum entanglement, superposition states, and vacuum fluctuation, for instance, are completely missing in the macroscopic world. But objects in the macroscopic world are indeed composed of a large number of microscopic particles, then what happens in the process that makes quantum effect disappear? Physicists are trying hard to figure it out. On the theoretical side, one of the most widely accepted explanations now is decoherence theory, but experimentally it requires a platform of continuous transition between macroscopic and microscopic worlds. The cavity optomechanical system exhibits an amazing potential in this respect. In addition, measurement accuracy, a fundamental requirement for the development of physics, drives the need for cavity optomechanics to move to the quantum region. As early as 30 years ago, physicists working on gravitational wave detection realized that the motion at the quantum mechanical scale of the suspended cavity mirror must be considered [11], due to the fact that measuring gravitational waves requires monitoring the movement of the cavity mirror to a very high precision. In the case of the LIGO system, for example, in spite of its 4-km long optical cavity, the accuracy of monitoring the movement of the cavity mirror needs to be at least on the order of 10^{-18} m to detect gravitational waves. This magnitude is already close to the SQL for position measurement. Finally, if cavity optomechanics, a combined system of laser and solid-state device, works in the quantum region, it has obvious advantages over other microscopic particle-based quantum systems in terms of quantum information transmission, storage, processing, etc. Quantum cavity optomechanics has recently become a hot research topic in quantum physics.

The interaction Hamiltonian of quantum cavity optomechanics can be derived from the analysis of the effect of cavity length change on the energy of the single-mode quantum optical field. According to the theory of quantum optics, the Hamiltonian of a single-mode quantum field in an optical cavity i $H = \hbar\omega\hat{a}^{\dagger}\hat{a}$, where $\hat{a}$ is the annihilation operator of photons, $\hbar$ is reduced Planck constant, and ω is the frequency of cavity mode, which has a dependence relationship with cavity length L. If a slight change x happens in cavity length, the Hamiltonian can be approximated as

$$H = \hbar\omega\hat{a}^{\dagger}\hat{a} + h\frac{\partial\omega}{\partial L}\bigg|_{L=L_0}\hat{a}^{\dagger}\hat{a}\hat{x} \tag{2.18}$$

where the second term describes the optomechanical interaction, the first-order derivative of ω at the original cavity length can be defined as the optomechanical coupling strength, while x is described in the quantum region using operators, representing the displacement of a quantum harmonic oscillator driven by the radiation pressure. While quantized optical fields can be easily generated in the laboratory, it is not an easy task to bring optomechanical oscillators into the quantum region. Specifically, the development of quantum cavity optomechanics can be divided into three lines of research: "top-down", "bottom-up", and "hybrid system", depending on the way to make optomechanical oscillators enter the quantum region.

As the name suggests, "top-down" is a way to gradually enter the quantum region by continuously reducing the scale of conventional cavity optomechanical system and suppressing its thermal noise. The main foundation for this is the fast development of Micro-Electro-Mechanical System (MEMS) and Nano-Electro-Mechanical System (NEMS) in recent years. On the one hand, the emergence of "high-finesse" optical microcavities has made it possible to accumulate massive radiation pressure even in ultra-small optical resonator structures. On the other hand, suspended cavity mirrors have been replaced by cantilevers used in atomic force microscopes, silicon nitride films, or whispering-gallery microresonators. These micron-scale structures, despite of a small area to withstand radiation pressure, can all demonstrate remarkable optomechanical effects under laser beams enhanced by high-finesse optical cavities by virtue of their smaller effective mass (~ng). In the case of a $1-\mathrm{W}$ laser beam, the radiation pressure generated by focusing it on an area of 100 μm^2 is about 7×10^{-9} N can give these miniature mechanical devices an acceleration of 10 m s^{-2}, similar to what we just estimated for the "solar sail".

Currently, the cavity optomechanical system demonstrates an amazing span in both mass and scale, with the mass of the suspended cavity mirror of the LIGO system on the order of kg, while that of the effective mechanical oscillator of the ultracold atomic gas coupled to the microcavity is about 10^{-20} kg; the optical cavity length of LIGO is 4 km, and the length of the optical microcavity is only 10^{-7} km. This makes the cavity optomechanical system an ideal platform for studying "quantum–classical transition" physics (Fig. 2.2).

Although the motion of mechanical components including micromirrors and cantilevers under the action of radiation pressure is complex and involves various modes of vibration and rotation, it has been found experimentally that their motion can be described by a single-mode damped oscillator as long as their mechanical quality factors are high enough. Furthermore, to show quantum mechanical effects, an oscillator needs to satisfy the following condition

$$k_{\mathrm{B}}T \leq \frac{\hbar\omega_{\mathrm{m}}}{2} \tag{2.19}$$

where k_B and T are the Boltzmann constant and the temperature of the oscillator, respectively. Therefore, if one wishes to observe quantum mechanical effects on macroscopic-scale mechanical oscillators, one needs very low temperatures or very high natural frequencies. For example, the typical natural frequency of a cantilever is in MHz, so it is possible to demonstrate quantum mechanical effects only at $T < 1\ \mu K$. If its frequency is increased to the GHz order of magnitude, the above relationship can be satisfied at 50 mK, a temperature that can be reached even in a cold storage. Considering that thermal noise will produce a random driving force that makes the

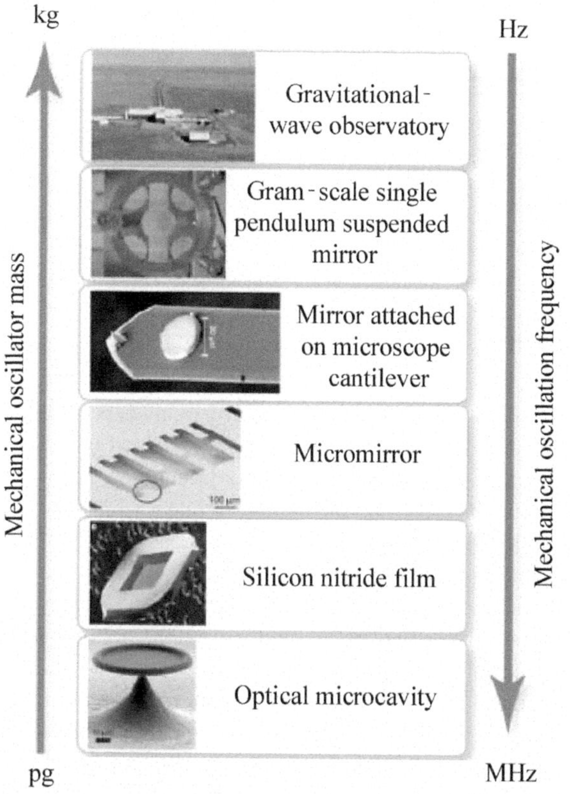

Fig. 2.2 A comparison of various cavity optomechanical systems now experimentally realized shows that their sizes are at both macroscopic and microscopic levels [12]

oscillator do Brownian motion and leads to the decoherence of the quantum system, physicists prefer to let the oscillator enter the quantum region by cooling. However, it is important to highlight that in recent years high frequency and low damping oscillators have been realized in solid state materials, and room-temperature quantum optomechanical systems are developing fast.

The usual means of cooling can at best bring the oscillator to the same temperature as its surroundings, which is far from the goal of μK, which is why the technology for laser cooling of solid mechanical components advances quickly. The technique cleverly exploits the special feedback mechanism of cavity optomechanics to cool mechanical oscillators and has two main branches of development. One is active feedback cooling, proposed by Stefano Mancini et al. in 1997 [13] and first implemented two years later by an experimental group in France [14], with the central idea of using radiation pressure to counteract the stochastic forces resulted from thermal noise. First, a laser beam with stable intensity and frequency is applied to drive the cavity optomechanical system, then the random vibration of the oscillator is monitored through "homodyne measurement" while the resulting signal is used to adjust the laser beam, thus changing the radiation pressure to stabilize the mechanical oscillator. This scheme of using feedback to control the radiation pressure to cool the

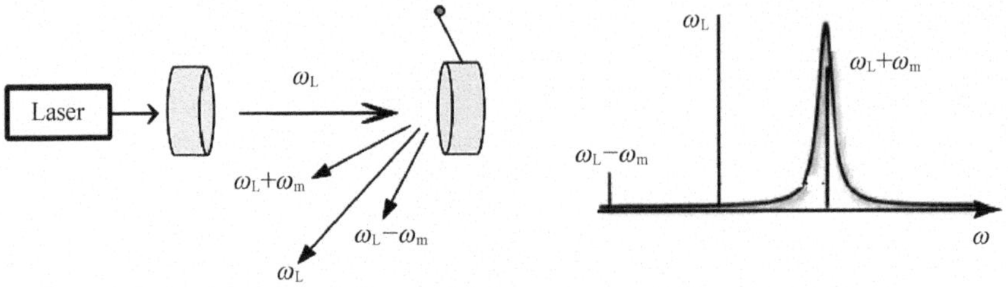

Fig. 2.3 The interaction between the laser and the cavity mirror leads to two frequency sidebands; to achieve the purpose of cooling the cavity mirror, the cavity needs to resonate with the blue-detuned sideband

oscillator has now been adopted by many experimental groups, allowing the temperature of the oscillator to drop to the order of μK. The other branch is modeled after the sideband cooling principle for laser cooling ions. Assuming that the incident laser frequency is ω_L and the natural frequency of the oscillator is ω_m, the light scattered by the oscillator produces two frequency sidebands, i.e., $\omega_L \pm \omega_m$. If the position of the oscillator is pre-adjusted so that the optical cavity resonates with the blue-detuned sideband, i.e., $\omega_L + \omega_m = \omega_c$, the energy of the oscillator is continuously transferred to the optical field, eventually achieving the purpose of cooling [15, 16]. This cooling mechanism is expected to cool the oscillator to its quantum–mechanical ground state, thus exhibiting vacuum fluctuation in the macroscopic world (Fig. 2.3).

The "bottom-up" approach is the exact opposite of the previous one, in which a cavity optomechanical structure is built up and amplified from an existing microscopic quantum system. A typical example is to put an ultracold atomic gas or atomic Bose–Einstein condensate (BEC) cooled to the quantum region into an optical cavity and use its atomic collective mode of motion to act as a quantum mechanical oscillator coupled to the in-cavity optical field to form an equivalent cavity optomechanical system. Atomic gases are much easier to cool than solid-state systems because of their low density and easy isolation from the surrounding environment. In particular, it undergoes quantum phase transitions at very low temperatures to produce atomic BECs, which can exhibit many macroscopic quantum effects. Although the mechanical effects of coupling a single atom to a cavity field are inapparent, the effects of coupling a collective mode of atomic motion to a cavity field is notable. Atomic BECs allow the coupling of the cavity field to the collective mode of atom motion and maintain obvious quantum effects similar to those of single atoms at the same time.

In an optical cavity with a fixed cavity length L, the many-body Hamiltonian of the atomic BECs coupled to the single-mode cavity field can be expressed as

$$H = \int \hat{\psi}^{\dagger}(x)\left[-\frac{\hbar^2\partial^2}{2m\partial x^2} + \hbar g_0 \hat{a}^{\dagger}\hat{a}\cos^2(kx)\right]\hat{\psi}(x)\mathrm{d}x \tag{2.20}$$

where g_0 is the single photon coupling strength of the cavity field and atom, $\cos^2(kx)$ describes the standing wave mode in the cavity with wave vector k, and $\hat{\psi}(x)$ is the field operator describing the atomic ensemble. The standing wave cavity field interacting with the atom will recoil the atom to a mode of motion with momentum $\pm 2\ hk$, so the atomic field operator can be expanded as

$$\hat{\psi}(x) \approx \frac{1}{L} c_0 + \frac{\sqrt{2}}{L} \cos(2kx)\hat{c}_2 \tag{2.21}$$

In this case, c_0 represents the annihilation operator of the atom that maintains the initial momentum mode, where most of the atoms are always in this mode due to the weak interaction with light, and so its quantum fluctuations can be neglected and expressed as a classical quantity, whereas $\hat{c}_2$ is the annihilation operator of the atoms recoiled whose quantum fluctuations are obvious due to a small number of atoms. If we plug the original Hamiltonian into the approximate expression, the interaction with the photon becomes $hg_0c_0\hat{a}^\dagger\hat{a}\left(\hat{c}_2 + \hat{c}_2^\dagger\right)/2\sqrt{2}$; as long as a quantum displacement operator $\hat{x} = \left(\hat{c}_2 + \hat{c}_2^\dagger\right)/\sqrt{2}$ is redefined, the form of this interaction is consistent with the Hamiltonian of quantum optomechanical interaction, and the interaction strength is greatly enhanced due to $c_0 \approx \sqrt{\mathrm{N}}$ (N is the total number of atoms). Thus, despite the fixed length of the optical cavity, the oscillation of an ultracold atom driven by light between the initial and recoil momentum modes can be equated to a quantum optomechanical oscillator. Since atomic BECs are substances with pronounced quantum effects and are well manipulated, quantum optomechanical systems based on them can exhibit many exotic quantum properties, e.g., quantum optomechanical measuring systems that manipulate the effective mass of an atomic BEC via an optical field and thus avoid quantum measurement backaction [17]; or, replacing a standing wave cavity with a toroidal cavity and combining it with an ultra-strong atomic bound potential well help realize a quantum cavity optomechanical system in which the roles of the atomic BEC and the optical field are swapped, and thus a vast spectrum of exotic quantum states of the atomic BEC can be prepared and measured by means of optomechanical manipulation [18]. However, compared to the "top-down" approach based on solid-state systems, the "bottom-up" approach is more advantageous in terms of principle exploration and less advantageous in terms of practical applications due to the characteristics of cold-atom experimental systems.

The third idea is to combine the strengths and weaknesses of the first two lines of research through a hybrid system whereby ultracold atomic oscillators are coupled to solid-state oscillators that act as cavity mirrors through an optical cavity. This hybrid system allows the use of ultracold atoms to co-cool solid-state oscillators into the quantum region, and it is less demanding for solid-state materials than the direct cooling method. Moreover, ultracold atomic systems have been researched as storage units of quantum information for many years, and such hybrid systems can transfer quantum information stored in cold atoms to solid-state devices. In addition, the manifestation of classical effects like multistability and chaos caused by nonlinearity in the hybrid systems in the quantum–classical transition region is also a research hotspot of decoherence theory. In addition to hybrid systems based on cold atoms, hybrid systems based on solid-state materials that can operate in different optical frequency domains are also of great interest. A number of optomechanical oscillators made of special solid-state materials can be coupled to both microwave and optical wave (visible frequency domain), and the hybrid-cavity optomechanical system built in this way is promising to become a signal processing node of composite acousto-optical quantum information network, which has a broad prospect in the applications of quantum information.

2.2.5 Summary and Prospect

Cavity optomechanics is originated from the observational study of macroscopic-scale cosmic phenomena, and the requirement for measurement accuracy pushes its study into the microscopic quantum region. Although the measurement of gravitational waves has not yet been successful (at the birth moment of the initial verion of this book), the study of cavity optomechanics is a unique part of quantum mechanics, and is developing rapidly and promisingly in applications such as quantum fundamental theory and quantum information processing. Non-classical states including superposition, entanglement, and squeezed states are invisible in the macroscopic world, but are very active in optical systems, and are central to the disciplines of quantum information, quantum computing, precision measurement, etc. Because the cavity optomechanical system couples the light field and the mechanical device, many physicists are trying to transfer nonclassical states from the light field to the mechanical device or use the light field to read out extremely weak nonclassical states that may already exist from the mechanical device in a bid to find traces of nonclassical states in the macroscopic world. For instance, Marshall et al. came up with an experimental scheme to prepare cavity mirror superposition states through the interaction of entangled state photons with cavity mirrors [19], and Knight et al. proposed to prepare a cavity field with multiple components in the Schrödinger cat state or entangled state by a cavity optomechanical system [20]. There are also more straightforward ideas to achieve entanglement between macroscopic objects using cavity optomechanical systems, such as the entanglement of two cavity mirrors with a cavity optomechanical system in which both mirrors are suspended [21], the multi-body (mirror-field-mirror) entanglement via a three-mirror cavity [22], and the entanglement between insulating films in a cavity [23]. However, according to decoherence theory, the stronger the coupling of a larger object to its environment, the faster its quantum mechanical wave packet collapses, so cooling techniques will remain key if entanglement between macroscopic objects is to be observed in accordance with these experimental scenarios. In addition to studying quantum–classical transition physics, the high sensitivity and easy manipulation of optomechanical cavities has given rise to hybrid cavity optomechanics. In other words, physicists couple various quantum mechanical systems, i.e., Bose–Einstein condensates, ultracold polar molecules, ions, single atoms, and the color centers of diamonds, with optomechanical cavities or their equivalent systems. In these systems, optomechanical cavities can be used to detect or manipulate quantum states or quantum effects in quantum systems, which is of great significance to the fields of precision measurement and quantum information. Cavity optomechanics has not only become a multidisciplinary research direction, but also will gradually develop into an ideal platform for exploring the practicality of quantum engineering and quantum technology.

2.3 Interactions of Ultracold Rydberg Atoms with Light Field

Recent years along with the development of laser cooling and atomic trapping techniques, the study of ultracold Rydberg atomic gases has become a new frontier of atomic physics, drawing the attention of an increasing number of scientists internationally. When the temperature of atoms reaches the micro-Kelvin region the center-of-mass motion of the atomic system can be well controlled, and scientists are able to focus on the internal interactions of these nearly "frozen" atoms. At this point, in order to describe the dynamics of them, it is usually sufficient to consider only the interactions between the atoms and the environmental effects. On the other hand, recent work has also suggested that the assumption of "frozen" atoms would not be valid if atoms are considered to be trapped in optical lattices or micro-potential wells, since the center-of-mass motion of the atoms would severely affect the excitation dynamics of atoms [24]. We will ignore this effect here and focus on the interaction between Rydberg atoms and their coupling effects with the environment.

2.3.1 Brief Introduction to Rydberg Atoms

In 1888, Swedish scientist J. R. Rydberg first experimentally studied the spectroscopic properties of atoms in highly-excited states, and these atoms are therefore called Rydberg atoms. Their major feature lies in that at least one of the outermost electrons should be excited to an energy level with a large principal quantum number. The outer electrons are far away from the nucleus and loosely bounded by it. Hence, Rydberg atoms are easily perturbed by an external electromagnetic field and their interactions with other neighboring atoms are also drastic. Earlier researches on Rydberg atoms were limited to the field of high-energy physics because the huge energy released during the production of Rydberg atoms could be used for plasma generation, detection of supernovae, etc. It was not until the invention of the laser in 1960 that the selective excitation of atoms was experimentally achieved, making it possible for preparing purer Rydberg atomic sources. In the following decades, many atomic gases at room temperature were used in the experimental researches of Rydberg atoms. Due to the existence of a relatively wider optical spectrum limited by the thermal motion of the atoms, it results in Doppler broadening, which greatly affected the sensitivity and resolution of measurements [25, 26]. It was only in the 1990s that three scientists S. Chu, C. C. Tannoudji, and W. D. Philips made a major breakthrough in the laser cooling and trapping techniques of atoms, leading to a significant step forward in cold atom experiments, and for which they were awarded the Nobel Prize in Physics in 1997. Subsequently, with the continuous development of atomic cooling technologies, scientists have been able to achieve high-density atomic gas sources at ultra-low temperatures, which also creates new opportunities for the development of ultracold Rydberg atomic physics. Currently, there are two main approaches to experimentally prepare ultracold Rydberg atoms: one is to achieve Stark deceleration of Rydberg atomic gases through a nonuniform electric field, since Rydberg atoms are highly polarized [27, 28]; the other is to generate atoms in the Rydberg state by means of continuous optical excitation with the help of ultracold atomic sources that have been successfully prepared (i.e., Bose–Einstein condensates) [29]. To our knowledge, the latter has two merits: first, it is not affected by the lifetime of Rydberg states, and second, the density of the initially prepared atomic sources in ground states is very high, e.g., the density of Bose–Einstein atomic condensate is up to 10^{20} m^{-3}. Thus, the center-of-mass motion of the atom in the ultra-low temperature environment is much smaller than the lifetime of the Rydberg-state atoms, which can be regarded as a "frozen" atom [30, 31].

2.3.1.1 Basic Properties of Rydberg Atoms

We explain some special properties of the Rydberg atoms based on the atomic Bohr model. If an electron charge *-e with* mass m moves circularly around the radius r, depending on the classical mechanics it is in agreement with Newton's second law of motion $F = ke^2/r^2 = mv^2/r$, where $k = 1/4\pi\varepsilon_0$, and ε_0 is vacuum permittivity. As the orbital angular momentum is quantized $mvr = nh$, the relationship between the electron's orbital radius r_{n} and the principal quantum number n can be obtained

$$r_n = a_0 n^2 \tag{2.22}$$

where the Bohr radius is $a_0 = 4\pi\varepsilon_0 h^2/me^2$. From Eq. (2.22) when $\mathrm{n} = 137$, the orbital radius of the Rydberg atom is about 1 μm and the collision cross-section is about 1 μm^2. The fact that the Rydberg atom has such a large atomic radius makes the outer electrons weakly bounded by its atomic kernel, giving rise to the easy ionization of atoms in environment. Here the Rydberg orbital energy E_{n} can be expressed as the sum of the kinetic and potential energies of the electrons outside the nucleus:

$$E_n = \frac{mv^2}{2} - \frac{kZe^2}{r} = -\frac{R_{\mathrm{m}}}{n^2} \tag{2.23}$$

where Z is the nuclear charge number (for neutral element, $Z = 1$), an $R_{\mathrm{m}} = k^2Z^2e^4m/2\hbar^2$ is the Rydberg constant of the element with mass m. For hydrogen atoms, $R_m = 13.6$ eV. Apparently, E_{n} is negative, because a negatively-charged electron always corresponds to a positively-charged electron, and the binding energy of the atomic nucleus to the electron is proportional to $1/n^2$. As can be seen in Eq. (2.23), the spacing of adjacent energy levels within the atoms should meet

$$E_{n+1} - E_n \approx \frac{2R_{\mathrm{m}}}{n^3} \tag{2.24}$$

From the above conclusion, we can see that when the principal quantum number n is large, the binding energy between the electron at the highly-excited Rydberg energy level and the nucleus is weak, which is proportional to $1/n^2$. In contrast, the orbital radius of the electron itself is large, which is proportional to n^2. Therefore, in a Rydberg atom, the outer electrons move with a large orbital radius, but the binding force with respect to the nucleus is very small.

Actually, any Rydberg energy level takes an energy of

$$E_{nl} = E_\infty - \frac{R_{\mathrm{m}}}{(n - \delta_{\mathrm{l}})^2} \tag{2.25}$$

where E_∞ is the ionization energy and δ_{l} is the quantum defect for a given angular momentum l energy level which is usually determined by empirical observation [32]. The quantum defect term arises because the orbit of the electron outside the nucleus is not absolutely circular, and even if the electron is located at a large energy level, it may be very close to the nucleus during its motion. The concept of quantum defect is introduced to describe the penetrating effect of outer electrons into the atomic nuclear region. Different angular momentum energy levels correspond to different quantum defects. In the case of rubidium atoms, $\delta_0 = 3.13$, $\delta_1 = 2.64$, $\delta_2 = 1.35$, and $\delta_3 = 0.016$. As the angular momentum quantum number l increases, the less time the electron stays around the nucleus and behaves more similarly to a hydrogen atom, the smaller the value of δ_{l} is. Table 2.1 describes dependence of some special properties of Rydberg atoms on the principal quantum number n, where $n \rightarrow n\text{-}\delta_{\mathrm{l}}$ is assumed by considering the effect of the quantum defect.

Table 2.1 Special properties of Rydberg atoms

Property	Correspondence	$n = 80$
Orbital radius	n^2	3 μm
Polarizability	n	10^3 GHz/(V/cm)^2
Electric dipole moment	n	10^4 ea_0
Spontaneous radiation lifetime	n^3	600 μs

2.3.1.2 Interactions of Rydberg Atoms

There are usually two types of interactions between Rydberg atoms: van der Waals interactions and resonant dipole–dipole interactions.

(1) The motion of electrons in an atom would generate an instantaneous dipole moment, leading to a transient polarization of neighboring atoms, which in return enhances the instantaneous dipole moment of the original atom. This process of coupling the instantaneous dipole moments to each other to produce attractive or repulsive forces is called "van der Waals interaction". Van der Waals forces are generally ubiquitous and nondirectional with its strength much smaller than the chemical bond energy; however, it can affect the properties of matter. For example, the melting point of nitrogen is 4 K whereas the melting point of radon gas is 221 K. Because radon has more electrons than nitrogen. Hence the resulting van der Waals forces exerted on it are larger, and the temperature to overcome the bonding of van der Waals forces and turn it into gas is also higher. In addition, heavier atoms bring larger instantaneous dipole moments to neighboring atoms, and thus have greater van der Waals interactions.

We introduce here a simple model to describe the atom–atom Rydberg interactions, with the Hamiltonian of the system expressed as

$$E_{nl} = E_{\infty} - \frac{R_{\mathrm{m}}}{(n - \delta_1)^2} \tag{2.25}$$

$$\frac{H}{\hbar} = \begin{pmatrix} 0 & V_{\mathrm{int}} \\ V_{\mathrm{int}} & \Delta \end{pmatrix} \tag{2.26}$$

Suppose there is a two-particle state $|AA\rangle$ and a two-particle state $|BC\rangle$ mutually coupled with a strength of V_{int} (V_{int} can be considered as a dipole–dipole interaction between Rydberg atoms, $V_{\mathrm{int}} = \mu_1\mu_2/r^3$), and the detuning between two energy states is Δ. In terms of Hamiltonian Eq. (2.26), the corresponding eigenvalues and eigenstates are

$$\lambda_{\pm} = \frac{\Delta}{2} \pm \sqrt{\left(\frac{\Delta}{2}\right)^2 + V_{\mathrm{int}}^2} \tag{2.27}$$

$$|\lambda_{\pm}\rangle = \frac{1}{N}\left\{\left(-\Delta \mp \sqrt{\Delta^2 + 4V_{\mathrm{int}}^2}\right)|AA\rangle + 2V_{\mathrm{int}}|BC\rangle\right\} \tag{2.28}$$

where N is a normalization coefficient. When the detuning Δ between energy levels is much larger than the interaction potential V_{int} ($\Delta \gg V_{\mathrm{int}}$, in the van der Waals region), due to the Rydberg interaction, V_{int} induces an intrinsic energy level shift $\Delta\left(\left(\frac{V_{\mathrm{imt}}}{\Delta}\right)^2 - \left(\frac{V_{\mathrm{int}}}{\Delta}\right)^4\right)$, which equals to $V^2{}_{\mathrm{int}}/\Delta$ if it is preserved to the second order.

From Table 2.1, the energy level spacing Δ is proportional to $1/n^3$, and the electric dipole moment is $\mu \sim n^2 m$; so, in the region of van der Waals force, this dipole–dipole interaction should be corrected to be $V_{VdW} \sim \frac{e^4 a_0^4 n^{11}}{r^6}$.

On the other hand, state $|BC\rangle$ has an energy level spacing Δ compared to $|AA\rangle$, and $|AA\rangle$ state can be considered as a low energy state or a ground state. The probability of Rydberg excitation can be estimated by studying the share of $|AA\rangle$ in the eigenstate $|\lambda\pm\rangle$ (the excited energy level caused by the Rydberg interaction V_{int}). Based on the eigenstate as displayed in Eq. (2.28), we assume that V_{int} is equal to 50 and Δ can be varied from 1 to 1000. The variation of Rydberg excitation rates with Δ is depicted in Fig. 2.4. The results show that when $\Delta \ll V_{\text{int}}$, the population of $|\text{AA}\rangle$ state on the two eigenstates $|\lambda\pm\rangle$ is identical; however when $\Delta \gg V_{\text{int}}$, the population on $|\lambda_-\rangle$ state tends to 0 whereas that on $|\lambda+\rangle$ state approaches 1. This indicates that in the van der Waals region away from resonance, the energy level shifts caused by Rydberg interactions can form only one energy band, which is determined by the symbol of van der Waals forces. Reversely, in the near-resonant dipole–dipole interaction region, two symmetrically-distributed energy bands are formed on both sides of the two-particle state, essentially arising from the process of coherent excitation of all atoms in a certain range [33].

(2) Dipole–dipole interactions are electrostatic interactions between atoms or molecules that have a permanent dipole moment or induced dipole moment. In the case of a neutral atom, although the entire atom is electrically neutral, it induces a permanent dipole moment under the influence of an external electric field. A strong dipole–dipole potential exists between atoms, as shown in Fig. 2.5.

We introduce a Hamiltonian to describe this electrostatic interaction between any two Rydberg atoms.

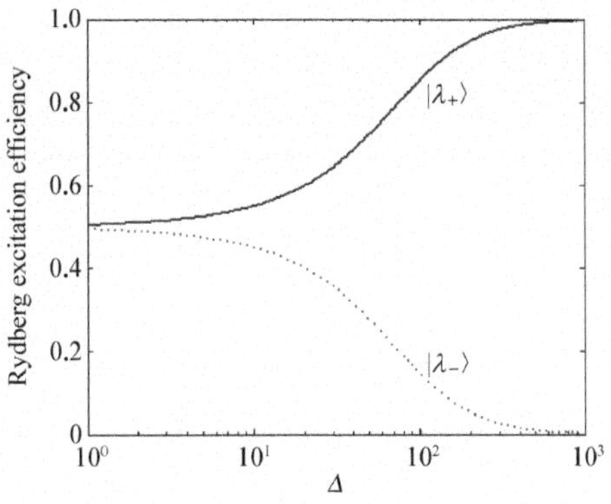

Fig. 2.4 The relationship between the detuning Δ and the population of $|\text{AA}\rangle$ state in eigenstates $|\lambda\pm\rangle$, where $V_{\text{int}} = 50$

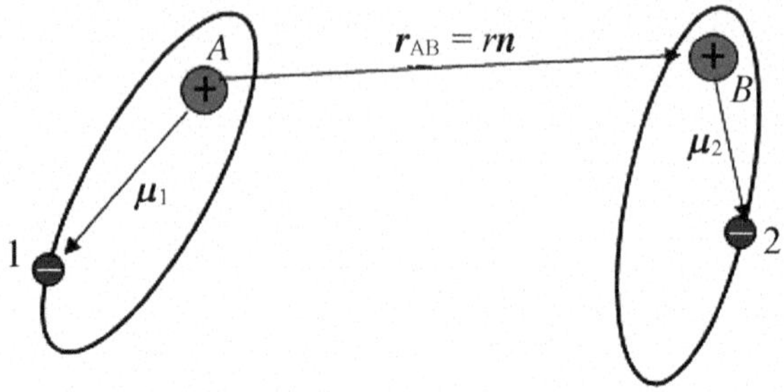

Fig. 2.5 For arbitrary atoms A and B there exists a dipole–dipole interaction between them

$$V_{12} = \frac{\mu_1 \cdot \mu_2 - 3(\mu_1 \cdot n)(\mu_2 \cdot n)}{4\pi \varepsilon_0 r^3} \tag{2.29}$$

where $\boldsymbol{\mu}_{1,2}$ denotes the dipole moment vector of the atom, n stands for the relative displacement between atoms A and B, and r means a relative distance.

When the applied electric field is much larger than the sum of the field strengths produced by all dipoles, the dipole moments of atoms are polarized and all oriented in the direction of external electric field. At this point, we define θ as an angle between the applied electric field and n. Equation (2.29) can be further reduced to

$$V_{12} = \frac{\mu_1 \mu_2}{4\pi \varepsilon_0 r^3}\left(1 - 3\cos^2\theta\right) \tag{2.30}$$

According to Eq. (2.30), the dipole–dipole interaction force between atoms is spatially dependent, a result that has been experimentally verified by Noel's group [34]. Different from the van der Waals force, the effective control of dipole–dipole interaction V_{12} can be achieved by changing the angle θ between the electric field direction and the relative displacement between atoms. When $\theta = \arccos(1/\sqrt{3}) \approx$ 54.7, $V_{12} = 0$. This particular angle is called "magic angle" and can be used to study the high-resolution spectral properties of solid-state NMR [35].

From the results of Eq. (2.30), we can already estimate the dipole–dipole interaction $V_{dd} \sim \frac{e^2 a_0^2 n^4}{r^3}$ for Rydberg atoms; in Eq. (2.26), when $\Delta \ll V_{int}$, the shift of the eigenenergy due to the Rydberg interaction potential V_{int} can be modified as $V_{int}\left(1 - \frac{1}{2}\frac{\Delta}{V_{int}} + \frac{1}{8}\left(\frac{\Delta}{V_{int}}\right)^2 + O\left[\frac{\Delta}{V_{int}}\right]^3\right)$; if only the zero-order term is retained, i.e., $V_{\mathrm{dd}} \sim V_{\mathrm{int}}$, the result is the same as that of Eq. (2.30).

2.3.2 *Light-Atom Interactions*

The interaction of laser field with atoms induces many fascinating physical phenomena. For example, the simplest model such as the interaction of a single two-level atom with a single-mode field of a specific frequency, is described in detail in the textbooks of quantum optics. The transition from the atomic ground state to the Rydberg state usually requires a laser of short wavelength and high frequency, e.g., an ultraviolet laser, so the Rydberg excitation has been realized experimentally by a multiphoton transition model [36]. Here, we will further explore the physical processes of three-level atoms interacting with two light fields, focusing on two important phenomena, i.e., Autler-Townes splitting and electromagnetically induced transparency.

2.3.2.1 Autler-Townes Splitting (ATS)

We first consider a single atom, as shown in Fig. 2.6a, where 5S, 5P and nS correspond to ground state $|g\rangle$, excited state $|e\rangle$ and Rydberg state $|r\rangle$, respectively. The laser field Ω_1 resonantly couples energy levels $|g\rangle$ and $|e\rangle$, Ω_2 couples energy levels $|e\rangle$ and $|r\rangle$, and the two-photon detuning between Ω_2 and $|r\rangle$ is \delta. The intermediate energy level $|e\rangle$ is unstable and its spontaneous radiation lifetime is γ, while the Rydberg state $|r\rangle$ has a long lifetime and is relatively stable with respect to the highly excited state $|e\rangle$. So here we neglect the decay of the Rydberg state.

In the Heisenberg picture, the motional equation for describing the single-atom dynamics can be written as

$$i\frac{d}{dt}\begin{pmatrix}\psi_g\\ \psi_e\\ \psi_r\end{pmatrix} = \begin{pmatrix}0 & \Omega_1 & 0\\ \Omega_1 & 0 & \Omega_2\\ 0 & \Omega_2 & \delta\end{pmatrix}\begin{pmatrix}\psi_g\\ \psi_e\\ \psi_r\end{pmatrix} - i\begin{pmatrix}0 & 0 & 0\\ 0 & \gamma & 0\\ 0 & 0 & 0\end{pmatrix}\begin{pmatrix}\psi_g\\ \psi_e\\ \psi_r\end{pmatrix} \tag{2.31}$$

In the case of $\Omega_1 \gg \Omega_2$, the $|r\rangle$ state can be regarded as being decoupled from $|g\rangle$ and $|e\rangle$. By diagonalizing Hamiltonian, we can obtain a set of new eigenstates $|\lambda_+\rangle$ and $|\lambda_-\rangle$ formed by the linear superposition of $|g\rangle$ and $|e\rangle$. The corresponding eigenvalues and eigenstates are

$$\lambda_\pm = \pm\Omega_1 \tag{2.32}$$

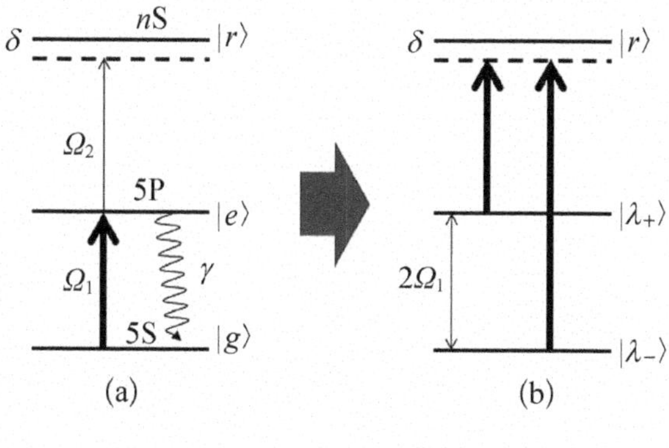

Fig. 2.6 **a** Energy level structure of a single three-level atom; **b** Representation of the atomic energy level structure in the dressed eigenstates when $\Omega_1 \gg \Omega_2$

Fig. 2.7 Autle-Townes splitting. The distance between two peaks is $2\Omega_1$, the Rabi frequency of the laser field is constant, and the pulse duration is as marked on the figure

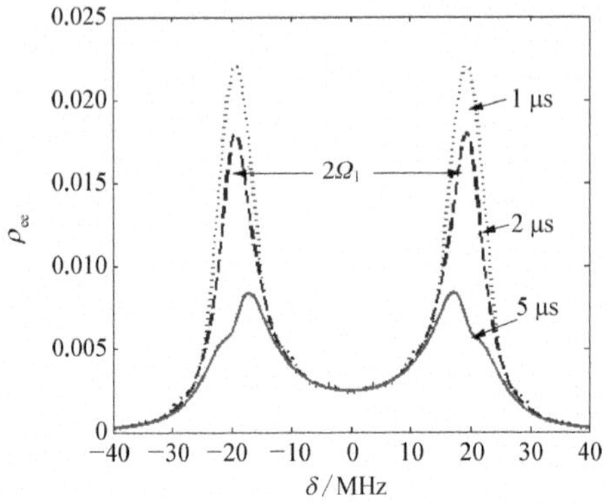

$$|\lambda_\pm\rangle = \frac{1}{\sqrt{2}}(|g\rangle \pm |e\rangle) \tag{2.33}$$

In the dressed-state picture, we assume $\Omega_1 = 20$ MHz, $\Omega_1 = 1.0$ MHz, $\gamma = 6.0$ MHz, and the two-photon vector Δ can be effectively tuned. The numerical results are shown in Fig. 2.7. We find that the population ρ_{rr} of Rydberg states $|r\rangle$ presents a typical double-peak structure with the detuning Δ, which is the famous Autler-Townes splitting. The peak positions locate at the resonant transitions of $|\lambda_+\rangle - |r\rangle$ and $|\lambda_-\rangle \rightarrow |r\rangle$, respectively.

2.3.2.2 Master Equation

Based on the principle of Autler-Townes splitting, the cooperative excitation effect of two Rydberg atoms is discussed in Fig. 2.8. Assuming that there are two three-level atoms A and B whose ground states $|g\rangle$ and excited state $|e\rangle$ are resonantly coupled by a strong laser field Ω_1. The spontaneous radiation rate of upper energy level excited state $|e\rangle$ is γ. The excited state $|e\rangle$ is coupled to the Rydberg state $|r\rangle$ by another weak laser field Ω_2 with detuning Δ. There exists a strong interaction V_{mt} between two Rydberg energy states $|r_1\rangle$ and $|r_2\rangle$.

Fig. 2.8 Model diagram for two interacting three-level atoms A and B

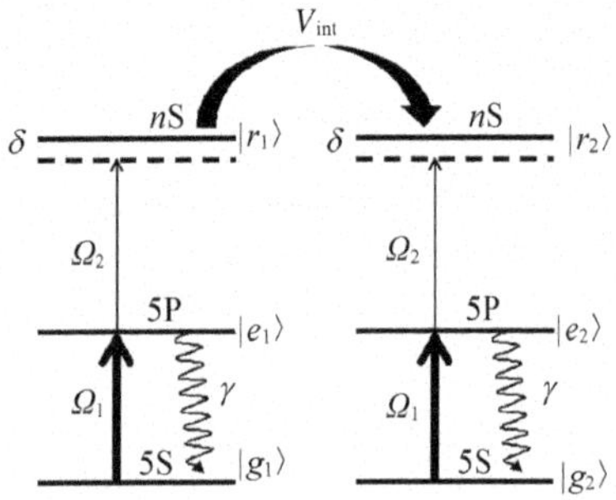

The two-body Hamiltonian of this system can be written as

$$H_{\mathrm{AB}} = H_{\mathrm{A}} + H_{\mathrm{B}} + H_{\mathrm{int}} \tag{2.34}$$

where H_{AB} denotes the Hamiltonian of a single atom (let h = 1), which in turn can be written as

$$H_{\mathrm{AB}} = \delta\left|r_{1(2)}\right\rangle\langle r_{1(2)}| + \left[\Omega_1\left|e_{1(2)}\right\rangle\!\langle g_{1(2)}| + \Omega_2|r_{1(2)}\rangle\langle e_{1(2)}| + H.c.\right] \tag{2.35}$$

H_{int} denotes the Hamiltonian of interactions between atoms, which is

$$H_{\mathrm{int}} = V_{\mathrm{int}}|r_1\rangle|r_2\rangle\langle r_2|\langle r_1| \tag{2.36}$$

A two-body density matrix is employed to describe the dynamical evolution of this system via a new set of diatomic basis vectors to expand the Hamiltonian. The standard complete basis vector should contain nine components: $|g_1g_2\rangle$, $|g_1e_2\rangle$, $|g_1r_2\rangle$, $|e_1g_2\rangle$, $|e_1e_2\rangle$, $|e_1r_2\rangle$, $|r_1g_2\rangle$, $|r_1e_2\rangle$, $|r_1r_2\rangle$. Considering the symmetry of Hamiltonian, these nine vectors can be reduced to six components that are orthogonal with each other. They are $|gg\rangle$, $|ee\rangle$, $|rr\rangle$, $(|ge\rangle+|eg\rangle)/\sqrt{2}$, $(|er\rangle+|re\rangle)/\sqrt{2}$, $(|gr\rangle+|rg\rangle)/\sqrt{2}$ where the subscripts $1, 2$ are omitted. Under this new set of two-particle basis vectors, Eq. (2.35) can be written as

$$H_{\mathrm{AB}} = \begin{pmatrix} 0 & 0 & 0 & \sqrt{2}\Omega_1 & 0 & 0 \\ 0 & 0 & 0 & \sqrt{2}\Omega_1 & \sqrt{2}\Omega_2 & 0 \\ 0 & 0 & 2\delta + V_{\mathrm{int}} & 0 & \sqrt{2}\Omega_2 & 0 \\ \sqrt{2}\Omega_1 & \sqrt{2}\Omega_1 & 0 & 0 & 0 & \Omega_2 \\ 0 & \sqrt{2}\Omega_2 & \sqrt{2}\Omega_2 & 0 & \delta & \Omega_1 \\ 0 & 0 & 0 & \Omega_2 & \Omega_1 & \delta \end{pmatrix} \tag{2.37}$$

Since the system contains spontaneous radiation loss γ from the intermediate state $|e\rangle$, we also introduce the Lindblad operator $L[\rho]$ to describe the role of this dissipation, and $L[\rho]$ is expressed in Eq. (2.38).

$$L[\hat{\rho}] = \gamma\left(\hat{A}\hat{\rho}\hat{A}^{\dagger} - \frac{1}{2}\hat{A}^{\dagger}\hat{\Lambda}\hat{\rho} - \frac{1}{2}\hat{\rho}\hat{\Lambda}^{\dagger}\hat{\Lambda}\right) \tag{2.38}$$

where $\widehat{p}$ is density matrix operator and $\hat{A} = |g\rangle\langle e|$ is the transition operator for spontaneous radiation. The field operator defining the system is $|\psi\rangle = c_1|1\rangle + c_2|2\rangle + c_3|3\rangle + c_4|4\rangle + c_5|5\rangle + c_6|6\rangle$, where $|1\rangle = |gg\rangle$, $|2\rangle = |ee\rangle$, $|3\rangle = |r\rangle$, $|4\rangle = (|ge\rangle+|eg\rangle)|\sqrt{2}$, $|5\rangle = (|er\rangle+, |re\rangle)/\sqrt{2}$, $|6\rangle = (|gr\rangle+|rg\rangle)/\sqrt{2}$, then the quantum master equation describing the evolution of the system can be expressed as

$$\dot{\rho} = -\mathrm{i}[H_{\mathrm{AB}}, \hat{\rho}] + L[\hat{\rho}] \tag{2.39}$$

where the density matrix operator $\hat{\rho}$ can be written as

$$\hat{\rho} = |\psi\rangle\langle\psi| = \begin{pmatrix} \rho_{11} & \rho_{12} & \rho_{13} & \rho_{14} & \rho_{15} & \rho_{16} \\ \rho_{21} & \rho_{22} & \rho_{23} & \rho_{24} & \rho_{25} & \rho_{26} \\ \rho_{31} & \rho_{32} & \rho_{33} & \rho_{34} & \rho_{35} & \rho_{36} \\ \rho_{41} & \rho_{42} & \rho_{43} & \rho_{44} & \rho_{45} & \rho_{46} \\ \rho_{51} & \rho_{52} & \rho_{53} & \rho_{54} & \rho_{55} & \rho_{56} \\ \rho_{61} & \rho_{62} & \rho_{63} & \rho_{64} & \rho_{65} & \rho_{66} \end{pmatrix} \tag{2.40}$$

From Eq. (2.39), we can obtain the time-dependent evolution of the system. In the following only some numerical results are given for analysis.

2.3.2.3 Numerical Results and Analysis

In numerical calculations, we assume the laser Rabi frequencies are $\Omega_1 = 20$ MHz and $\Omega_2 = 1$ MHz, and the spontaneous radiation rate of $|e\rangle$ is $\gamma = 6$ MHz. Figure 2.9 shows the Rydberg excitation probability ρ_{33} of the doubly-excited state $|3\rangle$ with the variation of the two-photon detuning Δunder different Rydberg interaction potentials.

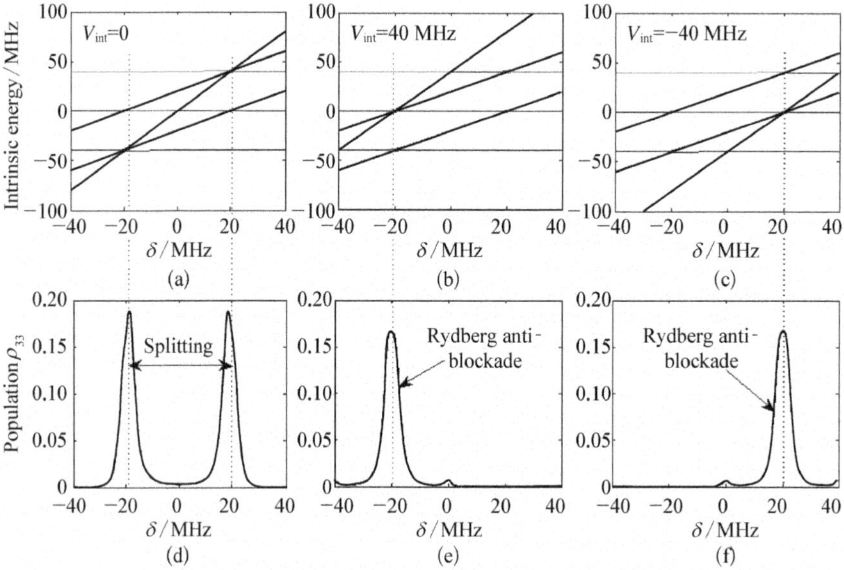

Fig. 2.9 Variations of eigenenergy (**a**)–(**b**) and the populations ρ_{33} along with the change of detuning \delta. When $V_{\text{int}} = 0$, ρ_{33} shows a double-peak structure, in agreement with the principle of AT splitting; when $V_{\text{int}} = 40$ MHz or -40 MHz, a high double Rydberg-state population appears only at $\backslash delta = -V_{\text{int}}/2$, and it is greatly suppressed in other regions due to the off-resonant excitation. Figures (**d**)–(**f**) represent the results measured at t = 10 μs

Figure 2.9a–c depict the variation of the eigenenergy of the system, and (d–f) show the corresponding doubly-excited Rydberg population ρ_{33} with detuning \delta. The first column exhibits the case of $V_{\text{int}} = 0$. When the interaction energy between two Rydberg atoms is neglected, the interaction model is equivalent to that contains two independent three-level atoms. If $\Omega_1 \geq \Omega_2$, the double Rydberg states $|rr\rangle$ till show a distinct double-peak structure, which is the well-known Aulter-Townes double-peak. Here (a) represents the distribution of eigenenergies, where the anti-crossing structure is observed between different energy levels when Δ is taken as -20 or 20. In these anti-crossing regions, the energies of the higher and lower levels are pretty close. At this point, only a little additional perturbation can realize the energy level transition easily. It is manifested as a Rydberg excitation.

Figure 2.9b–f describe the variation of eigenenergies and the population of state $|rr\rangle$ with Δ in the strong dipole–dipole interaction region. It is found that in the case of a large interaction strength $V_{\text{int}} \geq \Omega_1$, $\Omega 2$, the two-photon detuning can effectively compensate for the energy level shift due to strong interactions between Rydberg atoms as long as $\Delta = -V_{\text{int}}/2$ is satisfied [e.g., in Fig. 2.9e, $V_{\text{int}} = 40$ MHz, $\Delta = -20$ MHz and in Fig. 2.9f, $\text{V} = -40$ MHz, $\Delta = 20$ MHz]. Then, an effective double Rydberg excitation can be achieved even in the Rydberg blockade region, which is known as the Rydberg antiblockade effect. This effect was first predicted theoretically by the J. M. Rost group at the Max Planck Institute [37, 38] and then verified experimentally by the M. WeidemMer group in 2010 [39].

2.3.2.4 Rydberg Electromagnetically Induced Transparency (EIT)

To study the EIT effect of Rydberg atoms, a three-level atomic model is usually adopted (see Fig. 2.10). Unlike ATS, we consider that the coupling light Ω_2 is much larger than the pump laser Ω_1, and the detuning between the intermediate energy of $|e\rangle$ and the frequency of Ω_1 is denoted as Δ.

According to Eq. (2.31), the eigenenergies and eigenstates at $\Omega_2 \geq \Omega_1$ can be solved as:

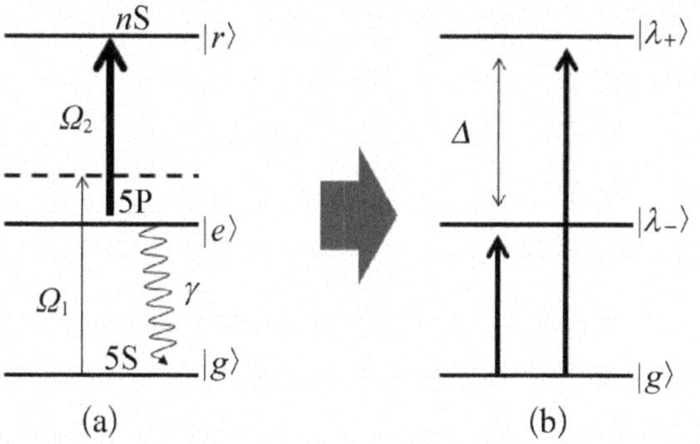

Fig. 2.10 **a** Energy level of a single three-level atom; **b** Representation of the atomic energy levels in the dressed-state picture when $\Omega_2 \geq \Omega_1$

$$\lambda_{\pm} = \frac{\Delta \pm \sqrt{\Delta^2 + 4\Omega_2^2}}{2} \tag{2.41}$$

$$\begin{aligned} |\lambda_+\rangle &= \sin\theta|e\rangle + \cos\theta|r\rangle \\ |\lambda_-\rangle &= -\cos\theta|e\rangle + \sin\theta|r\rangle \end{aligned} \tag{2.42}$$

where $2\theta = \arctan\left(-\frac{2\Omega_2}{\Delta}\right)$. The pump light Ω_1 is only mutually coupled to the $|e\rangle$-state component in the eigenstate $|\lambda_\pm\rangle$, while the symbol of the $|\mathrm{e}\rangle$ state population in the eigenstate $|\lambda_\pm\rangle$ is reversed as can be seen in Eq. (2.42). Therefore, two transition channels $|g\rangle \to |\lambda_+\rangle$ and $|g\rangle \to |\lambda_-\rangle$ will undergo destructive interference, forming a window for resonant transmission during the $|g\rangle \to |e\rangle$ transition, which is called electromagnetically induced transparency, or EIT for short. This effect was first discovered by the group of S. E. Harris at Stanford University in the 1990s [40, 41], and is now widely studied. Recently, the EIT effect in Rydberg atomic gas has also been greatly focused by scientists [42, 43]. The relationship between ρ_{ge} and the single photon detuning Δ can be calculated by master Eq. (2.39). According to the definitions of new basis vectors $|1\rangle, \ldots |6\rangle$, $\rho_{\mathrm{ge}} = \sqrt{2}(\rho_{14} + \rho_{42}) + \rho_{65}$, the numerical results are shown in Fig. 2.11. Let $\Omega_1 = 0.8$ MHz, $\Omega_2 = 1.2$ MHz, $\gamma = 6$ MHz and $\Delta = 0$. The variation of the transparent window on the EIT resonance-absorption spectrum is studied under different Rydberg interactions. It is obvious that the curve in the absence of interactions (see the solid line in Fig. 2.11) complies with the normal properties of EIT absorption spectrum, *i.e.*, there exists a dip with near-zero absorption near the area of $\Delta = 0$. It means that when single photon transition achieves resonance, the medium no longer absorbs photons to cause the $|g\rangle \to |e\rangle$ transition and can be considered to be transparent for the pump field. This is the classical EIT effect. As turning on the strong interaction between Rydberg atoms, the zero-absorption window with a full transmission probability becomes narrowed gradually. When the interaction energy V_{int} is increased to be comparable with the strength of laser Rabi frequencies, the EIT absorption spectrum becomes asymmetric (see the dashed line in Fig. 2.11). And when V_{int} is much larger than $\Omega_{1,2}$, the transparent window of EIT disappears and is replaced by a strong single absorption peak near $\Delta = 0$ (see the dotted line in Fig. 2.11). This is because the energy level shift caused by V_{int} destroys the condition for destructive interference between the transition channels of $|g\rangle \to |\lambda_+\rangle$ and $|g\rangle \to |\lambda_-\rangle$, so the pump field can still be greatly absorbed near the resonance point where $\Delta = 0$.

Also, we further consider that whether the EIT transparent window could be achieved via a similar method like Rydberg antiblockade. Because the two-photon detuning Δ can be used for compensating the Rydberg interaction potential. However, it was verified that this method is not feasible. Because both V_{int} and Δ themselves induce additional phases that destroy the condition for destructive interference letting the system reserve only one resonant absorption peak in this case.

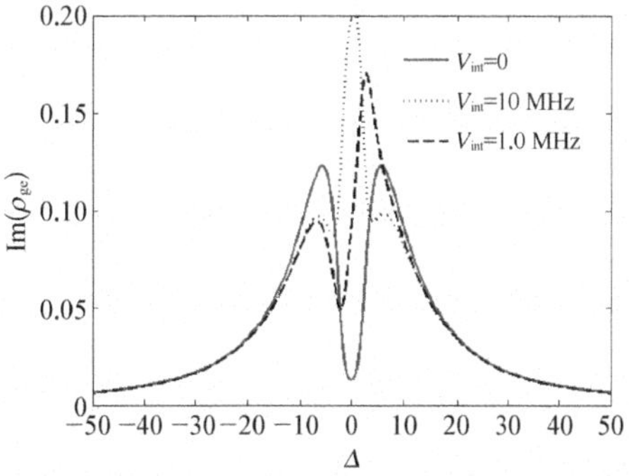

Fig. 2.11 EIT absorption rate between states $|e\rangle$ and $|g\rangle$ versus the variation of detuning Δ

2.3.3 Dipole Blocking Effect of Rydberg Atoms

Since 2000, when D. Jaksch et al. first theoretically proposed that fast quantum logic gates could be realized using ultracold Rydberg atoms [44], the applications of Rydberg atoms in the field of quantum information have been greatly extended. After that, more and more scientists have turned the research focus to the area of Rydberg atoms—a rising star in the atomic and molecular physics [45].

The essence of realizing quantum logic gates is to accomplish a controllable manipulation of the qubits, and the Rydberg dipole blockade effect can precisely satisfy this requirement. The strong dipole–dipole interaction between Rydberg states would give rise to the shift of atomic resonance levels. So when two atoms meet the resonant transition condition at the same time, only one of them can be excited to the Rydberg state, and the excitation of the other atom is strongly suppressed due to the shift of the energy level. By analogy, other nearby atoms are also affected by this Rydberg atom and cannot be excited. This effect are known as "Rydberg blockade" or "dipole blockade" [46]. The range of affected atoms can be measured by the blockade radius r_b (~μm), *i.e.*, the Rydberg excitation of any other atoms within a volume of radius ~ r_b^3 is greatly suppressed. Such an atomic ensemble which contains only single Rydberg excitation have named as "superatom" [47]. The Rabi frequency of the collective single excitation between the ground and excited states can be written as $\sqrt{N_b}\Omega$ in a "superatom" ensemble. Here N_b is the number of atoms involved in the "superatom" with radius ~ r_b^3, Ω is the effective Rabi frequency of a single atom. If in the case of a two-photon excitation, the effective Rabi frequency is $\Omega = \Omega_1\Omega_2/2\Delta$, where $\Omega_{1,2}$ is the Rabi frequency of single photon transition and Δ is the detuning with respect to the intermediate levels. The blockade radius r_b is defined as the atomic spacing when the strong dipole–dipole interaction energy of Rydberg atoms is equal to the excitation linewidth of the optical field. The value of r_b is usually influenced experimentally by the factors such as the laser power broadening and the detuning of the intermediate levels. An approximate estimation to r_b can be considered to be $r_b = \left(C_6/\hbar\sqrt{N_b}\Omega\right)^{1/6}$ when the van der Waals interaction of Rydberg atoms (assuming working in the van der Waals-interaction regime) is equal to the laser power broadening.

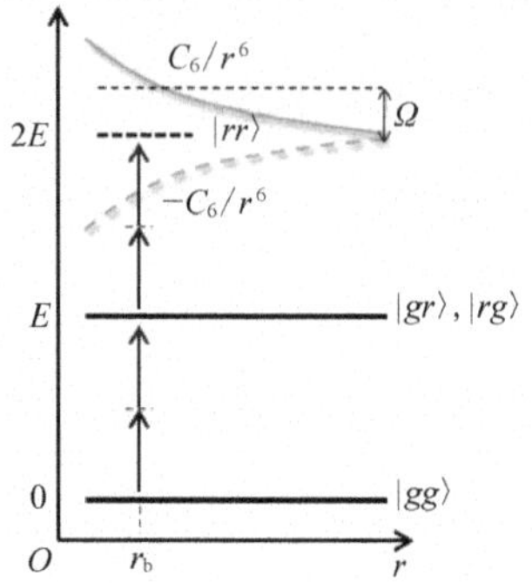

Fig. 2.12 Potential energy curve for Rydberg excitation of two two-level atoms, where r indicates the relative displacement between atoms and two arrows means the two-photon excitation process. In the van der Waals region, the van der Waals forces of attraction (shaded dashed line) or repulsion (shaded solid line) make the two-particle state $|r, r\rangle$ excitation blocked. The probability of double Rydberg excitation is largely reduced in the range of $r \ll r_b$

Figure 2.12 represents the energy level structure of the diatomic Rydberg excitation. $|g\rangle$ and $|r\rangle$ denote the single-atom ground state and the Rydberg state, respectively. So, $|gg\rangle$, $|gr\rangle$ and $|rr\rangle$ mean that two atoms in the ground state, one atom excited to the Rydberg state, and both atoms excited to the Rydberg state, respectively. In the case of van der Waals interactions, when two atoms are excited to the Rydberg state at the same time, the level shift due to the strong Rydberg interactions will form an energy band on one side of the state $|r, r\rangle$, preventing the excitation of other neighboring atoms. The dashed line in Fig. 2.12 represents the shift of the atomic resonance level $|r, r\rangle$, and the shaded dashed and shaded solid lines correspond to the van der Waals interactions of attraction and repulsion. If the laser power broadening Ω is identical to the level shift V_{int}, we can define the Rydberg dipole blockade radius as $r_b = (C_6/\hbar\Omega)^{1/6}$. If the "superatom" model involving N_b atoms is taken into account, Ω needs to be multiplied by a correction factor $\sqrt{N_b}$, leading to $r_b = \left(C_6/\hbar\sqrt{N_b}\Omega\right)^{1/6}$.

The Rydberg blockade effect has a wide range of applications, such as quantum information [48] and many-body physics [49]. In the following we discuss how to prepare entangled states by the blockade effect. In the case of two-atom systems, the Hamiltonian of system can be expressed as

$$H = \sum_{i=1,2} \left[-\Delta|r\rangle\langle r|_i + \frac{\Omega}{2}(|g\rangle\langle r|_i + |r\rangle\langle g|_i) \right] + V_{12}|r\rangle \langle r|_1 \otimes |r\rangle r|_2\rangle \tag{2.43}$$

where Δ is the detuning of laser field, and Ω is the Rabi frequency of single atom excitation. With a set of new two-particle basis vectors $\{|gg\rangle, |s\rangle, |a\rangle, |rr\rangle\}$, the Hamiltonian can be expressed in the matrix form, where states $|s\rangle$ and $|a\rangle$ represent symmetric vector $|s\rangle = (|gr\rangle + |rg\rangle)/\sqrt{2}$ and antisymmetric vector $|a\rangle = (|gr\rangle -$

$|rg\rangle)/\sqrt{2}$.

$$\tilde{H} = \begin{pmatrix} 0 & \frac{\Omega}{\sqrt{2}} & 0 \\ \frac{\Omega}{\sqrt{2}} & -\Delta & \frac{\Omega}{\sqrt{2}} \\ 0 & \frac{\Omega}{\sqrt{2}} & V_{12} \end{pmatrix} \tag{2.44}$$

Due to the symmetry of Hamiltonian, the antisymmetric vector $|a\rangle$ is found to be decoupled from other basis vectors, so any state $|\psi(t)\rangle$ can be expanded into a generalized form of $|\psi(t)\rangle = c_1(t)|gg\rangle + c_2(t)|s\rangle + c_3(t)|rr\rangle$, and the evolution of the population $c_i(t)$ on each eigen-basis should satisfy the motional equations.

$$\begin{aligned} i\dot{c}_1(t) &= \frac{\Omega}{\sqrt{2}}c_2(t) \\ i\dot{c}_2(t) &= \frac{\Omega}{\sqrt{2}}c_1(t) - \Delta c_2(t) + \frac{\Omega}{\sqrt{2}}c_3(t) \\ i\dot{c}_3(t) &= \frac{\Omega}{\sqrt{2}}c_2(t) + V_{12}c_3 \end{aligned} \tag{2.45}$$

If $V_{\text{int}} \geq 0$, we assume $C_3(t)\approx 0$ due to the Rydberg blockade effect (the Rydberg antiblockade is not considered here). Equation (2.45) can be simplified to

$$\begin{aligned} i\dot{c}_1(t) &= \frac{\Omega}{\sqrt{2}}c_2(t) \\ i\dot{c}_2(t) &= \frac{\Omega}{\sqrt{2}}c_1(t) - \Delta c_2(t) \end{aligned} \tag{2.46}$$

Given the initial preparation of atoms in the ground state, if $t = 0$, then $c_1(0) = 1$ and $c_2(0) = 0$ (let $\Delta = 0$). All the atoms are in the entangled state $|s\rangle$. If $V_{\text{int}} = 0$, we find this entangled-state preparation is not succeed, because state $|s\rangle$ is strongly coupled to state $|rr\rangle$.

In addition, the realization of the Rydberg blockade effect can be analyzed by using the motional Eq. (2.45) (considering the system of atoms A–B). We assume $\Delta = 0$, atom B is initially in the $|g\rangle$ state, and the external laser interacts with atom B only. If atom A is in $|g\rangle$, atom B will be excited to $|r\rangle$. However, if atom A is already pre-excited to $|r\rangle$, the Rydberg level of atom B will have a shift V_{int} due to the dipole–dipole interactions, and at this point, atom B will not be excited again but remain in the $|g\rangle$ state. In a word, whether atom B is excited to the Rydberg state or not, depends on the status of atom A.

To sum up, if two atoms are in the Rydberg state $|rr\rangle$ at the same time, the strong interaction between them leads to a level shift $V_{\text{int}} = \pm C_3/r^3$ (dipole–dipole interaction region) or $V_{\text{int}} = C_6/r^6$ or $V_{\text{int}} = -C_6/r^6$ (van der Waals interaction region). When the interaction energy is much larger than the Rabi frequency of the optical field as well as the laser linewidth, the doubly excited Rydberg state $|rr\rangle$ is moved away from the laser resonance frequency, so that only one atom can obtain a transition to the Rydberg state within the blockade radius at any moment.

2.3.4 Applications of Rydberg Atoms

Rydberg atoms with their outermost electrons excited to the higher energy level of principal quantum number, are weakly bound to the outer electrons. Thus they are susceptible to the influence of applied electromagnetic fields, promising for exhibiting a wealth of physical effects. Because of the uniqueness of the Rydberg atom (some properties are listed in Table 2.1), it has been an important research focus in the field of atomic physics. In recent years, along with the continuous development of experimental techniques for cold atoms, ultracold Rydberg atoms have been experimentally realized [50] and have again attracted a large number of researchers due to the virtues including powerful dipole–dipole interactions and quantum coherence.

The current research on Rydberg atoms involves many applications:

2.3.4.1 Rydberg Quantum Information

There are enormous dipole–dipole interactions between Rydberg atoms, which are usually proportional to the principal quantum number n^4. At the same time their lifetimes are very long, which are proportional to n^3. Therefore, as early as 2000, M. D. Lukin et al. proposed that these properties of Rydberg atoms could be exploited to realize fast two-qubit quantum logic gates [21]. Since then, Rydberg atoms have emerged with their unique advantages for many applications, such as two-qubit long-distance quantum logic gates, collectively encoded multi-qubit quantum registers, manipulating quantum interfaces between atoms and light, and simulating the effects of quantum many-body physics [22]. In 2009, H. Weimer et al. experimentally implemented the Rydberg quantum simulator. The most important feature of the Rydberg quantum simulation is the ability to manipulate many-body interactions between atoms in the spin regime. The system contains a more efficient manipulation than other conventional quantum simulation devices. Even for the system with higher-order interactions [51] it works efficiently. This research result brings new opportunity to the further realization of dissipative generation of quantum entangled states [52, 53], the study on the entanglement properties of high-fidelity multi-qubit quantum gates [54, 55], etc.

2.3.4.2 Rydberg Nonlinear Optics

Electromagnetic induced induction transparency is a bright spot in the domain of nonlinear optics, where the destructive interference between pump and probe fields can be used to enhance the optical nonlinear effect. The interaction time of atoms with light fields can be greatly enhanced due to the suppression of photon losses during the process [56]. Researches on the Rydberg EIT have also been performed for almost a decade. Ever since the first theoretical discovery by Friedler et al. that the long-range dipole–dipole interactions between Rydberg atoms could strengthen the nonlocal nonlinear interactions between single photon pulses, the EIT effect is led to the preparation of single photon pulses and the construction of photon phase gates [57]. Essentially, the nonlinear characteristic of the Rydberg atom originates from the third-order Kerr nonlinear coefficient induced by its huge polarizability [58]. On the experimental side one has been able to project the atomic nonlinearity onto the photon level in order to construct the nonlinearity of slow-light photons. The physical essence in this process is EIT. The above results have been used to generate Rydberg single-photon registers [59, 60], all-optical single-photon switches [61], efficient single-photon generators [62], and so on.

2.3.4.3 Rydberg Plasma

The ionization energy of the Rydberg atom is proportional to n^{-4}. When the principal quantum number n is extremely large, the Rydberg atom can be easily ionized to produce plasma. The method of direct-ionizing cold atoms is usually adopted to produce neutral plasma, which, however, is prepared in a non-equilibrium disordered quantum state that hinders subsequent researches of strong correlation properties. In contrast, when cold atoms are excited to a higher Rydberg state, there exists strong correlated interactions between Rydberg-state atoms. The kinetic properties of Rydberg-atom excitation are closely related to the coupling strength of the generated plasmas. At this point, the laser field will have the ability to independently manipulate both the temperature and density properties of the resulting plasmas. On the other hand these strong correlation properties of Rydberg atoms can also be exploited due to the studies of ultracold neutral plasmas [63].

2.4 Spin–orbit Quantum Gas and Gauge Potential

2.4.1 Spin–orbit Coupling

From the perspective of classical electrodynamics, spin–orbit coupling stems from the interaction of the electric field on the spin magnetic moment of motion. Assuming that in a system the electron's spin magnetic moment moves with a velocity v relative to the electric field E, the strength of the magnetic field at the position of the electron is

$$\mathbf{B} = -\frac{1}{c} v \times \mathbf{E} \tag{2.47}$$

The energy of the electron in this magnetic field is

$$-\mu \cdot \mathbf{B} = -\frac{e}{mc} s \cdot \mathbf{B} = -\frac{e}{m^2 c^2} s(\mathbf{E} \times p) \tag{2.48}$$

This term represents spin–orbit coupling. If we consider the motion of the electron in the electric field created by the nucleus,

$$\mathbf{E} = -\frac{1}{e}\frac{r}{r}\frac{dV}{dr} \tag{2.49}$$

The spin–orbit interaction energy becomes

$$-\frac{e}{m^2 c^2} s \cdot (\mathbf{E} \times p) = -\frac{e}{m^2 c^2} s \cdot \left(-\frac{1}{e}\frac{r}{r}\frac{dV}{dr} \times p\right) = \frac{1}{m^2 c^2}\frac{1}{r}\frac{dV}{dr}(s \cdot l) \tag{2.50}$$

The action of this term will produce a fine structure for the energy levels of the atom. As can be seen in the above derivation, spin–orbit coupling can only exist in charged particles such as electrons, and it has many applications in semiconductor spintronics. With the implementation of cold atoms and the improvement of laser technology, one can simulate the motion of electrons in electromagnetic fields and form spin–orbit coupling in neutral cold atomic gases through the interactions between light and atoms.

2.4.2 *Artificial Gauge Potential*

This section addresses how to simulate the motion of electrons in an electromagnetic field in a neutral atomic quantum gas based on the principle of creating artificial gauge potentials through light-atom interactions.

First, assuming the simplest case of a single two- level atom consisting of a ground states $|g\rangle$ and a excited states $|e\rangle$, which are coupled by an applied, space-dependent optical field.

In the Hilbert space spaned by $|g\rangle$ and $|e\rangle$, the single-particle Hamiltonian can be written as

$$H = \frac{p^2}{2m} + \frac{h\Omega}{2}\begin{pmatrix} \cos\theta & e^{-i\phi}\sin\theta \\ e^{i\phi}\sin\theta & -\cos\theta \end{pmatrix} \tag{2.51}$$

where m and p are the mass and momentum of the atom, respectively, Ω is the generalized Rabi frequency of light-atom interactions, $\Omega_{\cos\theta}$ represents the detuning between the frequency of the light field and the resonant frequency of the atom, and $\Omega_{\sin\theta}$ is the strength of the coupling between the atom and the light field, ϕ is the phase of the light field. It should be noted that both θ and ϕ are spatially dependent.

Diagonalizing the light-atom interactions Hamiltonian leads to the following two dressed states $|\chi_1\rangle$ and $|\chi_2\rangle$:

$$|\chi_1\rangle = \begin{pmatrix} \cos\frac{\theta}{2} \\ e^{i\phi}\sin\frac{\theta}{2} \end{pmatrix}, |\chi_2\rangle = \begin{pmatrix} -e^{-i\phi}\sin\frac{\theta}{2} \\ \cos\frac{\theta}{2} \end{pmatrix}$$

Corresponding eigenvalues are $\hbar\Omega/2$ and $-\hbar\Omega/2$. With dressed states, the wave function of atoms $|\Psi(\mathbf{r},t)\rangle = \psi_1(\mathbf{r},t)|\chi_1(\mathbf{r})\rangle + \psi_2(\mathbf{r},t)|\chi_2(\mathbf{r})\rangle$ can be constructed. By applying the momentum operator $\hat{p}$ on the state $|\Psi\rangle$, one can get

$$\hat{p}|\Psi\rangle = (\hat{p}\psi_1)|\chi_1\rangle + (\hat{p}\psi_2)|\chi_2\rangle - \sum_{j,l=1}^{2} A_{jl}\psi_l|\chi_j\rangle \tag{2.52}$$

where $\mathbf{A} = ih\langle\chi_j|\nabla\chi_i\rangle$. In the derivation of the above equation, the relationship $\nabla[\phi_j|\chi_j\rangle] = [\nabla\psi_j]|\chi_j\rangle + \psi_j|\nabla\chi_j\rangle$ is applied. According to the Schrödinger's equation, one can get $\hat{\boldsymbol{p}}$

$$i\hbar\frac{\partial\psi_1}{\partial t} = \left[\frac{(\mathbf{p}-\mathbf{A})^2}{2m} + \frac{\hbar\Omega}{2} + \mathbf{W}\right]\psi_1 \tag{2.53}$$

where $\boldsymbol{A}$ and $\boldsymbol{W}$ are the two artificial geometric gauge potentials, $\mathbf{A} = ih\langle\chi_1|\nabla\chi_1\rangle = \frac{h}{2}(\cos\theta - 1)\nabla\phi$ is an artificial vector potential which represents an effective magnetic field $\boldsymbol{B}$ acting on the atomic dressed state. When $\boldsymbol{B} \neq \boldsymbol{0}$, Eq. (2.53) describes the motion of an effective charged particle in a magnetic field and $W = \frac{h^2}{2m}|\langle\chi_2|\nabla\chi_1\rangle|^2$ is a geometric scalar potential. $\boldsymbol{A}$, $\boldsymbol{B}$ and $\boldsymbol{W}$ depend on the spatial variation of θ and ϕ, so they are called geometric potentials.

Next, in the model shown in Fig. 2.13, assuming an Λ-type three-level atom with two laser beams coupling the atomic ground state $|g_1\rangle$ and $|g_2\rangle$ to the excited state $|e\rangle$, respectively, the vectors of the two beams are both k_0 that counter-propagate in the x-direction. Carrying out adiabatic elimination for the excited state $|e\rangle$ under large detuning conditions, the effective Hamiltonian in the Hilbert space expanded by $|g_1\rangle$ and $|g_2\rangle$ are as follows

$$H = \frac{p^2}{2m} + \frac{\hbar}{2}\begin{pmatrix} \Delta & g e^{i2k_0x} \\ g e^{-i2k_0x} & -\Delta \end{pmatrix} \tag{2.54}$$

where Δ is two-photon resonance detuning and g is Raman transition strength. By comparison of Hamiltonian Eqs. (2.51) and (2.54), it can be seen that they have the same form. Hence, after the same steps, the effective Hamiltonian corresponding to the dressed state of this system can be deduced as

$$H_{\text{eff}} = \frac{(p-A)^2}{2m} \tag{2.55}$$

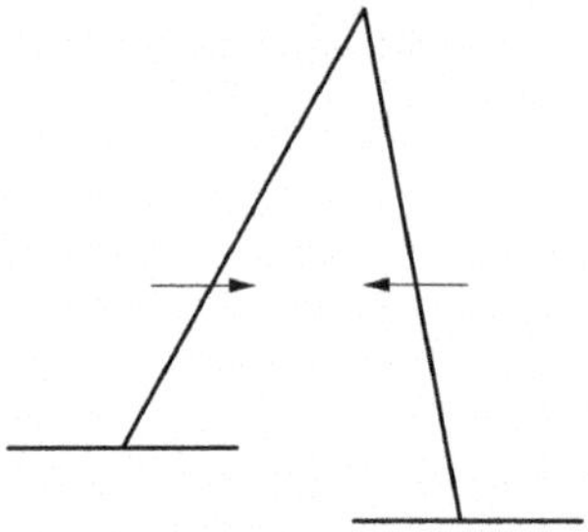

Fig. 2.13 Model for the generation of effective gauge potentials in a three-level atom

where $\mathbf{A} = (A_x, 0, 0)$, $A_x = \hbar k_0\left(\Delta/\sqrt{\Delta^2 + g^2 - 1}\right)$. Thus, an effective magnetic or electric field can be generated by modulating the detuning Δ. For example, Δ is spatially dependent on the y-direction, $\Delta = \Delta(y)$, and Ax is also spatially dependent, $A_x = A_x(y)$, leading $\mathbf{B} = \nabla \times \mathbf{A} = -\frac{\partial A_x}{\partial y} z$ to an effective magnetic field in the z-direction. Similarly, modulating Δ in time can produce an effective electric field $\mathrm{E} = -\frac{\partial \mathbf{A}}{\partial t} = -\frac{\partial A_x}{\partial t} x$ in the x-direction.

The above physical principle has been experimentally realized. In Ref. [64] people first prepared an atomic Bose–Einstein condensate of ^{87}Rb in the three spin states at $F = 1$, then apply a static magnetic field in the y-direction to cause energy shift, and then illuminate it with two Raman lasers counter-propagating along the z-direction, with the wave vector of light field denoted by k_r. The three spin ground states of the atom are then coupled by Raman transition (Fig. 2.14b). After a Raman transition, the atom acquires a recoil momentum $2k_r$ so that the Hamiltonian of the system can be expressed in the Hilbert space expanded by the states $\left\{\left|-1, \tilde{k}_x + 2k_r\right\rangle, \left|0, \tilde{k}_x\right\rangle, \left|+1, \tilde{k}_x - 2k_r\right\rangle\right\}$ as

$$\frac{H}{\hbar} = \begin{pmatrix} \frac{\hbar}{2m}\left(\tilde{k}_x + 2k_r\right)^2 - \delta & \Omega_R/2 & 0 \\ \Omega_R/2 & \frac{\hbar}{2m}\tilde{k}_x^2 - \varepsilon & \Omega_R/2 \\ 0 & \Omega_R/2 & \frac{\hbar}{2m}\left(\tilde{k}_x - 2k_r\right)^2 + \delta \end{pmatrix} \tag{2.56}$$

where k_x is quasi-momentum, Δ is the detuning of Raman transition, Ω_R is Raman Rabi frequency, and ε represents second-order Zeeman energy shift. By diagonalizing Hamiltonian (2.56), three energy eigenvalues can be obtained. If we take the smallest one of them (ground state), the dispersion relation of energy is

$$E = \frac{\hbar^2}{2m^*}\left[(k_x - k_{\min})^2 + k_y^2 + k_z^2\right] = \frac{1}{2m^*}(p - \mathbf{A})^2 \tag{2.57}$$

where m^* is effective mass and $k_{\min}$ is the minimum of ground state energy that relies on two-photon detuning Δ and Rabi frequency Ω_R. It can be seen that an effective vector potential $\mathbf{A} = \hbar k_{\min}\hat{\mathbf{e}}_x$ is generated in the system, and as introduced earlier, by adjusting the Raman light field δ to make Δ spatially dependent $\delta = \delta_0 + \delta' y$, one obtains an effective magnetic field in the z-direction. This principle has also been experimentally verified [65] and one obtains the effective magnetic field so that

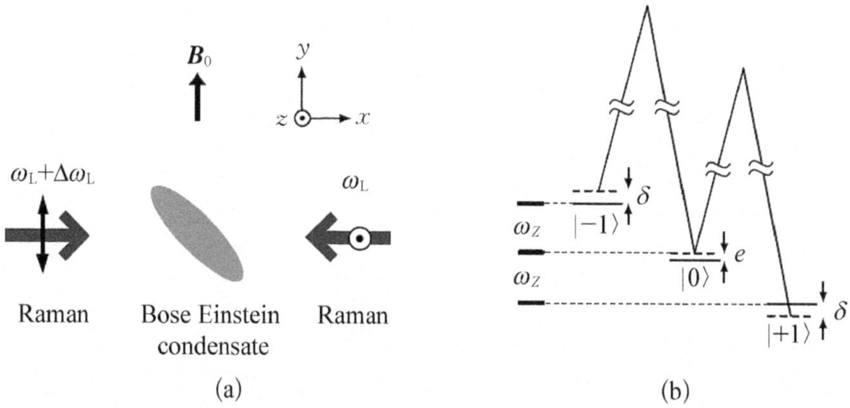

Fig. 2.14 Experimental schematic diagram for the generation of an effective vector gauge potential in an atomic ^{87}Rb Bose–Einstein condensate with $F = 1$ [64]. **a** Experimental schematic; **b** Energy level diagram

the quantized vortex can be generated without rotation. The results are shown in Fig. 2.15. If $\delta' = 0$, there are no vortices in the cold atomic gas, and as δ' increases, the effective magnetic field becomes stronger, and more and more quantized vortices are generated in the system. Likewise, the generation of an effective electric field in the Bose–Einstein condensate (BEC) has been experimentally verified [66].

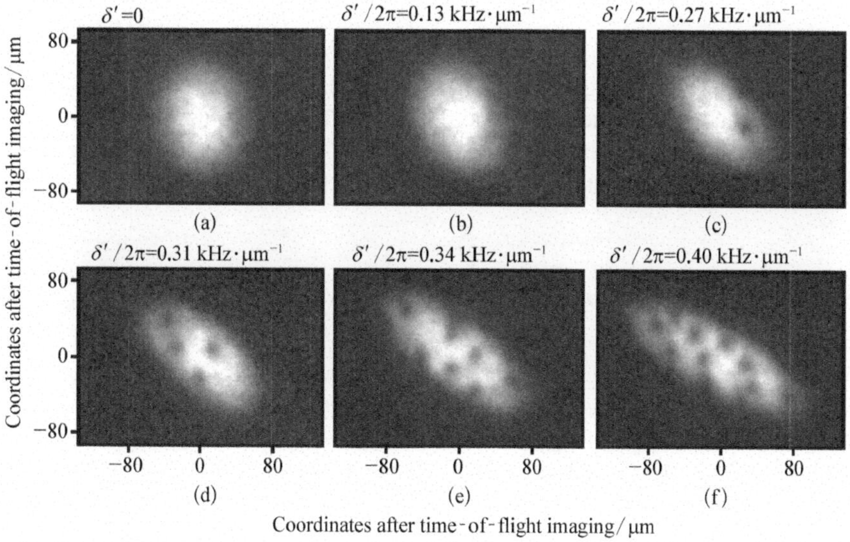

Fig. 2.15 Quantized vortex generation by effective magnetic field in Bose–Einstein condensate

The gauge potentials introduced above belong to the Abelian geometric gauge potential, i.e., after an arbitrary gauge transformation, any component of these potentials still maintains a commutation relation between each other. Next, we will present how to generate the non-Abelian gauge potential. Assuming an atom with $N + 1$ internal energy levels interact with a light field. The interactions between the atom and the light field are described by a time-independent $(N + 1) \times (N + 1)$-dimensional matrix, which ~~is~~ can be diagonalized to yield $N + 1$ dressed states $|\chi_{\mathrm{n}}(\mathrm{r})\rangle$, corresponding to eigenvalue $\varepsilon_n(r)$. The key to generating non-Abelian gauge potentials is that some of the dressed states are energy degenerate or nearly degenerated. We assume that q dressed states constitute a degeneracy subspace Z_{q} and that the initial state of the system is prepared in it, so the atomic wave function can be expanded as $|\Psi\rangle = \sum \psi_q(r)\big|\chi_q(r)\big\rangle$. After the same derivation process as above, the Schrödinger equation of the column vector $\tilde{\Psi} = \left(\psi_1, \cdots, \psi_q\right)^{\mathrm{T}}$ is satisfied is

$$i\hbar\frac{\partial\widetilde{\mathbf{W}}}{\partial t} = \left[\frac{(\mathbf{p} - \mathbf{A})^2}{2m} + \varepsilon + \mathbf{W}\right]\tilde{\Psi} \tag{2.58}$$

where ε is the energy of degenerate states. Equation (2.49) has the same form as Eq. (2.54), the difference being that both A and W are now $q \times q$ matrices, and the matrix element of the vector potential A is $A_{ij} = i\hbar\chi_i^\dagger(\mathbf{r})\nabla\chi_j(\mathbf{r})$, the corresponding effective magnetic field B is

$$B_i = \frac{1}{2}\varepsilon_{ikl}F_{kl},\ F_{kl} = \partial_k A_l - \partial_l A_k - \frac{i}{\hbar}[A_k, A_l] \tag{2.59}$$

$\frac{1}{2}\varepsilon_{ikl}[A_k, A_l] = (A \times A)_i$ is generally not equal to 0, because the components of the vector potential A are not necessarily commutative. Because of this term, the effective magnetic field B may not be 0 even if A is spatially homogeneous. It is the nature of the non-Abelian gauge potential, which also induces spin–orbit coupling.

2.4.3 Spin–Orbit Coupled Quantum Gases

A similar scheme to Fig. 2.13 was used experimentally to generate spin–orbit coupling [67], as shown in Fig. 2.16, with the difference that there is no applied magnetic field at this point. The atomic states $\{|-1, k_x + 2k_0\rangle, |0, k_x\rangle, |+1, k_x - 2k_0\rangle\}$ constitute a closed subspace in which the atomic Hamiltonian can be expressed as a 3×3 matrix.

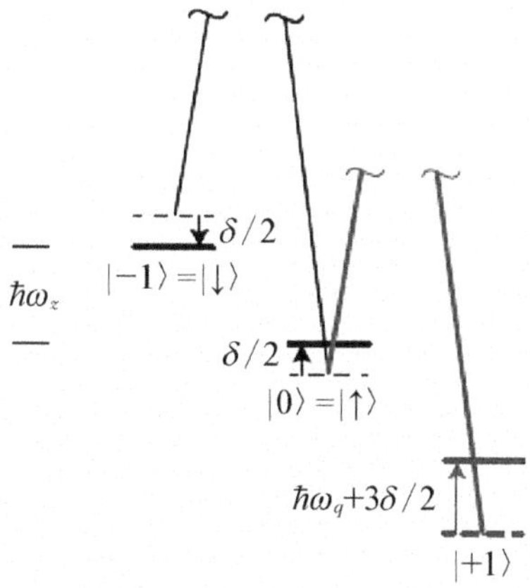

Fig. 2.16 Schematic diagram of the experimental scheme to generate spin–orbit coupling in a neutral atomic gas

$$H = \begin{pmatrix} \frac{\hbar^2 (k_x+2k_0)^2}{2m} + \delta_1 & \frac{g}{2} & 0 \\ \frac{g}{2} & \frac{\hbar^2 k_x^2}{2m} & \frac{g}{2} \\ 0 & \frac{g}{2} & \frac{\hbar^2 (k_x-2k_0)^2}{2m} - \delta_2 \end{pmatrix} \tag{2.60}$$

where $g/2$ is Rabi frequency and δ_1 and δ_2 represent the detuning of the Raman transition from the atomic state $|-1\rangle \rightarrow |0\rangle$ to $|0\rangle \rightarrow |1+1\rangle$. When $|\delta_1| \gg |\delta_2|$, adiabatic elimination can be made for $|+1\rangle$ state, Hamiltonian Eq. (2.60) can be reduced to a two-level system.

$$\begin{aligned} H_{\text{eff}} &= \begin{pmatrix} \frac{\hbar^2 (k_x+k_0)^2}{2m} + \frac{\delta}{2} & \frac{\delta}{2} \\ \frac{\delta}{2} & \frac{\hbar^2 (k_x-k_0)^2}{2m} - \frac{\delta}{2} \end{pmatrix} \\ &= \frac{\hbar^2 (k_x + k_0 \hat{\sigma}_z)^2}{2m} + \frac{g}{2}\hat{\sigma}_x + \frac{\delta}{2}\hat{\sigma}_z \end{aligned} \tag{2.61}$$

where the term proportional to $k_x \hat{\sigma}_z$ represents spin–orbit coupling.

Spielman's group at NIST in the United States first achieved Bose–Einstein condensation with spin–orbit coupling in ^{87}Rb [67], followed by Jing Zhang's group at Shanxi University [68] and Zwierlein's group at MIT [69], applying similar experimental principles to achieve spin–orbit coupling in Fermi ^{40}K and ^{6}Li atomic gases, respectively. Later, Jianwei Pan's group at the University of Science and Technology of China also achieved spin–orbit coupling in $^{87}R_b$ [70] and observed dipole oscillation phenomena.

Followings are some physical effects that can be brought about by spin–orbit coupling, and we first consider spin–orbit coupled Bose–Einstein condensates. When $\Delta = 0$ (i.e., Raman transition detuning is 0), by diagonalizing Eq. (2.56), we obtain the energy dispersion relation

$$\lambda_{\pm} = \frac{\hbar^2 k_x^2}{2m} \pm \sqrt{\frac{\hbar^4 k_x^2 k_R^2}{m^2} + \frac{g^2}{4}} \tag{2.62}$$

It can be seen that these two energy dispersion relations have an energy gap of Ω at $k_x = 0$. When $\Omega < 4E_L$ $\left(E_L = \frac{\hbar^2 k_R^2}{2m}\right)$, the lower one λ- has two energy minima located at $\sqrt{k_R^2 - \frac{m^2 g^2}{4h^4 k_R^2}}$, as shown in Fig. 2.17. After increasing Ω to be greater than $4E_L$, the spacing of these two energy minima becomes smaller and finally converges into a single minimum.

At this point, the collisional interactions of *s*-wave scattering between different components of BEC is then considered, and the phase diagram of this system can be obtained by calculation [5], as shown in Fig. 2.18.

In Fig. 2.18, $a = g_{10}/g$, $\beta = (g_1 - g_{00})/g$, $g = (g_{11} + {}_{00})/2$, and g_{ij} $(i, j = 1, 2)$ represent the strength of the collisional interactions between the *i* and *j* components of the atom, respectively. In region I, the system is in the superposition of two dressed states, while in regions II and III, the system is in two dressed states p_+ and p_-, respectively. In the presence of the resonance potential, the density distribution curves of the atoms have a stripe-like structure. The presence of different phases can be detected by turning off the potential well and measuring the displacement of the atomic cloud after expansion. For condensates with two dressed states, the

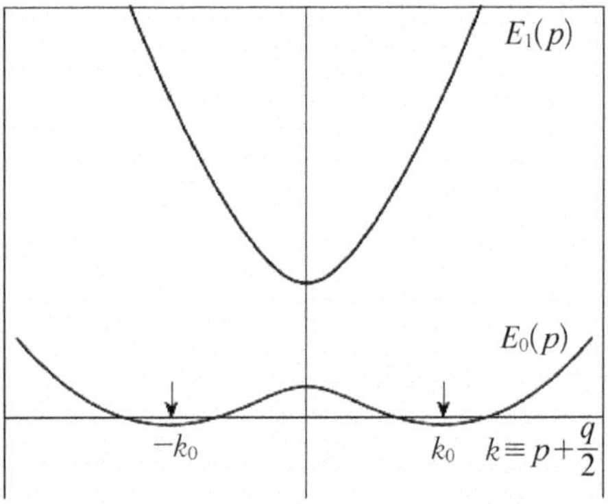

Fig. 2.17 Energy dispersion relation of spin–orbit coupled BEC [8]

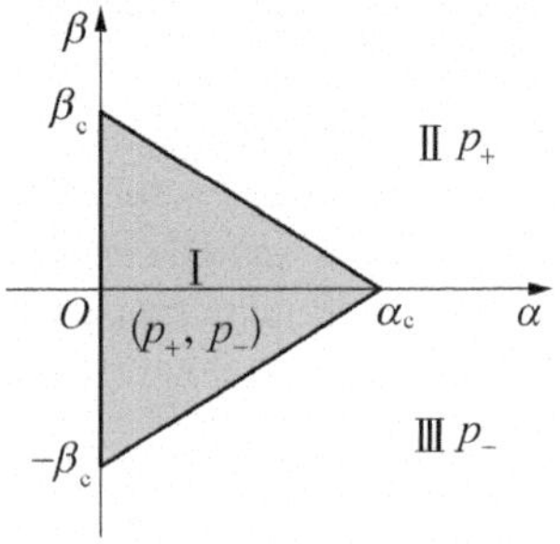

Fig. 2.18 Phase diagram of a bosonic gas with pseudo-spin 1/2 [5]

atomic cloud will split into two clouds with different momentum after expansion. On the contrary, regarding a condensate with only a single dressed state, the cloud will expand in only one direction, which depends on its momentum $p\pm$.

With respect to spin–orbit coupled fermionic atomic gases, collisional interactions between different components of the atom are studied, and it has been found that spin–orbit coupling can facilitate pairing between fermions [71, 72], which, together with the Zeeman effect, can produce topological superfluidity [73].

2.4.4 Spin–Orbit Coupling and Anderson Localization

Spin–orbit coupling is initially an important topic in condensed matter physics, which links electron spin with its orbital motion so that the propagation of electrons becomes spin-dependent. The spin–orbit interactions of the electron originate from the interactions between the electron spin and the magnetic field generated by its motion. As mentioned above, we also learned that spin–orbit interaction can be achieved in neutral atomic gases through the interaction between light and atoms.

Another intriguing issue in condensed matter physics is Anderson localization, a physical phenomenon discovered by the American physicist Anderson in 1958 [74]. He studied the motion of electrons without interactions in crystals with impurities and found that the electron spins would not diffuse because of crystal disorder, which is a result of quantum interference.

Since disorder is ubiquitous in nature, Anderson localization is universal and can occur in different physical systems, such as atomic matter waves and light waves. Because of the close connection between Anderson localization and conductor–insulator phase transition, it has always been a hot topic of research in physics. Based on these two research interests, in 2013, we studied the localization properties of spin-1/2 particles with spin–orbit coupling moving in a one-dimensional quasi-periodic potential [75]. The model can be implemented experimentally using cold atoms trapped in a quasi-periodic optical lattice potential plus an applied optical field (see Fig. 2.19). Considering that atoms with internal spin states $|\uparrow\rangle$ and $|\downarrow\rangle$ trapped in a spin-independent one-dimensional quasi-periodic optical lattice potential well, the potential well can be realized by superimposing two incommensurate optical lattice potentials. At the same time, a pair of counter-propagating Raman light couples atomic states $|\uparrow, k_x = q\rangle$ and $|\downarrow, k_x = q + 2k_{\mathrm{R}}\rangle$ to generate effective spin–orbit coupling [67].

In the Hilbert space expanded by the spin states $|\uparrow\rangle$ and $|\downarrow\rangle$ inside the atom, the single-particle Hamiltonian equation can be written as

$$\hat{h} = \left[\frac{p_x^2}{2m} + s_1 E_{\mathrm{R}1}\sin^2(k_1 x) + s_2 E_{\mathrm{R}1}\sin^2(k_2 x)\right]\hat{I} + \frac{\Omega}{2}\begin{pmatrix} 0 & \mathrm{e}^{2ik_{\mathrm{R}}x} \\ \mathrm{e}^{-2ik_{\mathrm{R}}x} & 0 \end{pmatrix} \quad (2.63)$$

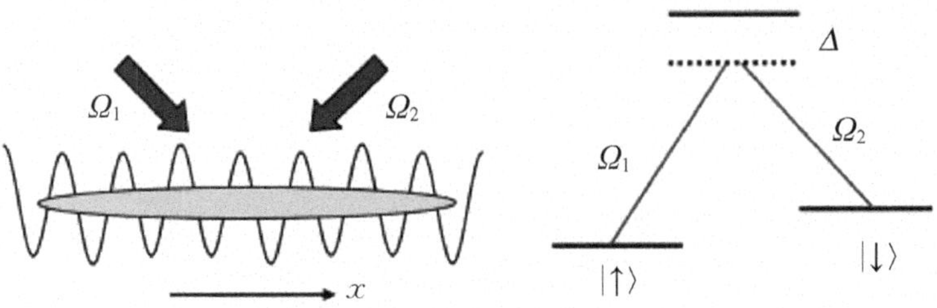

Fig. 2.19 Implementation of our studied system using cold atoms trapped in a quasi-periodic optical lattice and an applied optical field

where Ω is the strength of effective Raman coupling. Unitary transformation is performed for Hamiltonian Eq. (2.63) by the operator $\hat{R}(k_{\mathrm{R}}) = \exp(-\mathrm{i}k_{\mathrm{R}}x\hat{\sigma}_z)$, which is equivalent to transforming the system into the Hilbert space expanded by the spatially modified pseudospin atomic states $|\hat{\uparrow}\rangle' = |\uparrow\rangle \mathrm{e}^{-ik_{\mathrm{R}}x}$ and $|\downarrow\rangle' = |\downarrow\rangle \mathrm{e}^{ik_{\mathrm{R}}x}$ and making pseudospin rotation $\hat{\sigma}_z \to \hat{\sigma}_y, \hat{\sigma}_y \to \hat{\sigma}_x, \hat{\sigma}_x \to \hat{\sigma}_z$. Under this transformation, the Hamiltonian Eq. (2.63) can be rewritten as

$$\hat{h} = \frac{\left(p_x - \hat{A}\right)^2}{2m} + \frac{\Omega}{2}\hat{\sigma}_z + s_1 E_{\mathrm{R1}} \sin^2(k_1 x) + s_2 E_{\mathrm{R1}} \sin^2(k_2 x) \tag{2.64}$$

where the vector potential $\hat{A} = -\hbar k_R \hat{\sigma}_y$ then represents the effective spin–orbit interaction in the system. Hamiltonian Eq. (2.64) describes the motion of spin-1/2 particles with spin–orbit coupling in a one-dimensional quasi-periodic potential, which can be used to model the motion of corresponding condensed matter systems, for instance, electrons in a one-dimensional semiconductor nanowire.

Under the tight-binding approximation, the wave function of the particle can be written as $|\psi\rangle = \sum \psi_{j,\sigma}\hat{c}^{\dagger}_{j,\sigma}|0\rangle$, where j represents the j_{th} lattice point, defined by the relatively deep main optical lattice. Starting from the Hamiltonian Eq. (2.64), the corresponding stationary Schrödinger equation can be written as

$$\begin{aligned}
&- J\cos(\pi\gamma)\left(\phi_{j+1,\uparrow} + \phi_{j-1,\uparrow}\right) - J\sin(\pi\gamma)\left(\psi_{j+1,\downarrow} - \phi_{j-1,\downarrow}\right)\\
&- \Delta\cos\left(2\pi\beta_j\right)\psi_{j,\uparrow} + \frac{\Omega}{2}\psi_{j,\uparrow} + E\psi_{j,\uparrow}\\
&- J\cos(\pi\gamma)\left(\phi_{j+1,\downarrow} + \psi_{j-1,\downarrow}\right) + J\sin(\pi\gamma)(\psi_{j+1,\uparrow} - \phi_{j-1,\uparrow})\\
&- \Delta\cos\left(2\pi\beta_j\right)\psi_{j,\downarrow} - \frac{\Omega}{2}\psi_{j,\downarrow} = E\psi_{j,\downarrow}
\end{aligned} \tag{2.65}$$

where $\gamma{=}k_r/k_1$, J represents the tunneling amplitude of the atom between neighboring lattice sites, and Δ describes the strength of disorder. In the absence of spin–orbit coupling, i.e., $\gamma = 0$ and $\Omega = 0$, the two spin states of the atom will decouple and Eq. (2.65) becomes

$$-J\left(\psi_{j+1}+\psi_{j-1}\right)-\Delta\cos(2\pi\beta j)\psi_j=E\psi_j \tag{2.66}$$

Equation (2.66) satisfies Aubry duality, i.e., the discrete Fourier transform $\psi_j = \sum_m \psi_m e^{im(2\pi\beta_j)}$ of particle wave function, substitute it into Eq. (2.66) and we can obtain

$$-\frac{\Delta}{2}\left(\tilde{\psi}_{m+1}+\tilde{\psi}_{m-1}\right)-2J\cos(2\pi\beta m)\tilde{\psi}_m=E\tilde{\psi}_m \tag{2.67}$$

By comparison of Eqs. (2.66) and (2.67), it can be seen that the they are identical when $\Delta = 2J$. Since Fourier transform represents the transformation between the extended and local states of the system, $\Delta = 2J$ becomes a critical point. When $\Delta < 2J$, the single-particle wave function of the system is in the extended state, while when $\Delta > 2J$, the single-particle wave function of the system is in the localized state.

Now considering the effect of spin–orbit coupling on the localization of the system, we give the phase diagram of the system in the parameter space of the disorder strength Δ and the spin–orbit coupling strength Ω (see Fig. 2.20). The phase diagram is verified by multi-branch analysis of the atomic wave function and numerical simulations of diffusion dynamics. The spin–orbit coupling is found to be able to lead to a mixing of energy spectra (i.e., the co-existence of extended and localized states) and the appearance of mobility edges.

To verify the correctness of the phase diagram, the diffusion dynamics of the system is simulated, i.e., a Gaussian wave packet is prepared in the extended phase, critical phase, and local phase, respectively. The evolution of the wave packet with time is shown in Fig. 2.21.

We found that when the system is in the extended phase, the wave packet spreads rapidly with time evolution and eventually expands almost to the whole optical lattice; while when the system is in the local phase, the wave packet hardly spreads; when the system is in the critical phase, along with the spread of the wave packet, there is always a soliton-like structure at its center, accompanied by a diffuse sideband structure. These phenomena can be experimentally observed by time-of-flight imaging.

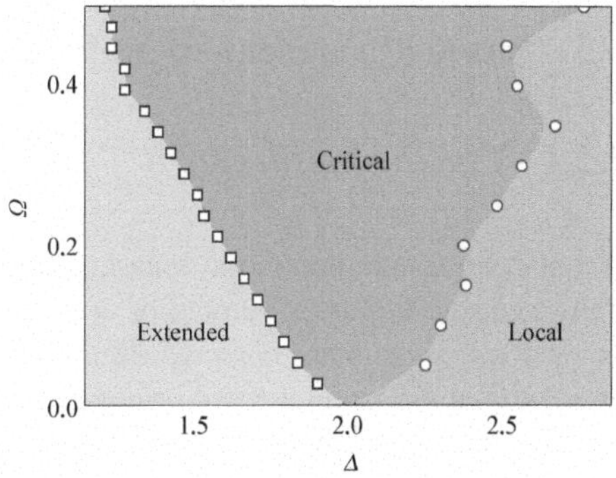

Fig. 2.20 Phase diagram of the system, extended represents the particle in the extended state and local the particle in the local state, coexisting in the critical phase

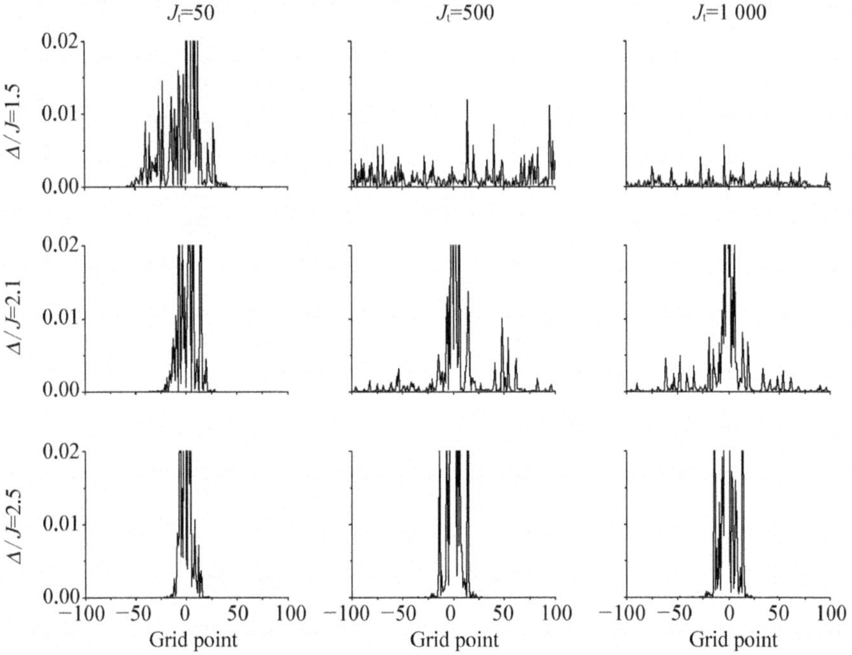

Fig. 2.21 Time evolution of the wave packet, from top to bottom, corresponding to the extended phase, critical phase, and local phase

In short, as an emerging interdiscipline research field between condensed matter physics and atomic and molecular physics, the research of spin–orbit interactions and gauge potentials in quantum gases is still under fast development. A series of new physical phenomena are being studied and discovered, especially in the areas of quantum optics, ultracold atomic physics, and quantum simulation, of which only a small fraction of the results can be covered by this subject.

2.5 Phase Measurement and Quantum Noise

2.5.1 Phase Measurement

In optical measurements, the measurement of physical quantities can usually be translated into the measurement of optical field phase. For example, weak gravitational wave signals are detected through the stretching and contracting effects of gravitational waves on the two orthogonal arms of an interferometer that will cause a subtle change in length, which in turn leads to a phase shift of the optical field in the interferometer. By measuring the phase of the optical field, the length variation caused by the gravitational waves can be determined, and thus the gravitational wave signal can be finally determined. The phase mentioned in this section refers to relative phase, i.e., the variation of phase or simply phase shift. Theoretically, discrete variables can usually be read accurately to their precise values. Phase, however, is a continuous variable, and for continuous variables, there are always unavoidable errors in reading its value. These errors come not only from experimental imperfections, but also, and more importantly from the limitations of the uncertainty principle in quantum mechanics by nature, which has a lower limit on its measurement accuracy.

In general, the phase measurement consists of 3 processes as shown in Fig. 2.22: first, the preparation of the probe system sensitive to the measured phase in the initial state $\rho(0)$; second, the evolution to $\rho(\Phi)$ after the probe system acts; third, the measurement of the probe system to extract the quantity reflecting the phase information. In order to reduce the influence of errors, the classical method of measurement is to repeat the measurement under exactly the same conditions (repeating the whole process in Fig. 2.22) and obtain the average result. But quantum mechanics sets limits to the parametric accuracy of such a measurement process. The fluctuation in the measurement result is proportional to a scale factor $1/\sqrt{\mathrm{N}}$, and N is the number of repeated measurements. This scale factor defines the so-called standard quantum limit (SQL) [76, 77], which has been confirmed to be an unbreakable limit by classical repeated measurement methods. This SQL is not unbreakable in principle; it can be broken by carefully designing quantum processes, such as changing the classical independent N-time measurements into one-time measurement of a strongly correlated N-body system (entangled or squeezed states) with the highest accuracy up to the $1/N$ Heisenberg limit (HL) [78]. In phase measurement, the $1/\sqrt{\mathrm{N}}$ SQL can be pushed via quantum process measurement accuracy, but it is still difficult to reach the 1/N Heisenberg limit. In order to get better phase measurement results, the whole process should be optimized. The theoretical analysis of phase measurement accuracy includes quantum fisher information (QFI) and error propagation formula (EPF). An EPF is taken to analyze how to optimize the measurement process. The linear error equation of phase measurement is

$$\delta\phi = \frac{\Delta\hat{O}}{\left|\partial\langle\hat{O}\rangle/\partial\phi\right|}$$

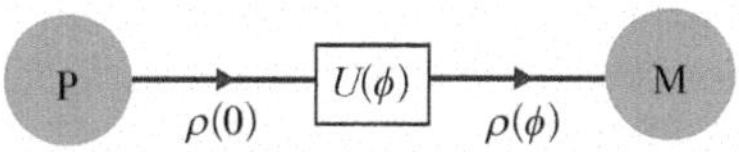

Fig. 2.22 Parameter (phase) measurement process. Process P preparation system is in the initial state $P(0)$, after the evolution of the phase $U(\Phi) = \mathrm{e}^{-\mathrm{i}\Phi\mathrm{H}}$ the state is $P(\Phi)$. The information about Φ is obtained by measuring process M

where $\Delta\hat{O} = \sqrt{\langle\hat{O}^2\rangle - \langle\hat{O}\rangle^2}$. In the calculation, the average value of the observable measurement $\hat{O}$ as well as the standard deviation are given under optimal measurement conditions. According to the above equation, the measurement process of quanta provides two paths to break this limit: ① preparing special detection states, i.e., changing the quantum statistical properties of the input probe state to reduce the quantum noise of the observable measurement to improve the accuracy of measurement. An example is that the input probe light is changed from Poisson statistics coherent light to sub-Poisson statistics compressed light, thereby reducing the noise of observable measurement and improving the measurement accuracy; ② designing special measurement interactions, i.e., changing the form of the coupled Hamiltonian to improve the slope of the output signal. For instance, changing linear interaction to nonlinear interaction an enhancing the output intensity of the probe light to increase the measurement accuracy, which will be described in Sect. 2.8. Both of these methods, alone or in combination, can help break the SQL of measurement accuracy.

High-accurate phase valuation is beneficial to multiple scientific fields, such as quantum precision measurement, imaging, sensing, and information processing. The accuracy of phase measurement depends on the states of the system and observable measurements; figuring out how to design and improve the measuring process to get the maximum accuracy of phase measurement has gradually bred the much-concerned frontier domain of phase precision measurement. Optical interferometers are useful and flexible measuring tools, among which the Mach–Zehnder interferometer, and its other derivatives, has been used for many years as the standard equipment for phase measurement. In the past decades, researchers devoted great efforts to improve the resolution and sensitivity of interferometers, investigated different physical mechanisms to improve the resolution and sensitivity of phase measurement [76–78]. In this section, we take interferometer phase measurement as an example to introduce how to push the SQL of measurement using quantum processes or quantum enhancement effects and recent progresses.

2.5.2 Phase Measurement by Mach–Zehnder Interferometer

Mach–Zehnder interference is shown in Fig. 2.23, consisting of two beam splitters, two plane mirrors, and a phase shifter. After the beam is shot on the beam splitter, the reflected beam has $\pi/2$ more phases than the transmitted beam, that is, the amplitude of the reflected light is preceded by an additional factor *i*. If two input ends a and b are injected into coherent states $|\alpha\rangle_a$ and $|\beta\rangle_b$, respectively, the input state is a product state $|\alpha\rangle_a \otimes |\beta\rangle_b$. For a 50:50 beam splitter, there is an input–output transformation relation for the following states

$$|\alpha\rangle_a \otimes |\beta\rangle_b \rightarrow^{\mathrm{BS}} \left|\frac{\alpha + \mathrm{i}\beta}{\sqrt{2}}\right\rangle_a \otimes \left|\frac{\beta + \mathrm{i}\alpha}{\sqrt{2}}\right\rangle_b \tag{2.68}$$

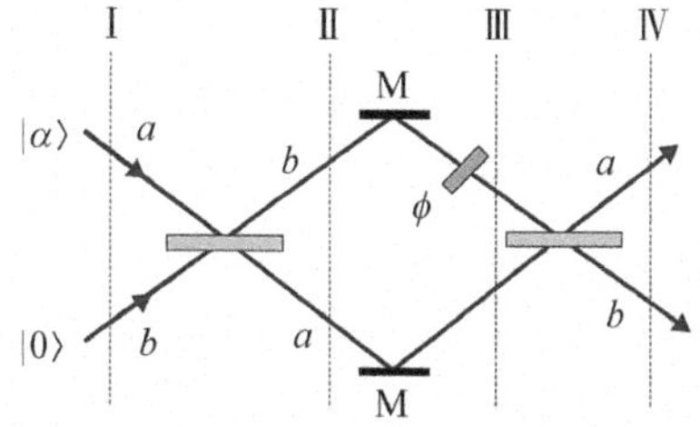

Fig. 2.23 Mach–Zehnder interferometer. *a* and *b* refer to the two channels of the interferometer, inside which channel *b* acquires a relative phase Φ with respect to channel a. I, II, III, and IV stand for the different phases of the interferometer, and *M* is a reflector. The coherent state $|a\rangle$ and the vacuum state $|0\rangle$ denote the input state of a certain combination. If there is no phase difference ($\Phi = 0$), all photons of coherent light come out from the port of channel b; if there is a π phase difference ($\Phi = \pi$), all photons of coherent light come out from the port of channel a. If the phase difference is in the middle, photons come out from both ports, and the phase difference or phase shift Φ is obtained by measuring the output light intensity of both ports. In the case of coherent light incidence, the statistical error of obtaining the phase shift Φ is proportional to $1/\sqrt{\mathrm{N}}$

The interferometer can have many types of input states and can inject coherent states, Fock states, squeezed states, etc. In the following, we consider a coherent state $|a\rangle$ and a vacuum state $|0\rangle$ injected into the interferometer (see Fig. 2.23), such as the b-port injected into the vacuum state, then $\beta = 0$ in the above expression. After passing the first beam splitter BS_1, the input state $|\psi\rangle_I = |\alpha/_a \otimes |0\rangle_b$ will be transformed by Eq. (2.68) to $|\psi\rangle_{\text{II}} = |\alpha/\sqrt{2}\rangle_a \otimes |\mathrm{i}\alpha/\sqrt{2}\rangle_b$. Subsequently, the beam passes through the plane mirror and phase shifter. The phase of the plane mirror to the beam is a negligible global phase. The b-channel phase shift ϕ is unitary operator $\hat{U}_{\text{PS}} = \exp(\mathrm{i}\phi\hat{n}_b)$, where $\hat{n}_b$ is the photon number operator. According to $\exp(\mathrm{i}\phi\hat{n}_b)|n\rangle_b = \exp(\mathrm{i}n\phi)|n\rangle_b$, state $|\psi\rangle_{\text{II}}$ undergoing a phase shift process can be described as $|\psi\rangle_{III} = \hat{U}_{\text{PS}}|\psi\rangle_{\text{II}} = |\alpha/\sqrt{2}\rangle_a \otimes \left|\mathrm{i}\alpha\mathrm{e}^{\mathrm{i}\phi}/\sqrt{2}\right\rangle_b$. After passing the second beam splitter BS_2, $|\psi\rangle_{III}$ is again transformed by Eq. (2.68), and the output state is $|\psi\rangle_{IV} = \left|\alpha(1-\mathrm{e}^{i\phi})/2\right\rangle_a \otimes \left|\mathrm{i}\alpha(1-\mathrm{e}^{i\phi})/2\right\rangle_b$. Since the intensity of the quantized field is proportional to the average photon number of the quantum states of the optical field, i.e., $I \propto \langle\hat{n}\rangle$. The intensities of the two output ends *a* and *b* of the interferometer are

$$I_{a\text{ - out}} = \langle\hat{n}_a\rangle = \overline{n}(1-\cos\phi)/2 = \overline{n}\sin^2(\phi/2) \tag{2.69}$$

$$I_{b\text{- out}} = \langle\hat{n}_b\rangle = \overline{n}(1+\cos\phi)/2 = \overline{n}\cos^2(\phi/2) \tag{2.70}$$

where $\overline{n} = |\alpha|^2$. From Eqs. (2.69) and (2.70), it is clear that regarding the coherent light incoming from the port of channel *a*, in the absence of phase difference ($\Phi = 0$), all photons of coherent light will come out from the port of channel b; in the presence of π phase difference ($\Phi = \pi$), all photons of coherent light come out from the port of channel a. The intensity at the outgoing end and $I_{a\text{ - out}} + I_{b\text{ - out}} = \overline{n}$

are obviously independent of the Φ phase. Therefore, using the intensity difference $I_D = I_{b\text{-out}} - I_{a\text{-out}}$ at the outgoing end as the detection variable, it is easy to obtain the measured signal

$$I_D = \bar{n}\cos\phi \tag{2.71}$$

The accuracy of phase measurement is first described below from the perspective of a classical physical image. The variation of intensity difference I_D with phase shift ϕ is shown in Fig. 2.24. According to $\phi = kx = 2\pi x/\lambda$ (x is the length difference between the two arms of the interferometer), the difference between the two arms can be studied by measuring the phase to study gravitational waves, which are relatively weak and require a particularly high accuracy of phase measurement, and this aspect is described specifically in Sect. 2.2. Next is the analysis of the measurement accuracy of phase shift in the interferometer. When $\phi = \pi/2$, the intensity output by two ports is the same, the intensity difference is 0, and the slope of this point is the maximum, as shown in Fig. 2.25a, b. If the phase has a small shift $\Delta\phi$, the corresponding intensity also has a slight variation ΔI_D, which, according to Eq. (2.71), can be approximated as such a differential relationship as $\Delta I_D = \bar{n}\sin\phi(\Delta\phi)$ or written as

$$\frac{\Delta I_D}{\Delta\phi} = \frac{\partial I_D}{\partial\phi} = \bar{n}\sin\phi \tag{2.72}$$

The above equation can also be written as

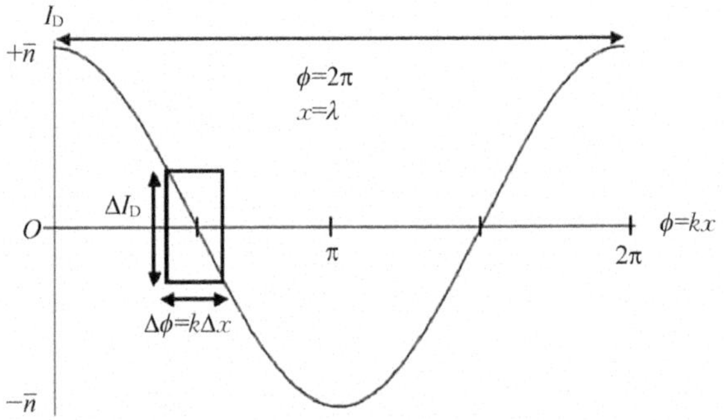

Fig. 2.24 Plot of the intensity difference I_D with phase shift with a period of 2π. If there is a little shift $\Delta\phi$ in the phase, then there is also a small change in intensity ΔI_D. According to quantum mechanics, the intensity of the optical field can never be measured with infinite accuracy, so ΔI_D will always have a finite value, and $\Delta\phi$ will have a lower limit. The minimum detectable phase shift $\Delta\phi$ is limited by optical field fluctuation, the magnitude of which is determined by the statistical nature of the optical field. For coherent light detection, the lower limit of phase measurement is $1/\sqrt{n}$

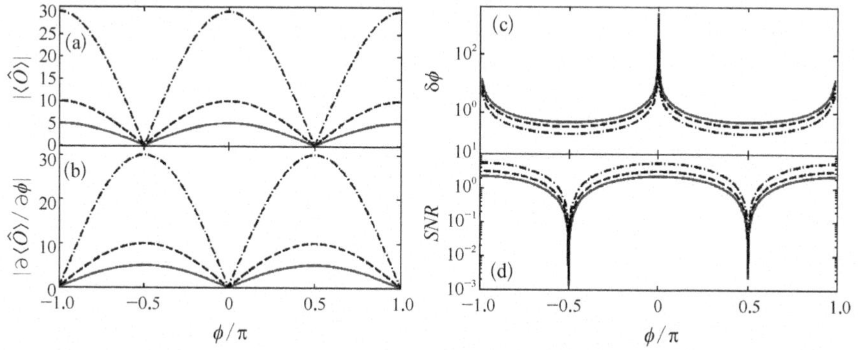

Fig. 2.25 The various properties of the output signals changed with phase under the different input strengths. **a** Average $|\langle\hat{O}\rangle|$; **b** slope $|\partial\langle\hat{O}\rangle/\partial\phi|$; **c** phase sensitivity $\delta\phi$; **d** SNR, change relation with phase ϕ, where $\overline{n}=5$ (solid line), $\overline{n}=10$ (dotted line), $\overline{n}=30$ (chain line)

$$\Delta\phi = \frac{\Delta I_{\rm D}}{\partial I_{\rm D}/\partial\phi} = \frac{\Delta I_{\rm D}}{\overline{n}\sin\phi} \tag{2.73}$$

$\partial I_{\rm D}/\partial\phi$ is slope and at $\phi=\pi/2$ and $\sin(\pi/2)=1$, the slope is maximum at this phase point, i.e., the smallest detectable phase shift $\Delta\phi$ that can be detected at this phase point. If we can measure the intensity shift $\Delta I_{\rm D}$ to infinite accuracy ($\Delta I_{\rm D}=0$), we can measure the phase shift to infinite accuracy ($\Delta\phi=0$) too. If this is the case, any interferometer can measure weak gravitational waves. Obviously, such analysis of classical physical images as above is not correct. Because it does not take into account quantum effects and that the intensity of the light field is not a constant but about the fluctuation of some averages, which originate from the vacuum fluctuation of the electromagnetic field and are quantum in nature. According to quantum mechanics, the intensity of light field can never be measured infinitely accurate, so $\Delta I_{\rm D}$ in Fig. 2.24 will always have a finite value and $\Delta\phi$ is a lower limit. For coherent light detection, this lower limit is $1/\sqrt{\overline{n}}$, i.e., SQL. It is based on this that Caves [11] first noted that changing the quantum statistical properties of the detection field could improve the measurement accuracy and break the SQL of measurement. As the statistical properties of the detection field are changed, that is to say, $\Delta I_{\rm D}$ is actually changed, which in turn leads to the reduction of the minimum detectable, and accordingly the measurement accuracy can be improved.

The accuracy of phase measurement is now described from the perspective of quantum physical images. According to the previous analysis, we choose the population difference operator $\hat{O}=\hat{n}_b-\hat{n}_a$ as the detection variable, then the intensity difference at the output end is $|\langle\hat{O}\rangle|=\langle\hat{n}_b-\hat{n}_a\rangle=\overline{n}\cos\phi$. Phase shift measurement involves two quantities: one is the magnitude of phase shift, and the other the accuracy of phase shift, also known as phase sensitivity. For the measurement of the phase shift size, it is detected by modulating the intensity difference $\langle\hat{O}\rangle$ at the output end of the interferometer. As shown in Fig. 2.25a, different line types correspond to different input intensities. Here, the oscillation frequency of the detected variable depends only on the phase ϕ, having nothing to do with the optical field intensity of input detection. With respect to the measurement of the interferometer's phase sensitivity, that is, the minimum detectable phase shift, according to linear error equation $\delta\phi = \Delta\hat{O}|\partial\langle\hat{O}\rangle/\partial\phi|^{-1}$, there is $\Delta^2\hat{O} = \langle\hat{O}^2\rangle - \langle\hat{O}\rangle^2 = \overline{n}$ in the case of input coherent state case, and the phase sensitivity can be easily derived from

$$\delta\phi = \frac{1}{\sqrt{\overline{n}}|\sin\phi|} \tag{2.74}$$

When $\phi = \pi/2$, the detectable min phase shaft is $\delta\phi_{\min} = 1/\sqrt{n}$. This is the usual standard quantum limit (SQL), also known as shot-noise limit (SNL). By comparison with the classical image, the molecular ΔI_{D} in Eq. (2.73) is a minor change while in the quantum image $\Delta\hat{O}$ is a mean square error, reflecting the quantum noise at work.

In the measurement of phase shift sensitivity, it is sometimes necessary to measure the *SNR* of the output signal, which is defined as $SNR = |\langle\hat{O}\rangle|/(\Delta\hat{O})$. In the current measurement scheme, $SNR = \sqrt{\overline{n}}|\cos\phi|$, the larger the better; as for sensitivity $\delta\phi = \Delta\hat{O}|\mathrm{d}\langle\hat{O}\rangle/\mathrm{d}\phi|^{-1}$, the smaller the better. So, by comparing the definitions of these two quantities, it shows that they both require the uncertainty $\Delta\hat{O}$ to be as low as possible, but $\Delta\hat{O}$ will never be zero, and its magnitude is determined by the quantum fluctuation of the input signal. The SNR and sensitivity measurement require $|\langle\hat{O}\rangle|$ and $|\partial\langle\hat{O}\rangle/\partial\phi|$ to be as large as possible. As can be seen in Fig. 2.25a, b, the maximum value of signal$|\langle\hat{O}\rangle|$ and signal slope $|\partial\langle\hat{O}\rangle/\partial\phi|$ are sometimes impossible to obtain at the same phase point. This leads to a low SNR in this case when the sensitivity is high and vice versa, as shown in the curves in Fig. 2.25c, d. In some measurement processes like sensitivity measurements that are independent of the phase ϕ, this allows both optimal sensitivity and optimal SNR to be obtained when measurements are made at some optimal phase point.

From $\delta\phi_{\min} = 1/\sqrt{n}$, it is simple to assume that by increasing the intensity of the probe field for interferometer input, one could always enhance the sensitivity of phase measurement. However, as pointed out by Caves [11], increasing the input intensity also increases the noise of the radiation pressure on the interferometer mirror, thus leading to the same error in the measurement process. As shown in Fig. 2.26, in the region of low input intensity, the shot noise predominates and the radiation pressure noise is negligible; in the region of high input intensity, the radiation pressure noise plays a dominant role. Hence, there is a saddle point where either increasing or reducing the input intensity will lead to a growth in quantum noise. To avoid measurement errors caused by the increase in radiation pressure noise, researchers prefer to use the quantum effect to improve the sensitivity of the interferometer at low intensity input rather than simply enhancing the input intensity.

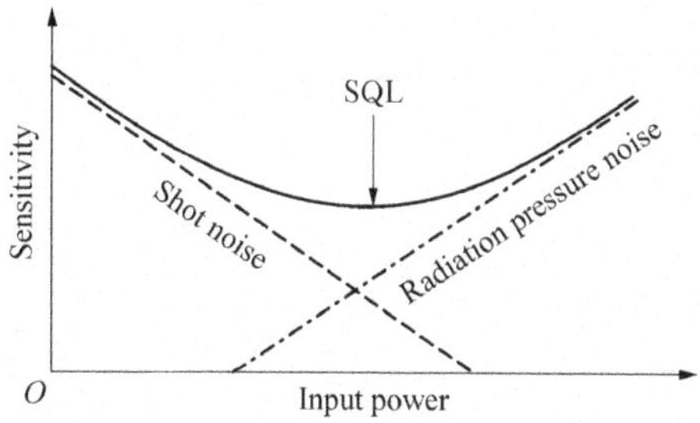

Fig. 2.26 Relation graph of the shot noise and radiation pressure noise of coherent light incidence interferometer varying with the coherent light incident power

2.5.3 Phase Measurement of Quantum Effect Enhancement

The quantum enhancement effect can be leveraged to achieve phase valuation to break the SQL and thus reach the Heisenberg limit (HL). The approach to improve the accuracy of phase measurement with quantum effect will be set forth in the ensuring paragraphs from the perspective of the linear error equation, from which, the accuracy can be enhanced by lowering the uncertainty $\Delta\hat{O}$ of $\hat{O}$ be measured or increasing its slope.

The first person to show interest in reducing the uncertainty of the detection quantity $\Delta\hat{O}$ was Caves [11], who first pointed out that the SQL of phase measurement can be broken with nonclassical compressed light. As the coherent light is incident on the interferometer, the detection quantity $\langle\hat{O}\rangle$ can be simply expressed as $\langle\hat{O}\rangle = A_p\cos(\phi_0+\phi)$, as indicated by the solid lines in Figs. 2.27 and 2.28. In view of improving the slope to enhance measurement accuracy, two paths can be taken: one is to increase the oscillation frequency of measurement, e.g., NOON state of input entanglement $\left(|NOON\rangle = (|N,0\rangle + |0,N\rangle)/\sqrt{2}\right)$ [79]. The average of detection quantity is $\langle\hat{O}\rangle = A_p\cos(\phi_0+N\phi)$, as shown by the dashed line in Fig. 2.27, and the oscillation frequency is improved due to the input *NOON* state with respect to the coherent light incidence (see the solid line in Fig. 2.28). Second is to enhance amplitude Ap. The intensity of the input signal is amplified by a nonlinear process, as shown by the dashed line in Fig. 2.28, where the amplitude of the output signal is enhanced, and the solid line in this figure indicates coherent light input, but not amplified. One effective approach is to replace the linear beam splitter in the Mach–Zehnder interferometer with a parametric process or a four-wave mixing process, which is also called SU(1,1) interferometer and was first proposed theoretically by Yurke et al. [80]. This new type of interferometer has now been experimentally implemented by our group [81, 82].

The ways to enhance phase valuation using quantum effects are summarized in Table 2.2. Of course, it is also possible to simultaneously lower the fluctuation $\Delta\hat{O}$ of mechanical quantity $\hat{O}$ as well as increase its slope $|\partial\langle\hat{O}\rangle/\partial\phi|$, together to improve the phase measurement accuracy [83]. Two examples of enhanced phase measurements are presented below to specify the quantum enhancement effect.

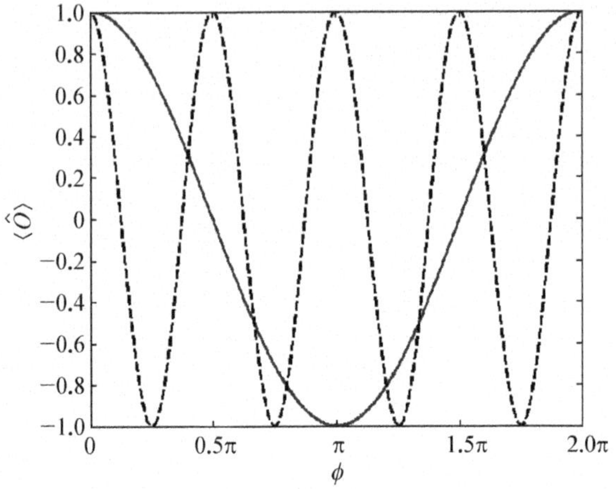

Fig. 2.27 NOON state input enhancing slope to improve the accuracy of phase measurement, where the solid line indicates the measurement signal of the coherent state input and the dashed line indicates the measurement signal of the NOON state input. The increase in the oscillation frequency of dashed line has actually enhanced the slope of output signal, thus improving the sensitivity of phase measurement

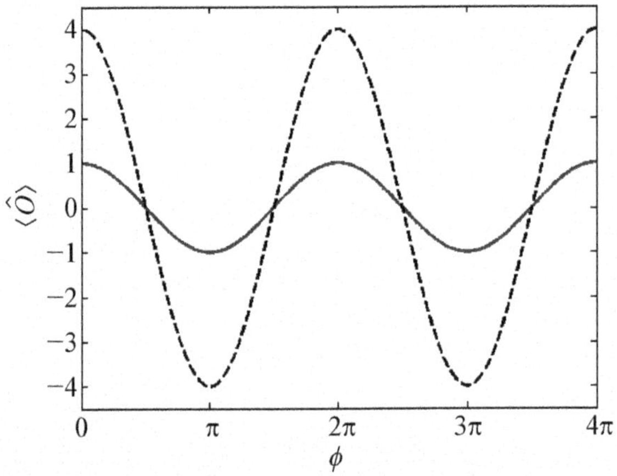

Fig. 2.28 Amplifying the strength of the detection signal to increase the slope and enhance phase measurement accuracy, where the solid line indicates the measurement signal of the coherent state input and the dashed line indicates the amplified measurement signal of the coherent state input. The increased amplitude of the dashed line means that the slope of the output signal and the sensitivity of phase measurement are enhanced as well

Table 2.2 Increase of phase measurement sensitivity by quantum effect

Ways to increase sensitivity	Means adopted
Lowering noise $\triangle\hat{O}$	Inputing squeezed state
Increasing the slope of detection quantity $\lvert\partial\langle\hat{O}\rangle/\partial\phi\rvert$	Improving the oscillation frequency of detection quantity
	Enhancing the output amplitude of detection quantity

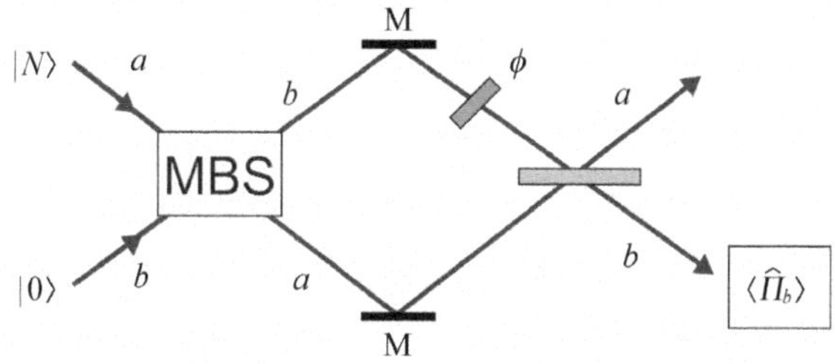

Fig. 2.29 The first beam splitter in the conventional Mach–Zehnder interferometer, replaced by a special beam splitter (MBS) that is capable of producing a NOON state for initial input states $|N\rangle_a|0\rangle_b$. $\hat{\Pi}_b = (-1)^{\hat{n}_b}$ means that the output signal of port b is measured by the method of parity detection

(1) Phase measurement of oscillation frequency enhancement

In this section, the enhancement of phase measurement sensitivity by increasing the oscillation frequency of the output signal is presented, specifically through the *NOON* state of entanglement [79]. As shown in Fig. 2.29, in a conventional Mach–Zehnder interferometer, the first beam splitter is replaced by a special beam splitter MBS with magic. For the initial input states $|N\rangle_a|0\rangle_b$, a *NOON* state is generated after passing this special beam splitter MBS:

$$|NOON\rangle = \left(|N\rangle_a|0\rangle_b + e^{i\Phi_N}|0\rangle_a|N\rangle_b\right)/\sqrt{2} \tag{2.75}$$

The following analysis shows how the above quantum state evolves through the second beam splitter. For a 50:50 beam splitter (see Fig. 2.30), the transformation relation of the operator is $\hat{a}' = (\hat{a}+\mathrm{i}\hat{b})/\sqrt{2}$, $\hat{b}' = (\hat{b}+\mathrm{i}\hat{a})/\sqrt{2}$. Their inverse transformation is $\hat{a} = (\hat{a}'-\mathrm{i}\hat{b}')/\sqrt{2}$, $\hat{b} = (\hat{b}'-\mathrm{i}\hat{a}')/\sqrt{2}$. The input vacuum state $|0\rangle_a|0\rangle_b$ leads to the output of vacuum state $|0\rangle_{a'}|0\rangle_{b'}$, meaning that for this linear beam splitter there is $|0\rangle_a|0\rangle_b|0\rangle_{a'}|0\rangle_{b'}$. Considering the input of population state $|N\rangle_a|M\rangle_b$, the population state of this input can be written as $\||N\rangle_a|M\rangle_b = \left(\hat{a}^\dagger\right)^N\left(\hat{b}^\dagger\right)^M|0\rangle_a|0\rangle_b/\sqrt{N!M!}$. According to the input transformation relation of the vacuum state and the substitution of inverse transformation, the output state of the population output state $|N\rangle_a|M\rangle_b$ after passing the beam splitter is as follows

where Φ_{N} is a N-dependent phase, which is related to specific generating devices. Applying the phase-shifted unitary operator $\hat{U}_{\mathrm{PS}} = \exp(\mathrm{i}\phi\hat{n}_b)$ of channel b acting on the above NOON state, we have

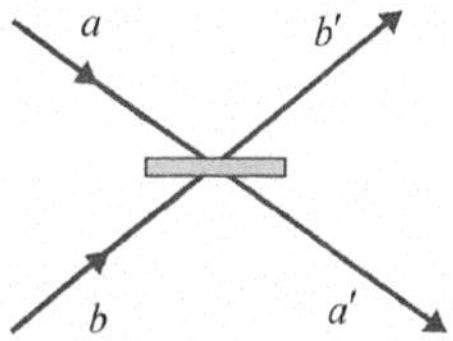

Fig. 2.30 50:50 beam splitter

$$\hat{U}_{\mathrm{PS}}|NOON\rangle = \left(|N\rangle_a|0\rangle_b + e^{i(N\phi+\Phi_N)}|0\rangle_a|N\rangle_b\right)/\sqrt{2} \tag{2.76}$$

$$|N\rangle_a|M\rangle_b \to \left(\frac{\hat{a}'^{\dagger} + \mathrm{i}\hat{b}'^{\dagger}}{\sqrt{2}}\right)^N \left(\frac{\hat{b}'^{\dagger} + \mathrm{i}\hat{a}'^{\dagger}}{\sqrt{2}}\right)^M |0\rangle_{a'}|0\rangle_{b'}/\sqrt{N!M!} \tag{2.77}$$

Therefore, the state $\left(|N\rangle_a|0\rangle_b + \mathrm{e}^{i(N\phi+\Phi_N)}|0\rangle_a|N\rangle_b\right)/\sqrt{2}$ undergoing phase shift evolves as follows after passing the second beam splitter

$$|\mathrm{out}, N\rangle = \frac{1}{\sqrt{2^{N+1}N!}}\left[\left(\hat{a}^{\dagger} + \mathrm{i}\hat{b}^{\dagger}\right)^N + \mathrm{e}^{i(N\phi+\Phi_N}\left(\hat{b}^{\dagger} + \mathrm{i}\hat{a}^{\dagger}\right)^N\right]|0\rangle_a|0\rangle_b \tag{2.78}$$

Regarding the output state $|\mathrm{out}, N\rangle$, when $N > 1$, the average of population difference operators is 0, namely $\langle\hat{O}\rangle = 0$, and it is impossible to get any phase information by measuring the population difference. The photon number parity measurements introduced by C. C. Gerry et al. is introduced below, which allows a good analysis of phase measurement sensitivity [84]. The parity operator is defined as $\hat{\Pi} \equiv (-1)^{\hat{n}}$, and the result is $+1$ when the photon number of the output mode is even and -1 when the photon number of the output mode is odd. In this example, the parity measurement of channel b for interferometer output is considered (Fig. 2.29), and the parity operator for this mode is written as $\hat{\Pi}_b = (-1)^{\hat{n}_b} = \exp(i\pi\hat{n}_b)$. With respect to output state $|\mathrm{out}, N\rangle$, the average of the parity operator

$$\langle\hat{\Pi}_b\rangle = \frac{\mathrm{i}^N}{2}\left[\mathrm{e}^{\mathrm{i}(N\phi+\Phi_N)} + (-1)^{\mathrm{N}}\mathrm{e}^{-\mathrm{i}(N\phi+\Phi_N)}\right] \tag{2.79}$$

or according to the odd and even of N

$$\langle\hat{\Pi}_b\rangle = (-1)^{N/2}\cos(N\phi + \Phi_N), \quad \text{N is an even number} \tag{2.80}$$

$$\langle\hat{\Pi}_b\rangle = (-1)^{(N+1)/2}\sin(N\phi + \Phi_N), \quad \text{N is an odd number} \tag{2.81}$$

It is clear that the oscillation frequency of average parity operator $\hat{\Pi} \equiv (-1)^{\hat{n}_b}$ is proportional to N. Also, the slope of the average value $\left|\partial\langle\hat{\Pi}_b\rangle/\partial\phi\right|$ is proportional to N and becomes larger as N increases. Substituting for the linear error equation, the sensitivity of phase measurement is higher than the SQL and same as the Heisenberg limit

$$\delta\phi = \sqrt{\frac{1 - \langle\hat{\Pi}_b\rangle^2}{\left(\partial\langle\hat{\Pi}_b\rangle/\partial\phi\right)^2}} = \frac{1}{N} \tag{2.82}$$

where $\hat{\Pi}_b^2 = \hat{I}$. About using the NOON state and performing parity detection, the *SNR* of phase measurement is

$$SNR = \left|\frac{\cos(N\phi + \Phi_N)}{\sin(N\phi + \Phi_N)}\right|, \quad \text{N is an even number} \tag{2.83}$$

$$SNR = \left|\frac{\sin(N\phi + \Phi_N)}{\cos(N\phi + \Phi_N)}\right|, \quad \text{N is an odd number} \tag{2.84}$$

When it is measured near $\Phi \approx 0$, it is able to get a high SNR by letting $\Phi_N = 0$ for an even number of N, and a high SNR can be obtained by letting $\Phi_N = \pi/2$ for an odd number of N.

In terms of the schemes to enhance the sensitivity of phase measurements by increasing the oscillation frequency of the output signal, in addition to the *NOON* state, there are also input pairs of Fock states $|N\rangle_a|N\rangle_b$ or of coherent states in a Mach–Zehnder interferometer [84]. Performance of parity detection in experiments: first, direct measurement; second, indirect measurement. Direct measurements, i.e., by photon counting. Then, +1 or −1 is given depending on odd or even numbers. To achieve such a purpose, this requires detectors with resolvable photon numbers. The current detectors described in the literature have photon number resolution in single digits. To achieve a large number of photon resolutions, no such detector is currently available. Indirect measurement, i.e., through the relationship $W(\alpha) = 2\langle\hat{D}^\dagger(\alpha)\hat{\Pi}\hat{D}(\alpha)\rangle/\pi$ between the parity operator and the Wigner function $W(_\alpha)$. $\hat{D}(\alpha) = \exp(\alpha\hat{a}^\dagger - \alpha\hat{a})$ is the translation operator for single-mode field and α is a complex number in phase space. At the origin of phase space ($\alpha = 0$), there is $\langle\hat{\Pi}\rangle = \pi W(0)/2$. Therefore, we can obtain the expectation value of the parity operator by simply measuring the value of the Wigner function at the origin of the phase space. The Wigner function of the field can be reconstructed by a series of homodyne detections using quantum tomography. In this way, the expected value of the parity operator can be obtained without direct measurement.

(2) Phase measurement by signal amplification

Increasing the amplitude of the output signal to enhance the sensitivity of phase measurement is discussed in this section. The SU(1,1) interferometer introduced by Yurke et al. [80] (Fig. 2.31), substitutes for the optical parametric amplifier (OPA) of the linear beam splitter in the conventional Mach–Zehnder interferometer. As already described, the input signal will be amplified after passing through this interferometer compared to the same input signal put into the Mach–Zehnder interferometer. In the case of the same output noise, the sensitivity of phase measurement will be greatly improved by linear error equation $\delta\phi = \Delta\hat{O}\left|\mathrm{d}\langle\hat{O}\rangle/\mathrm{d}\phi\right|^{-1}$. Our recent work considers not only the use of OPA output signal to increase the slope $\left|\mathrm{d}\langle\hat{O}\rangle/\mathrm{d}\phi\right|$, but also the use of input squeezed states to reduce the noise $\Delta\hat{O}$ of the output signal [83]. Therefore, at the two input ends of the interferometer, a coherent state $|\beta\rangle_a (\beta = |\beta|e^{i\theta}\beta)$ is input at one port and a squeezed vacuum state $|0, \xi\rangle (\xi = re^{i\eta})$ is input at the other port. Both enhancements are applied to improve the sensitivity of phase measurement.

Based on Fig. 2.31, in the SU(1, 1) interferometer, $\hat{a}(\hat{a}^\dagger)$ and $\hat{b}(\hat{b}^\dagger)$ are the annihilation (creation) operators of the two modes of the interferometer. The transformation between the input operator and output operator of the interferometer in the

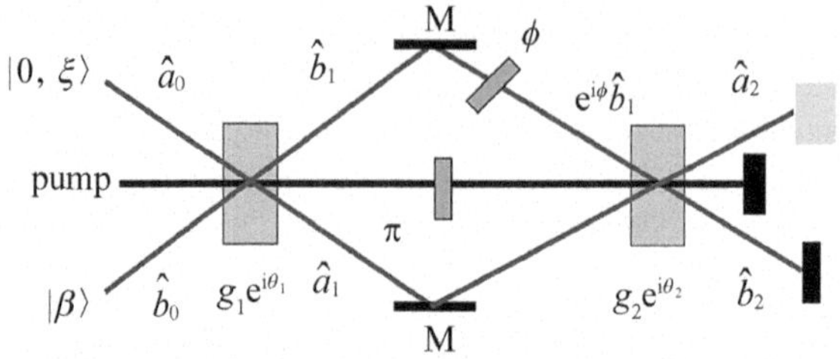

Fig. 2.31 SU(1, 1) interferometer. It is the replacement of both beam splitters in a conventional Mach–Zehnder interferometer with an optical parametric amplifier (OPA). Here, two ports of the interferometer are studied, one with input coherent state $|\beta\rangle$ and the other with input squeezed vacuum state $|0, \xi\rangle$, and the output field $\hat{a}_2$ is obtained by homodyne measurement

Heisenberg representation is

$$\begin{aligned} \hat{a}_2 &= u\hat{a}_0 - v\hat{b}_0^\dagger \\ \hat{b}_2 &= e^{i\phi}\left(u\hat{b}_0 - v\hat{a}_0^\dagger\right) \end{aligned} \tag{2.85}$$

where $u = \cosh g_1 \cosh g_2 + e - {}^{i\phi}e^{i(\theta 2—\theta 1)}\sinh g_1 \sinh g_2$, $\upsilon = e^{i\theta 1} \sinh g_1 \cosh g_2 + e - {}^{i\phi}e^{i\theta 2} \cosh g_1 \cosh g_{2,}$ and $|u|^2 - |\upsilon|^2 = 1$. $g_1(g_2)$ and $\theta_1(\theta_2)$ are the intensity and phase of two parametric processes, respectively. ϕ is the phase shift difference between the two arms inside the interferometer, which is the quantity we need to measure. If $\theta_{2.}\theta_1 = \pi$, $g = g_1 = g_2$, then $u = \cosh^2 g\text{-}e^{-i\phi} \sinh^2 g$, $\upsilon = e^{i\theta 1} \sinh g \cosh g\ (1 - e^{-i\phi})$ correspond to the equilibrium geometries of the interferometer. In equilibrium, $u = 1$ and $\upsilon = 0$ when $\phi = 0$, that is, the second parametric process is the retroaction of the first one.

Here, the measured quantity of the output signal is the quadrature amplitude component $\widehat{\text{X}}$ other than the sum of photon numbers $\hat{N} = \hat{n}_a + \hat{n}_b$. The quadrature amplitude component $\widehat{\text{X}}$ is defined as $\widehat{\text{X}} = \left(\hat{a}_2 + \hat{a}_2^\dagger\right)/\sqrt{2}$, and the linear error equation is $\left(\Delta_\phi\right)^2 = \langle(\Delta\hat{X})^2\rangle/\left|\partial\langle\hat{X}\rangle/\partial\phi\right|^2$. The output signal of the interferometer in the equilibrium geometry and the slope of the output signal are $\widehat{\text{X}}$

$$\langle\hat{X}\rangle = -\frac{1}{\sqrt{2}}|\beta| \sinh(2g)\left[\cos\left(\phi + \theta_\beta - \theta_2\right) - \cos\left(\theta_\beta - \theta_2\right)\right] \tag{2.86}$$

$$\frac{\partial\langle\hat{X}\rangle}{\partial\phi} = \frac{1}{\sqrt{2}}|\beta| \sinh(2g) \sin\left(\phi + \theta_\beta - \theta_2\right) \tag{2.87}$$

The two equations above show that due to the introduction of two nonlinear amplification processes, the output signal and its slope are enhanced by a factor of sinh(2 g) relative to the conventional Mach–Zehnder interferometer in the case of

$\sinh(2g) > 1$. If $\sinh(2g) < 1$, the output signal is not amplified and its slope is reduced instead.

The two equations above show that due to the introduction of two nonlinear amplification processes, the output signal and its slope are enhanced by a factor of $\sinh(2g)$ relative to the conventional Mach–Zehnder interferometer in the case of $\sinh(2g) > 1$. If $\sinh(2g) < 1$, the output signal is not amplified and its slope is reduced instead.

In the equilibrium case, regarding the two input ends of the interferometer, the coherent state is input at one port and the squeezed vacuum state at the other port, the noise of the output signal is

$$\left\langle(\Delta\hat{X})^2\right\rangle = \frac{|v|^2 + |u|^2[\cosh(2r) - \sinh(2r)\cos(\Theta)]}{2} \tag{2.88}$$

where $\Theta = \eta + 2\theta u$, and θu is given by $u = |u|\exp(i\theta_u)$. When the port for inputting squeezed vacuum state is changed to input vacuum state, the noise is $\left\langle(\Delta\hat{X})^2\right\rangle = 1/2 + |v|^2$. The noise of the output signal is larger than the noise of the coherent state due to the presence of ϕ, resulting in υ not being 0; only when $\phi = 0$, $\left\langle(\Delta\hat{X})^2\right\rangle = 1/2$ is the noise of the coherent state. In this particular case, the interferometer does nothing and the output signal is the same as the input signal. Thus, the noise of output signal is determined by that of input noise. If the input is a squeezed state, the noise of the output signal is compressed. As shown in Fig. 2.31, the squeezed vacuum state is injected into one of the modes; when $\phi = 0$ and $\eta = 0$ (in this case, $\theta_u = 0$), the noise at the output end is $\left\langle(\Delta\hat{X})^2\right\rangle = e^{-2r}/2$, which is lower than the vacuum noise.

The characteristics of the output signal of the SU(1, 1) interferometer are shown in Fig. 2.32. Compared with the previously introduced coherent state injected into the Mach–Zehnder interferometer, the same thing is that the magnitude of the signal and its slope vary with phase shift ϕ as depicted in Fig. 2.32a, b; the difference is that the noise of the output signal given by Fig. 2.32c is phase dependent, as given by Eq. (2.88), and the noise remains phase dependent even when only coherent states are injected. This leads to the fact that when measurements are made at these points of maximum slope, the best phase sensitivity will not be achieved at all these points. As shown in Fig. 2.32d for phase sensitivity, when the phase shift ϕ is the odd number of π, the signal slope is the largest but the signal noise is also large, making the phase sensitivity not the best. The SNR of the output signal is also different. As indicated in Fig. 2.32e, the best measuring position is not at the maximum signal, but near the lowest noise of the output signal, despite that the signal intensity at this position is also low. Therefore, in the SU(1,1) interferometer, the noise factor plays a dominant role when performing phase measurements.

In conclusion, when phase sensitivity is measured at the optimal point $\phi = 0$, the optimal sensitivity is

$$(\Delta\phi')^2 = \frac{1}{N_\beta}\frac{1}{e^{2r}}\frac{1}{N_{OPA}(N_{OPA} + 2)} \tag{2.89}$$

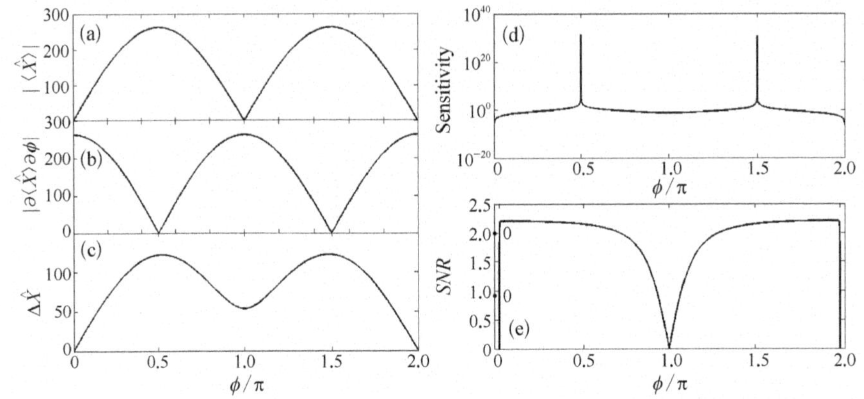

Fig. 2.32 Plots of **a** average $|\langle\hat{X}\rangle|$; **b** slope $|\partial\langle\hat{X}\rangle/\partial\phi|$; **c** noise $\langle(\Delta\hat{X})^2\rangle$; **d** SNR $= |\langle\hat{X}\rangle|/(\Delta\hat{X})$; **e** sensitivity Δ_ϕ versus phase shift ϕ for the output signal of SU(1, 1) interferometer in equilibrium geometry, where $\theta_\beta - \theta_2 = \pi/2$, $g = 2.5$; $r = 1.5$, $|\beta|$=5

Here, it requires $\eta = 0$ and θ_β-$\theta_2 = \pi/2$, where $N_{\text{OPA}} = 2\sinh^2 g$. The factor $1/e^{2r}$ is the result of inputting the squeezed vacuum state, and the factor $1/N_{\text{OPA}}$ $(N_{\text{OPA}} + 2)$ is due to the introduction of the parametric amplification process.

The Heisenberg limit for the phase measurement of the SU(1, 1) interferometer is $\Delta\phi_{\text{HL}} = 1/N_{\text{Tot}}$. Note that $N_{\text{Tot}}\left(\equiv \left\langle \hat{a}_1^\dagger \hat{a}_1 + \hat{b}_1^\dagger \hat{b}_1 \right\rangle\right.$ is the number of photons inside the interferometer, i.e., the number of phase-sensitive photons, not the number of input photons. The N_{Tot} is calculated as

$$N_{\text{Tot}} = (N_{\text{OPA}} + 1)\left(|\beta|^2 + \sinh^2 r\right) + N_{\text{OPA}} \tag{2.90}$$

where $|\beta|^2 + \text{smh}^2 r$ is the number of input photons. If it is used as the number phase-sensitive photons, the so-called "super-Heisenberg limit" will be obtained. When $N_{\text{OPA}} \geq 1$ and $|\beta| \simeq e^r/2 \simeq \sinh r \geq 1$, the photon number inside the interferometer can be written as $N_{\text{Tot}} \simeq 2N_{\text{OPA}}\ N_\beta = 2N_{\text{OPA}}\ \text{Sinh}^2 r$. The optimal phase sensitivity can be written as

$$(\Delta\phi\prime)^2 = \frac{1}{2N_{\text{OPA}}N_\beta}\frac{1}{2N_{\text{OPA}}\sinh^2 r + e^{2r}} \simeq \frac{1}{N_{\text{Tot}}^2} \tag{2.91}$$

That is to say, in this case, $\Delta\phi' \simeq \Delta\phi_{\text{HL}}$, the optimal phase sensitivity is close to the Heisenberg limit. If we directly let $\Delta\phi' \simeq \Delta\phi_{\text{HL}}$, it requires $|\beta|=\tanh(2g)e^r/2$. Therefore, the condition for the optimal phase sensitivity approaching the Heisenberg limit is $|\beta| \simeq e^r/2 \simeq \sinh r$, and $\tanh(2g) \approx 1$.

Figure 2.33 numerically shows the phase sensitivity breaking SQL as well as approaching HL. Figure 2.33a numerically calculates the variation of the phase sensitivity with the parametric process coefficient g (see the dashed line in Fig. 2.33). In the case of $|\beta| \simeq e^r/2 \simeq \sinh r$, the phase sensitivity approaches HL when $g \geq 2$[tanh

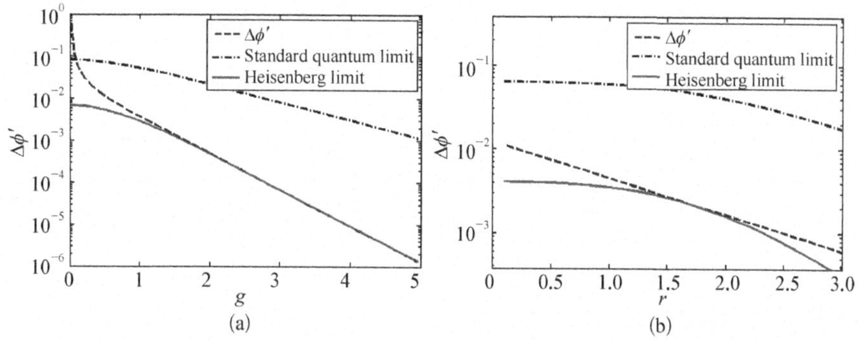

Fig. 2.33 **a** Plot of the phase sensitivity of SU(1, 1) interferometer varies with the the gain g of optical parametric process; **b** Variation with the squeezed parameter r of the squeezed vacuum state input

(2 g) ≈1. And as g becomes larger, the optimal phase sensitivity is closer to HL. Figure 2.33b depicts the variation of the phase sensitivity with the squeezed parameter r of the squeezed state input. For a certain number of other parameters, it can be easily derived from the figure that there is an optimal input squeezed parameter r, complying with $|\beta| \simeq e^r/2 \simeq \sin r$. In other words, the average photon number at both input ends can reach the optimal value only when they are in equilibrium, which is the same requirement as that of the conventional Mach–Zehnder interferometer, as in the theoretical analysis above. However, regarding the measurement of SNR, since the noise of the output signal is determined only by the squeezed state of the input in certain measurement situations, the greater the intensity of the input coherent field, the better the SNR for specific output noise.

2.5.4 Summary and Prospect

Quantum based measuring processes can break the SQL of measurement, even up to the Heisenberg limit, and these are obtained in an environment without additional noise. For open environments with noise, the degree of measurement accuracy improvement is definitely compromised by making use of quantum properties such as entanglement. Still, the sensitivity of phase measurement can exceed the SQL, but the Heisenberg limit is yet to be reached [85, 86]. A large number of researchers are currently focusing on new theory, experiment, and development for the study of phase measurements in open environments.

As quantum information is an exciting and rapidly growing area of scientific interest and development, it has attracted world-leading theoretical and experimental researches. Its rapid development has given rise to many novel quantum applications, excitingly opening new horizons in numerous areas of measurement. The combination of quantum information, quantum computing and precision measurement is a possible trend for the future development of quantum precision measurement. Currently, many researchers are trying to explore some possibilities of multi-particle correlation to challenge a number of quantum limits, and the combination of multi-body correlation with quantum precision measurement is also a possible direction for future development. This book only briefly outlines phase measurement, phase measurement limit, and phase measurement by quantum enhancement. For more details, interested readers may consult the review articles and monographs associated with precision measurement.

2.6 Two-Dimensional Magneto-Optical Trap and Heralded Single-Photon Source

2.6.1 Two-Dimensional Magneto-Optical Trap (2D MOT)

From the 1980s, laser cooling and trapping techniques have become a new favorite of atomic-molecular physics. Laser cooling and trapping as well as Bose–Einstein condensation that is closely associated with it won the Nobel Prize in physics in 1997 and 2001, respectively. As early as 1970 [87], Ashkin noted that laser beams can produce a considerable scattering force on atoms if their frequency resonate with atomic level transition. At the same time, he also theoretically suggested the possibility of implementing laser cooling techniques. The resonant radiation pressure available for optical cooling can be divided into two categories: spontaneous scattering force and optical dipole–dipole force. The latter is caused by the intensity gradient of the laser beam and can drag a macroscopic particle with a diameter between 1 and 100 μm into the high-intensity region of the laser beam by means of a transverse gradient force. The "optical tweezer" manipulates macromolecules or biological cells by taking advantage of the optical field is an important application of optical dipole–dipole forces. At the same time, the spontaneous scattering force of laser is widely used to cool atoms, which was first proposed by Hänsch and Schawlow [88]. The physical process can be described simply as follows: assuming that the laser wavelength is λ and its frequency comes near a specific transition frequency of the atom. After an atom absorbs a photon close to the transition frequency, the momentum of this photon will be transferred to the atom. Since the direction of the photon momentum subsequently released by this atom in the excited state as a result of spontaneous radiation is random, the average momentum change ends up along the momentum direction of the originally absorbed photon. Just as hitting one billiard against another on a pool table can cause it to change direction, so do photons in a laser beam exert radiation pressure on an atom to change its motion. If the incident laser is red-shifted (the laser frequency is lower than the transition frequency) for atomic transition, it follows from the Doppler effect that the atom has a better chance of absorbing an oncoming photon than a photon moving in the same direction as the atom. When there is a pair of red-shifted lasers acting on the atom at the same time, the thermal motion of the atom will be weakened by this non-equilibrium radiation pressure. Thus, it can be imagined that the place where six laser beams converge must form "photon adhesive" and the atoms enter it as if they were deep in the mud. The atoms are subjected to a damping force $F = -\alpha \upsilon$, where α is the damping factor, regardless of the direction of motion. The first experiment using "photon adhesive" for laser cooling was completed by Steven Chu et al. [89].

The laser field can cool atoms but fails to provide a stable binding point. Just as a small ball is bound to a position where the spring restoring force is zero, atoms need such a restoring force to keep them confined at some point in space. By splitting the atomic magnetic energy level in a magnetic field, Raab et al. experimentally implemented the first magneto-optical trap (MOT) in 1987 [90], combining a laser field with a gradient magnetic field and thus trapping the atom at the magnetic field zero. Assuming that a hypothetical atom is composed of a ground state $J = 0$ and an excited state $J = 1$ $(m = -1, 0, +1)$. The Zeeman energy level $m = -1, 0, +1$ of excited state degeneracy appears splitting in a non-uniform magnetic field $B = \beta z$ (β stands for magnetic field gradient) generated by a pair of anti-Helmhotz coils, and the spacing of the split levels is $\Delta E = mg\mu_B B$ (g stands for the Lande g-factor, which we can tentatively consider as a positive number; μB is Bohr magneton). As shown in Fig. 2.34a (the positive direction of z-axis is set in Fig. 2.34), if the frequencies of a pair of incident lights propagating in opposite directions are red detuned for transition $J = 0 \rightarrow J = 1$, the atoms at the $z > 0$ end will absorb more σ^- polarized light while the atoms at the $z < 0$ end more σ^+ polarized light. As a result, if the polarization of the laser field and the direction of light propagation are shown in Fig. 2.34a, the atoms are subjected to a restoring force pointing to the origin $z = 0$. By the same principle, if the beam pairs in the x- and y-axes are set in Fig. 2.34b, the atoms will bear a restoring force pointing to the magnetic field zero in all six directions in space, making the edge of magnetic field zero a stable point for trapping the atoms.

The magnetic field zero of a 3D MOT is only a point in space (trapping point), so the cooled atomic clusters are spherical in shape. It is not difficult to imagine that if the trapping point of the magneto-optical trap can be extended to a line, the cooled atoms will be lined up along the line and the confinement area of the trap will be effectively increased. Thus, the two-dimensional magneto-optical trap developed on the basis of the 3D MOT adopts a magnetic field gradient that is distributed in two dimensions (transverse) whereas the magnetic field is uniform in the third dimension (longitudinal). Theoretically, such a magnetic field can be obtained by extending the anti-Helmholtz coil in Fig. 2.34 along the x-axis to infinity (with respect to the y- and z-axes). In practice, however, the magnetic field coil cannot be infinitely long, and thus the ends of the finite length coil have the undesirable effect of the residual magnetic field on the trapping region. Because of this, two pairs of anti-Helmholtz coils are generally used to generate such a magnetic field gradient in a two-dimensional magneto-optical trap, as shown in Fig. 2.35a. As the longitudinal magnetic field gradient is zero, the 2D MOT does not have a longitudinal binding effect, the 2D MOT formed under the action of four trapping beams is generally used to generate atomic flows. To make stable atomic clusters, a pair of trapping light is

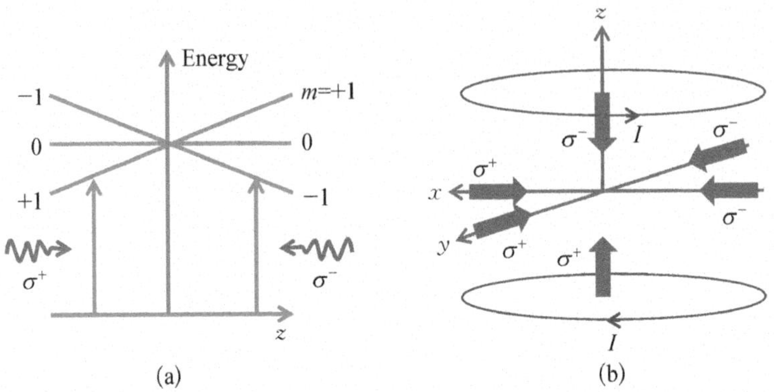

Fig. 2.34 3D MOT schematic diagram. **a** The energy levels of a hypothetical atom include ground state $J = 0$ and excited state $J = 1$. Assuming Lande factor $g > 0$ and spatial magnetic field $B > 0$ at $z = 0$ (spatial magnetic field $B < 0$ at $z < 0$), the Zeeman magneton level $m = +1$ is shifted up at $z > 0$ while energy level $m = -1$ is shifted down at $z > 0$; $m = + 1$ is shifted down at $z < 0$ and $m = -1$ is shifted up at $z < 0$; **b** Six-way trapping light of 3D MOT. The magnetic field is generated by a pair of anti-Helmholtz coils with currents passing in opposite directions

also required in the longitudinal direction to exert a photon adhesive effect on the atomic clusters. In a 2D MOT, the atomic clusters end up in a long cigar shape because the magnetic field is uniformly distributed in one of the dimensions and the trapping region is linear along this dimension [91]. We are going to introduce in particular a 2D MOT configuration published in Scientific Instrument [92]. As shown in Fig. 2.35, the length of the two pairs of anti-Hemholtz coils along the z-axis is absolutely greater than the geometric length in other directions to ensure a uniform distribution of magnetic fields in the z-axis. The inhomogeneous magnetic field distribution of the coils on the xy plane is revealed in Fig. 2.35b, while the magnetic field along the z-axis has zero magnitude or a weak offset. The polarization of the six laser beams is set up in Fig. 2.35, where Fig. 2.35b lists only the beams on the xy plane, and the laser beams along the z-axis are shown in Fig. 2.35a. Unlike the configuration of a typical magneto-optical trap, this configuration has two dimensions (y-axis and z-axis directions) in which the trapped light field has a 45° angle with the actual laser field and is an equivalent synthesis of the laser field. This configuration has effectively improved space utilization and avoided occupying the axial direction of the atomic cluster because this direction is the orientation of the highest optical depth of the atomic cluster.

The key to laser cooling technology is to select a suitable atomic level structure for the cooling process, and to achieve effective cooling, it must be an ideal two-level structure, i.e., a structure within which atomic transition can proceed in an infinite cycle. Actually, very few atoms meet this requirement. Therefore, in operation, in addition to the cooling light field, we need pump light to bring the atoms back into the cooling process. Take the atomic level transition of $^{85}R_b$ as an example, $S_{1/2}$, $F = 3 \rightarrow P_{3/2}$, and $F' = 4$ are closer to the ideal two-level system transition. Thus, the cooling light adopts this transition process as the effective action process, and the

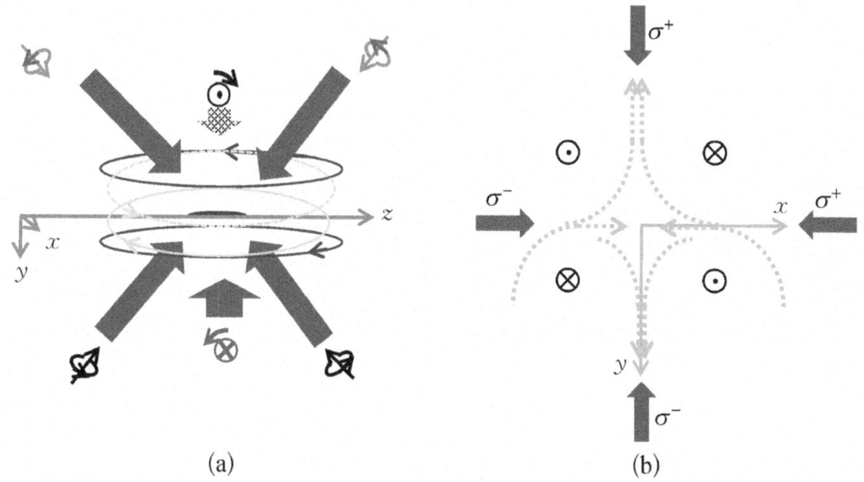

Fig. 2.35 Setup diagram of magnetic field coil and trapped optical path. **a** Polarization setup of six beams. A pair of Anti-Hemholtz coils is placed along each of the *y*-axis and *x*-axis directions. The arrows represent the current direction of each coil; **b** A sketch of magnetic field distribution, where the dotted lines and arrows indicate the distribution of magnetic field lines and the direction of magnetic field. According to the magnetic field direction on the *xy* plane shown in Fig. 2.33b, the two pairs of laser beams on the *yz* plane should be levorotation while the laser on the *x* direction corresponds to dextrogyration

redshift of cooling light frequency is $\delta = 20$ MHz (which can be adjusted slightly up and down to get the best cooling effect). Although the atoms on the $P_{3/2}$, $F' = 4$ energy level in an ideal two-level system do not transition directly to $S_{1/2}$, $F = 2$ due to the transition rule, the trapping light in fact may spur the $S_{1/2}$, $F = 3$ atoms to pump to $P_{3/2}$, $F' = 3$ or $P_{3/2}$, $F' = 2$. From another point of view, $S_{1/2}$, $F = 2$ is not the lowest energy ground state and the atoms will gather on the stable ground state $S_{1/2}$, $F = 2$, thus opening the atomic two-level system used for cooling. This two-level system will eventually lose all the atoms and make the cooling process unsustainable. Repumping light ($S_{1/2}$, $F = 2 \rightarrow P_{3/2}$, $F' = 2$) is used precisely to pump the atoms that fall on the ground state $S_{1/2}$, $F = 2$, returning it into the two-level system for cooling. Without the action of the repumping light, the cold atomic cluster cannot be formed either. A pair of back-pumped light acts in the x-axis direction, coinciding with the trapped light pair in that direction. When the ratio of the window for measuring OD to the whole trapping-measurement process is 1:10 and the power of the repumping light is adjustable (up to 13 mW with a speckle diameter of 1 ft), we can obtain an optical depth from 0 to 45° on this system [93]. The fluorescence image of cold atomic clusters is shown in Fig. 2.36, with a longitudinal length of about 15 cm and an aligned distribution along the *z*-axis.

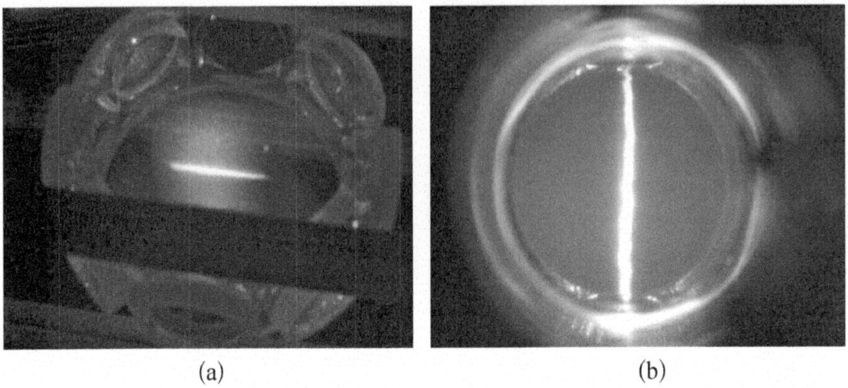

Fig. 2.36 Cold atom fluorescence images in a 2D MOT, **a**, **b** observed along the x- and y-axis directions. The longitudinal length of the atomic cluster is about 1.5 cm

2.6.2 Increasing Optical Depth

To characterize the strength of interactions between atoms and light fields, we need to define the optical depth (OD) of the atomic ensemble. In general, we can define the OD of an atomic ensemble using the intensity of the incoming and outgoing light.

$$I_0 = I_i e^{-OD} \tag{2.92}$$

For example, when the OD = 1, the intensity of the outgoing light decays to 1/e that of the incoming light. therefore, and OD expresses the ability of the medium to absorb light. Of course, it is not difficult to infer that the optical depth so defined must be related to the detuning of the incident light frequency with respect to atomic transition. The smaller the frequency detuning, the greater the absorption capacity of the medium. Therefore, the optical depth is generally defined in terms of transition resonance. Under resonance conditions, OD $= N_{\alpha 0} L$, where N is atomic number density, α_0 is the absorption cross section of the medium at resonance, and L is the effective length of the atomic cluster. Increasing the optical depth of a cold atomic ensemble helps to enhance the coupling between atoms and light as well as nonlinear effect. For example, the group velocity of light propagation under the EIT slow light effect in atomic clusters of high optical depth is approximated by $v_g \approx \frac{2L}{OD}\frac{\Omega_c^2}{4\gamma_{13}}$, i.e., the larger the OD, the more pronounced the slow light effect. The photon pairs generated through this EIT medium have long coherence time.

According to the definition of optical depth, the approach to increase OD leads to an increase in the atomic density and the effective length of the atom. On the one hand, increasing the length L of atomic clusters has a significant effect on increasing the optical depth of media. Due to the high effective length of atomic clusters in 2D MOT compared to 3D MOT, the optical depth measured in 2D MOT can usually reach 2–4 times that of 3D MOT under the same condition. On the other hand, the enhancement of atomic density is more obvious for OD, but laser cooling and trapping techniques have proven in principle the existence of a density limit for grabbing atoms. To increase the atomic density, we first need to understand the reason for the existence of an upper density limit for atoms in magneto-optical traps. Primarily, the atomic density is restricted by two processes [94]. First, the collisions between ground and excited atoms caused by the thermal motion of atoms have the effect of accelerating the atoms, forcing them to leave the trapping region. Second, because the trapped atoms still undergo spontaneous radiation, their radiant energy is reabsorbed by neighboring atoms and converted into kinetic energy. This corresponds to the repulsive force within the atoms having a strongly fluorescent cluster. When a certain limit density is reached, this repulsive force between atoms balances with the trapping force in the optical trap so that the atomic density cannot increase any further. The first to adopt the "dark-spot" technique (DSPOT) was the D. E. Pritchard's research group [95]. They introduced parameter ρ to represent the distribution of atoms on the two ground states in the atomic hyperfine structure. They made a prediction of the parameter ρ in terms of the total number of atoms captured by the experimentally measured magneto-optical trap ($\sim 10^{10}$) and suggested that $\rho \sim 0.01$ is able to maximize the atomic density. The conclusion is actually not difficult to understand. As mentioned in Sect. 2.5.1.1, the spontaneous radiation of excited atoms causes them to stay in the ground state $S_{1/2}$, $F = 2$, which can be called "dark state", because the atoms in this state can hardly be re-pumped to the excited states of $P_{3/2}$ without the effect of pumping back light. As a result, the atoms in this ground state do not fluoresce or exert a repulsive force on other atoms due to the pressure of light radiation. So, the distance between atoms in the dark region of the repumping light can be effectively shortened to further increase the atomic density. Naturally, the "bright region" at the periphery of the back-pumped light is still an important part of the magneto-optical trap for trapping atoms, while the "dark spot" must be at the center of the atomic cluster to receive the successfully trapped atoms. The group of D. E. Pritchard used an atomic flow source to efficiently load sodium atoms into a 3D MOT, where this "dark spot" technique was employed to increase the atomic density in the dark region to nearly $10^{12}/\text{cm}^3$. The optical depth measured in this cluster of atoms reached 135. In a recent study of quantum information networks, A. G. Radnaev et al. applied this technique to obtain an atomic ensemble with extremely high optical depths (OD = 152) [95]. S. Zhang et al. also constructed a long-biased dark region on a 2D MOT to increase its optical thickness from 40 to 160 [92].

Another effective solution to increase the atomic density is the compressed magneto-optical trap (CMOT). Regardless of the total number of atoms in the potential well, the atomic density always increases with the increase of the equivalent restitution coefficient of the potential well; on the other hand, the increase of this coefficient slows down the loading rate of the magnetic trap. Accordingly, in CMOT technology, two processes, separate atomic loading and magnetic trap compression, are usually used to minimize the loss of atoms. Moreover, the linear increase of the magnetic field gradient is accompanied by a corresponding increase of the trapped light detuning to extend the action of the photon adhesive and to cut the atomic loss in magnetic field compression. B. M. Sparkes [96] et al. compressed the magnetic field within 20 ms after the magneto-optical trap was loaded with atoms, increasing the magnetic field gradient linearly from 16 G/cm at the time of loading to 40 G/cm, and in the meantime, the detuning of the trapping light and repumping light is manipulated (trapping line detuning: 30 MHz $\rightarrow$ 70 MHz; repumping detuning: 0 MHz $\rightarrow$ 8 MHz). The compressed magneto-optical trap is capable of trapping approximately 4×10^9 atoms at a temperature of about 200 μK.

2.6.3 Controllable Heralded Single-Photon Source

In quantum mechanics, the Fock state $|n\rangle$ of the light field is the eigenstate of the particle number operator $\hat{n} = \hat{a}^\dagger \hat{a}$, and the eigenvalue n corresponds to a determined number of photons. A coherent light source (e.g., laser) that gradually decays to make the average number of photons per light pulse less than or equal to one, thus constituting a single-photon level. However, the light pulse from a coherent source, even if weak, still obeys Poisson distribution and is a coherent state not a Fock state. That is, even if the average of photons per pulse is 1, there are still pulses with more than one photon. However, the single photon in Fock state makes it clear that there is only one photon in each light pulse. According to the uncertainty principle, the number of photons and the phase of the light field cannot be determined with infinite accuracy at the same time. Therefore, the phase information of a single photon in Fock state cannot be determined. How to prepare a single photon in Fock state is an important research direction. At present, it can be generated from quantum dots, single molecules, single atomic cavities, and atomic clusters.

The entangled photon pairs generated from parametric effect conversion are the preferred solution for single photons. However, the emerging time of the photon pair obtained from parametric effect conversion is completely random, and the two photons are also generated almost simultaneously. Single-photon systems like this cannot accurately predict when a single photon will be emitted. To produce on-demand single-photon sources, various schemes have been proposed, for example, single-atom cavities or highly nonlinear cavities. Single-photon gate devices based on mesoscopic dual-potential barrier p-i-n heterojunctions have been reported to generate periodic single-photon streams [9, 97] with controllable repeatability. Single-photon gate devices can also be realized by single quantum dots embedded in high quality factor microcavities. The interval time jitter of single photon generating events can be reduced by tuning the frequency to achieve the resonance of the quantum dots and the microcavity. On the other hand, single photons can also be generated by spontaneous radiation from single molecules in solids excited to fluorescent excited states by a fast adiabatic process. The transition frequency of a single molecule is a sinusoidal function of time and its sweep frequency is determined by the electric field acting on the electrode on the molecule. When the molecular transition frequency crosses the resonant frequency of the excited optical field, the molecule is excited and fluorescent photons are emitted within a certain time delay [98]. Lounis and Moernerthe [99] demonstrated experimentally that laser pulses pumping a single molecule can induce the molecule to produce single photons during relaxation even at room temperature.

Let us take a look at the commonly accepted measurement methods for identifying single photons in Fock state today. As a single photon source emits only one photon at a time, it can be verified using a beam splitter. Unlike classical light waves, a photon is a non-subdividable energy packet that either transmits or reflects as it passes through the beam splitter, and it is impossible for both receiving terminals to receive photons at the same time. As shown in Fig. 2.37, if the photon counting rate at D1 (transmission end) is labeled N_T and the photon counting rate labeled N_R at D2 (reflection end), the composite D1 and D2 photon counting rate is NTR, the second-order autocorrelation function can be expressed as $g_{auto}^{(2)} = N_{TR}/N_T N_R$. The classical light field satisfies $1 \leq g(2) \leq 2$. A light source meeting $g(2) < 1$ can no longer be explained by the classical electromagnetic field theory. Therefore, it is called non-classical light source. The ideal Fock state single photon passing through the beam splitter can transmit or reflect only, both of which cannot occur at the same time. It is therefore known that $N_{TR} = 0$, and $g_{auto}^{(2)} = 0$. Regarding Fock photon state $|2\rangle$ satisfying $g_{auto}^{(2)} = 0.5$, it can be calculated: in the predictive counting method for example, two photons in the Fock photon state are injected in front of the beam splitter in Fig. 2.38a, and they can be temporarily numbered 1, 2. Also, each time the pair is emitted, it is forecast by the photon namely gate signal. In this counting method, the autocorrelation coefficient $g_{auto}^{(2)} = \frac{N_{TR} N_G}{N_{TG} N_{RG}}$, where N_G represents the average number of heralded photons, and N_{TR}, N_{TG}, and N_{RG} stand for the average composite photon counts at the transmitting end T, the reflecting end R, and the gate signal end G, respectively. There are four different photon distributions, as shown in Fig. 2.38b, c. Let $N_G = 4$, and one can get $N_{TR} = 2$, comprising the two cases shown in Fig. 2.38b. Similarly, N_{TG} (RG) $= 1 + 1 + 2 \times 1$ can be calculated, where the case in Fig. 2.38c contributes two photons to the counts of transmitting or reflecting ends, so the average photon number in this case should be recorded as 2. Substituting the above items into the autocorrelation coefficient formula, the autocorrelation coefficient of the Fock state with photon number 2 is $g_{auto}^{(2)} = 0.5$, and this value is the basis for determining single photons $g^{(2)} < 0.5$.

Another scheme for generating single photons is single photons obtained from correlated photon pairs, which are called heralded single photons and also have

Fig. 2.37 Measurement of the second-order autocorrelation coefficient of a single photon in Fock state

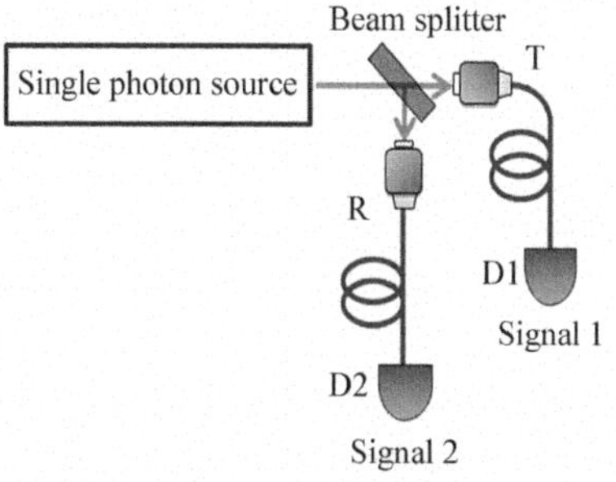

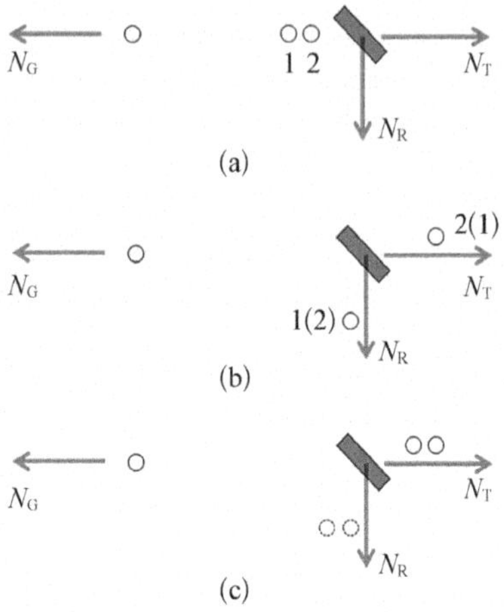

Fig. 2.38 Calculation of autocorrelation coefficient for Fock state light source with photon number 2. **a** Two photons injected into the beam splitter in the heralded counting mode; **b** The case where one photon transmits and the other reflects, photon 1 transmits and photon 2 reflects or vice versa; **c** The case where both photons transmit or reflect

important applications in quantum information manipulation. Since a pair of photons in a correlated photon pair has a very high correlation, the receiving of one photon forecasts the corresponding photon with a heralding accuracy of 100% (under electromagnetically induced tunneling effect). Thus, after receiving the heralded photon, the other photon in the pair constitutes a good single-photon source. The coherence time of the photon pair limits the time resolution of this heralded single photon. The two-photon bandwidths formed in the parametric conversion process in conventional crystals are extremely wide, reaching the THz magnitude. Hence, the coherence time of photon pairs is fairly short, usually only a few to tens of femtoseconds. Even the use of narrow-bandwidth filters can only increase the two-photon coherence time to picoseconds, a time scale that places stringent requirements on both the modulator and the photodetector. Moreover, the use of filters inevitably results in the loss of frequency-dependent features of the photon pair. The most direct way to solve this problem is to extend the coherence time of the photon pair such that the arrival time difference between the two photons can be easily detected. Compared to solid media with multiple rotations, vibrations and strong atomic collisions, atomic gases are a more desirable and controllable media system. Further, the cooled and trapped atomic system in the magneto-optical trap more effectively reduces atomic thermal motions and collisions, and thus it is suitable for generating narrow linewidth two-photon pairs. From this, controllable heralded single-photon sources are produced.

2.6.4 *Single Photon with Controllable, Narrow Linewidth, and Wide Time Domain Wave Function*

Due to the narrow linewidth of ground-state and excited-state transitions, the two-photon pair generated by the third-order nonlinear coefficient of cold atoms has a narrow linewidth that effectively prolongs two-photon coherence time. On this basis, the slow-light effect of electromagnetically induced transparency (EIT) is used to further extend and regulate the delay time of the anti-Stokes photon relative to the Stokes photon. The coherence time of correlated photon pairs generated by cold atomic systems based on high optical depth can be extended to 100 ns–1 μs, thus constituting a single photon linewidth that can be less than 1 MHz. The first time-domain controllable two-photon source was first generated by Harris's group [100]. Figure 2.39 is the schematic diagram for generating narrow-linewidth entangled photon pairs. The far-detuned pumping light and the resonant coupling light are two control lights of four-wave mixing process. The photon pairs generated by the pump and coupled light in the continuous state, i.e., Stokes and anti-Stokes photons, are a pair of photons with time–frequency entanglement, because they need to meet the phase-matching condition. Put another way, their time or frequency parameters cannot be expressed independently. Stokes photons are resulted from a spontaneous Raman process, while anti-Stokes photons are generated by an excited Raman process assisted by the EIT effect. The narrow-bandwidth high-transmission window induced by coupled light allows anti-Stokes photons to pass through almost without loss, thereby improving the pairing success rate. At the same time, the slow-light effect of the EIT allows the delay time of the anti-Stokes photon to grow and be controlled with respect to the Stokes photon, so that the coherence time of the time-domain wave function of the generated photon pair can reach 1/μs, corresponding to a two-photon bandwidth of the order of MHz or even lower (see Fig. 2.39 for the decoherence rate limit of the ground state $|1\rangle$ and sub-stable state $|2\rangle$ forbidden by transition). The generation of this two-photon source opened up the possibility of later modulating the time-domain wave function of heralded single photons arising from double photons.

In this heralded single-photon system, modulating the wave function of a photon is as easy as modulating a light wave pulse. A modulator placed in the outgoing photon beam can serve to modulate the photon wave function [101, 102]. The simplest modulation is to insert an electro-optical modulator with a modulation bandwidth of 10 GHz or more into the outgoing optical path of the anti-Stokes photon. At this

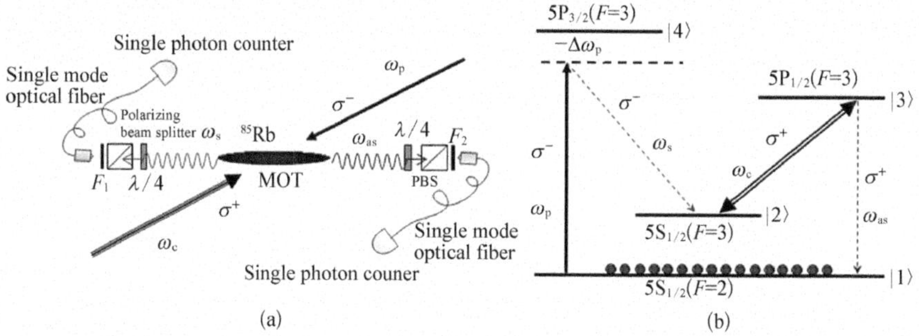

Fig. 2.39 Generation of narrow-linewidth wide-entanglement photon pairs via four-wave mixing. **a** Schematic diagram of optical path during four-wave mixing; **b** Level structure diagram of rubidium atom

point, the Stokes photon works not only as a trigger of the single-photon signal to define and herald the time starting point of single-photon wave function, but also as a triggering signal for the waveform generator controlling the modulator. In the wave function modulation of this heralded single photon, we are not concerned with the absolute arrival time of the photon pair, but are more interested in the relative time of the two photons $\tau = t_{as} - t_s$. The amplitude modulation function applied to the anti-Stokes photon can be represented by $m(\tau)$, and the single-photon wave function can be expressed as

$$\psi_0(\tau) = \left\langle 0|m(\tau)\hat{a}_{as}(t_s + \tau)\hat{a}_s(t_s)|\Psi_{s,as}\right\rangle\Big|_{t_s=0} = \frac{1}{2\pi}\int \Phi_0(\omega)e^{-i\omega\tau}d\omega \tag{2.93}$$

where $\Phi_0(\omega)$ denotes the spectrum of single photons. A similar single-photon modulation scheme is the simultaneous modulation of Stokes and anti-Stokes proposed by Bethangady [18] et al. A pair of modulators is placed in the optical path of the two generated coherent photons to modulate them sinusoidally in amplitude, and controlled by the two output terminals of the same function generator with the same modulation frequency and random phase difference. In this scheme, they use a slow detector, i.e., the time difference that can be resolved by a detector relative to the coherence time length of the photon pair, and the coincidence counting of the photon pair it detects are equivalent to the integration of the coincidence count over the entire relative time axis τ. Theory shows that this modulation scheme can yield the Fourier discrete cosine transform function pair of the original photon pair wave function. Both of the above modulation schemes employ the modulator directly to manipulate the generated photons, and thus the modulator unavoidably introduces ambient noise into the single-photon source. Also, since the modulation of single-photon waveform is in fact realized by controlling the passage of photons, it means that a large number of the generated photons are annihilated in the modulator and wasted. To improve the utilization of photons, the scheme of modulating two classically controlled light with four-wave mixing was proposed by Du et al. and demonstrated experimentally by Chen et al. [103, 104]. They demonstrated that under certain conditions, the arbitrary time-domain waveform of the input optical field of the four-wave mixing can be transcribed in the two-photon wave function, i.e., the two-photon wave function (and also the heralded single-photon wave function) can be manipulated by designing the waveform of the classical light wave input.

2.6.5 *Application*

In future quantum networks and quantum information, the communications media must avoid low loss and remain stable during transmission. Compared to other particles (e.g., electrons, ions, atoms), photons have low losses in transmission with the help of today's optical fiber communication technology and, most importantly, the quantum state of a single photon can be kept intact during transmission. Therefore, single photons are important tools in the present stage of quantum communication [105–107]. Ultra-narrow linewidth photon pairs or single-photon sources feature high brightness, i.e., the photon production rate is concentrated within a particular frequency linewidth. Moreover, their corresponding time length of single-photon time-domain wave function can reach the μs order, capable of greatly improving the controllability and scalability of single-photon sources. It is worth mentioning that quantum key distribution, which has been successfully applied, has been widely proven to effectively improve the security of key information transmission, and has broad application prospects for future network security and military. The non-subdivision principle of individual photons and the non-cloning theorem of quantum states are the core of quantum key distribution. Narrow-bandwidth heralded single-photon sources can effectively improve the key generation rate because their wave functions have particularly long coherence time constants, and furthermore, their single-photon nature can greatly lower the risk of being eavesdropped [108].

In fundamental research of single photons, the single photon sources with controllable time-domain wave function have galvanized experimental studies on the propagation speed of individual photons in a medium. Recent experimental research on the propagation speed of single photons [109] have used this technique to generate controllable single-photon sources and make them pass through cold atomic clusters to produce optical precursor waves, thus confirming that the speed limit proposed by Einstein still holds for single photons. The emergence of single-photon light precursor waves clearly disproves the conjecture that individual photons may have superluminal velocities under the quantum tunneling effect. Although we cannot yet explain the origin of this limiting speed that manipulates the motion of the entire universe, the research on single-photon quantum light sources is bound to advance human understanding of the quantum world.

2.7 Quantum Correlation in Raman Scattering

2.7.1 Raman Scattering in Atomic Ensemble

Raman scattering is a kind of inelastic scattering. When a strong pump field, whose frequency is off-resonant, passes through a medium and interacts with the molecules or atoms in the medium, the scattered field can be generated by Raman scattering process. We called the scattered field as Stokes or anti-Stokes field whose frequency is smaller or larger than the pump field. The frequencies of the scattered light and the pump light should satisfy two-resonant condition. The frequency difference is only related to the energy levels of the medium, so Raman scattering provides an important way for the study on the structure of crystal and molecular. Since its discovery by the Indian physicist Raman in 1928 [110], many kinds of Raman scattering effects have been widely used in a wide range of fields such as chemistry, physics, biology, mineralogy, materials science, and even archaeology and industrial product quality control. In the past twenty years, with the rapid developments of quantum optics and precision measurement, Raman scattering in atomic ensemble has also been extensively researched due to the long lifetime of the atomic level. In addition to the conventional Raman scattering processes, such as spontaneous process, several new types of Raman scattering have been developed, i.e., enhanced Raman scattering and phase-correlation Raman scattering. Because of the long coherent time of the atoms, Raman scattering can generate atomic spin excitation, originating from the phase correlation between atomic internal states, and it is easier to manipulate the light-atom interaction process based on atomic excitation to enhance the coupling strength of light-atom interactions and to improve the signal-to-noise ratio (SNR) of precision measurement.

2.7.2 Atomic Spin Excitation

When a coherent electromagnetic field couple two energy levels of an atom by Raman scattering, a defined phase relationship can be built between two levels, which can be written as a coherent superposition states of two atomic levels. In the three-level atom in Fig. 2.40, for a single atom, the coherent superposition state can be written as $|\psi\rangle = \cos\theta|g\rangle + e^{i\varphi}\sin\theta|m\rangle$; for an atomic ensemble, it can be written as $\hat{S}_a = 1/\sqrt{N_a}\sum|g\rangle_i\langle m|$. When there are only a few atoms in the |m⟩ state, it is generally called atomic spin wave which is an analogy with the spin excitation in ferromagnets. When there are a lot of atoms in the |m⟩ state, it is called atomic coherence. Atomic spin excitation can greatly affect the absorption and scattering behaviors of the atomic ensemble to the light field, has reveal lots of new physical phenomena, including Hanle effect, quantum beat, quantum echo, electromagnetic induced transparency (EIT), electromagnetic induced absorption, etc.; it also can significantly improve the coupling strength between light and atomic systems and also the SNR. So atomic spin excitation and its modulation technologies are important for quantum optics, quantum information and precision measurement, etc. The preparation of atomic spin excitations is mostly achieved by Raman scattering [98], stimulated adiabatic Raman passage (STIRAP) [111], or EIT [112]. We focus here on Raman scattering because it has a major difference from STIRAP and EIT. Raman scattering, while STIRAP and EIT require two light fields to interact with the atomic ensemble simultaneously, converting one of the light fields as atomic spin excitation, and the corresponding light field vanishes when the atomic spin excitation is generated. Raman scattering only requires one pump field. The Stokes field and atomic spin excitation are generated simultaneously, and these two have the intensity and phase correlation. Based on the correlation between the Stokes field and the atomic spin excitation, we can accurately control the light field and the atomic system.

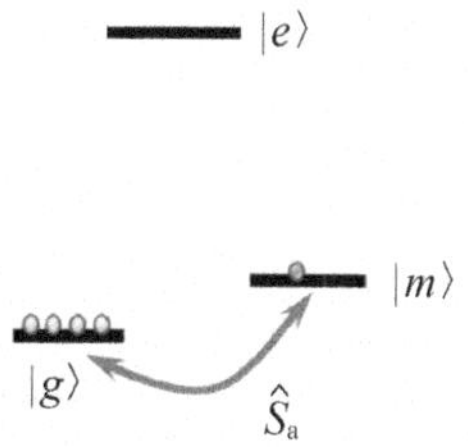

Fig. 2.40 Schematic diagram of atomic level. *g*: atomic ground state; *m*: sub-stable state; *e*: excited state

2.7.3 Correlation Between Atomic Spin Excitation and Light

There are three kinds of Raman scattering: spontaneous Raman scattering, stimulated Raman scattering [113, 114] and enhanced Raman scattering [115–117], respectively corresponding to (a)–(c) in Fig. 2.41. In experiments, regardless of whatever Raman scattering, almost all atoms are initially populated on the $|g\rangle$ level by a resonant pumping light to enhance the interaction efficiency of Raman scattering and decrease the loss of the Stokes light as shown in Fig. 2.40. The atomic ensemble can be written as $|g_1, \ldots g_{\mathrm{j}}, \ldots, g_{\mathrm{N}}\rangle$.

Then a strong pump beam P, whose amplitude can be considered as constant A_{P}, enters into the atomic ensemble and interacts with the atoms by Raman scattering. The annihilation of one pump photon is accompanied by the creation of an atomic spin excitation $\hat{S}_{\mathrm{a}}$ and a Stokes photon $\hat{a}_{\mathrm{Sout}}$, whose Hamiltonian is [106]

$$\hat{H} = i\hbar\eta A_p \hat{S}_a^{+} \hat{a}_{\mathrm{Sut}}^{+} - i\hbar\eta^{*} A_p^{*} \hat{S}_a \hat{a}_{\mathrm{Sut}} \tag{2.94}$$

η is the coupling strength. Equation (2.94) is similar to the Hamiltonian of three-wave mixing process in a crystal, where three waves are optical waves. While the two

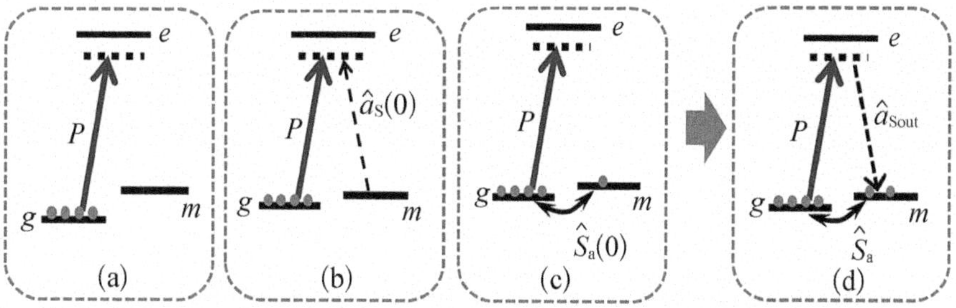

Fig. 2.41 **a–c** *P* is the pump light. Initial conditions of the atomic ensemble and light field for Raman scattering: **a** Spontaneous Raman scattering process, where all atoms are populated on the ground state $|g\rangle$; **b** Stimulated Raman scattering process, where all atoms are on the ground state $|g\rangle$, $\hat{a}_s(0)$ is the Stokes seed; **c** Enhanced Raman scattering process, where the atoms are initially prepared as atomic spin wave; **d** Atomic ensemble and light field after the Raman scattering, $\hat{a}S_{\mathrm{out}}$ is the output Stokes field, and $\hat{S}_{\mathrm{a}}$ is the final generated atomic spin wave

waves in Raman scattering are optical waves and the other wave is a matter wave-atomic spin wave. Raman scattering satisfies the two-photon resonant condition $\omega_P - \omega_s = \omega_{gm}$. The solution to Eq. (2.94) can be written as

$$\begin{aligned} \hat{a}_{Sut}(t) &= \hat{a}_S(0)\cosh\left|\eta A_p\right|t + \hat{S}_a^+(0)\sinh\left|\eta A_p\right|t \\ \hat{S}_a(t) &= \hat{S}_a(0)\cosh\left|\eta A_p\right|t + \hat{a}_S^+(0)\sinh\left|\eta A_p\right|t \end{aligned} \tag{2.95}$$

For spontaneous Raman scattering as shown in Fig. 2.41a, the Stokes seed is vacuum state, and the atomic initial state $\hat{S}_a(0)$ is the ground state $\left|g_1, \ldots g_j, \ldots, g_N\right\rangle$. The output Stokes field S_{out} is generated from vacuum fluctuation, For stimulated Raman scattering (SRS) in Fig. 2.41b, the initial state of atomic ensemble is $\left|g_1, \ldots g_j, \ldots, g_N\right\rangle$, and the difference compared with spontaneous process is that there is a Stokes seed $\hat{a}_s(0)$. The Stokes seed and the pump field are required to satisfy the two-photon resonance condition, which is achieved by phase-locking between two independent lasers or frequency shifting using AOM. As given in Eq. (2.95), the Stokes seed is amplified to the output Stokes field by stimulated Raman scattering. The intensity of the Stokes $\hat{a}_{Sout}$ field depends on the parameters including the atom number, the intensity of the pump light, and the light-atom coupling strength. For enhanced Raman scattering process, as shown in Fig. 2.41c, the initial state of atomic ensemble is atomic spin wave $\hat{S}_a(0) \equiv 1/\sqrt{N_a}\sum_i |g\rangle_i\langle m|$ but the Stokes seed is vacuum state. From Eq. (2.94), when the atomic initial state $\hat{S}_a(0)$ is not the ground state $\left|g_1, \ldots g_j, \ldots, g_N\right\rangle$, the effect of initial atomic spin wave on the Stokes field $\hat{a}S_{out}$ is equivalent to the Stokes seed $\hat{a}_s(0)$, which can greatly improve the efficiency of Raman scattering and the enhancement strength is also comparable to that of SRS [118].

As can be seen in Eq. (2.95), the spontaneous Raman scattering has the following correlation function [119]

$$\left\langle \hat{a}_{Sout}(t)\hat{S}_a(t)\right\rangle = \cosh\left(\left|\eta A_p\right|t\right)\sinh\left(\left|\eta A_p\right|t\right) \tag{2.96}$$

or expressed as a normalized correlation function

$$\gamma_{a_{Sut}S_a} = \frac{\left\langle \hat{a}_{Sout}(t)\hat{S}_a(t)\right\rangle}{\left\langle \hat{S}_a^+(t)\hat{S}_a(t)\right\rangle\left\langle \hat{a}_{Sout}(t)\hat{a}_{Sout}^+(t)\right\rangle} = 1 \tag{2.97}$$

From the above equation, it is clear that the Stokes light field phase $\varphi S_{ou}t$ and the atomic spin wave phase φS_a meet the inverse correlation $\varphi_{Sout} + \varphi_{Sa} = \text{const}$, which has been experimentally verified [119].

2.7.4 Application of Spin Excitation-Photon Modulation

Raman scattering has been an indispensable process in quantum information and precision measurements, e.g., quantum repeaters, generation of single-photon sources and photon pairs, and atomic clocks, due to its quasi-three-wave-mixing properties. We give several examples of applications as following.

(1) Improvement of interaction strength

The scattering efficiency of Raman scattering is determined by transition matrix element and the intensity of the pump field [109]. In general, the transition matrix element for Raman scattering between a single atom and the pump field is very small, which results in a small efficiency and a small SNR for spontaneous Raman scattering, limiting its applications. To improve the coupling efficiency, many methods have been developed in the past decades. The first method is to put atom into a high-Q optical microcavity. The light field can be reflected back and forth in the cavity. A good optical cavity has a Q value of about 10^6, i.e., the optical cavity can be employed to increase the pump intensity by a factor of nearly 10^6 to improve the interaction strength; the second method is to initially prepare the atom into the Rydberg state before the pump field enters into the atom and then enlarge the dipole transition matrix element of the atom via large atomic volume; the third method is the utilization of atomic ensemble replacing single atom to increase the coupling efficiency of Raman scattering. The experimental implementation of the third method is much simpler than the first two methods. The efficiency improvement by using atomic ensemble is due to the increase of interaction atoms and the collective effect between atoms in the atomic ensemble. Based on the atomic ensemble, the interaction efficiency can be further enhanced, for example, initially-prepared atomic spin wave before the atom-light interaction or introducing feedback mechanism during the interaction process to increase the nonlinear interaction coefficient. These methods have been reported experimentally and theoretically. Scully [120] prepared the maximum coherence between the vibrational energy levels of molecular to enhance the Raman efficiency by femtosecond laser pulses, also known as FAST-CARS (femtosecond adaptive spectroscopic techniques to CARS), which is useful for detecting viruses and other biomolecules. Jian prepared the maximum coherence between two atomic levels by EIT in a rubidium atomic ensemble [111]. On the basis of this atomic spin wave, they converted pump light (425 nm) to Stokes light (295 nm) with 40% Raman efficiency. The conversion efficiency was later increased to nearly 100% by further increasing the atomic density.

(2) Photon conversion driven by spin wave

Atomic spin waves can be used to manipulate two different frequency modes of photons to produce a Rabi-like oscillation effect [117]. A strong atomic spin wave, which can be considered as constant S_a, is initially prepared in the atomic ensemble, and then two weak light fields, namely, the Raman pump field P and the Stokes field S, are injected into the atomic ensemble. The Hamiltonian of the light-atom interaction can be written as

$$\hat{H} = ih\eta S_a^* \hat{a}_p \hat{a}_S^+ - ih\eta^* S_a a_p^+ \hat{a}_S \tag{2.98}$$

its solution is

$$\begin{aligned}
\hat{a}_S(t) &= \hat{a}_S(0)\cos|\eta S_a|t + \hat{a}_p(0)\sin|\eta S_a|t \\
\hat{a}_p(t) &= \hat{a}_p(0)\cos|\eta S_a|t - \hat{a}_S(0)\sin|\eta S_a|t
\end{aligned} \tag{2.99}$$

It can be seen that the strong atomic spin wave drives two weak optical fields with different frequencies to convert each other in an oscillatory manner. This oscillation of photons between different frequency modes is quite similar to the famous "Rabi oscillation" of the atoms driven by a strong laser in quantum optics. The difference is that the atoms and the photons reverse the roles and here the atoms are used to drive the photons. The conversion efficiency of the optical fields between two frequencies is experimentally up to 80% with the assistance of strong atomic spin wave. This process has potential application in quantum information and precision measurement technology, such as efficient photon conversion, preparation and control on optical quantum states, and wavelength division multiplexing of single-photon frequency entangled states and photon-atom quantum states.

(3) Quantum repeater

Quantum repeater is an essential element for quantum communication and quantum computer. The fundamental process in the famous DLCZ quantum repeater scheme [113] is Raman scattering. Based on the Raman scattering in the atomic ensemble, the entanglement of two atomic ensembles L and R is achieved, and then quantum communication can be implemented via entanglement exchange and entanglement purification. The scheme of the DLCZ scheme is shown in Fig. 2.42a, where a strong write beam is split into two and interacts with two atomic ensembles simultaneously. The forward Stokes photon emitted from the two ensembles are mixed using a beam splitter. When one and only one of the two detectors clicks a photon, it means that which-way atomic excitation is automatically created and two ensemble are entangled, which can be written as

$$|\psi_\phi\rangle_{LR}^{\pm} = \frac{1}{\sqrt{2}}\Big(|0_a\rangle_L|1_a\rangle_R \pm e^{i\beta\phi}|1_a\rangle_L|0_a\rangle_R\Big) \tag{2.100}$$

ϕ is the phase difference between two ensembles L and R. Equation (2.100) is the maximum entangled state. Many experiments of this parallel scheme [121] have been reported. However, it requires the phase locking technology because the environment will cause the micrometer-level random jitter on the optical elements, thus breaking the phase correlation between two ensembles. This problem has been solved by the cascade scheme [122], which is shown in Fig. 2.42b. A strong pump beam passes through two atomic ensemble in serial, then generates one Stokes photon via spontaneous Raman scattering detected by detector. Similar as the DLCZ proposal, a

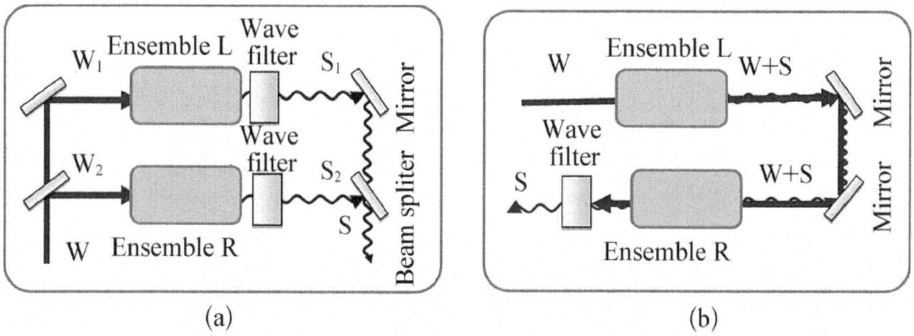

Fig. 2.42 **a** Entanglement generation scheme with two parallel ensembles, where the write beam W is divided into W_1 and W_2 and injected into two atomic ensembles L and R at the same time via spontaneous Raman scattering. The generated Stokes photons S_1 and S_2 are mixed into a total Stokes field S at the beam splitter; **b** entanglement generating scheme with two ensembles in serial, where the write beam W passes through two atomic ensembles L and R sequentially and interacts with the atoms by spontaneous Raman scattering. As the atomic ensemble is pencil-shaped, the probability of forward Stokes photon S is much larger than that in other azimuthal angles, and the Stokes photon S flies out of the ensembles with the write beam together

which-way atomic excitation is automatically created and two ensemble are entangled. The write beam and generated Stokes photon pass together to stabilize the phase correlation between two ensembles.

(4) Quantum light source

Photon source and quantum memory are very important for quantum information. Generally, the bandwidth of the single photon source generated by crystals or quantum dots is extremely wide, about several GHz. The crystals and quantum dots is not suitable as an information memorizer due to the short coherent time. On the other hand, the absorption bandwidth of the atomic ensembles is about 1 MHz and the coherence time of atomic system is longer than the μs level. So the atomic ensemble is suit to act as the memorizer for the single photon source generated by atomic system but not for that by the crystal and quantum dot. Generally, the single photon sources in the atomic ensemble is generated by spontaneous Raman scattering in the atomic ensemble [123]. The generation of the Stokes photon corresponds atomic spin wave in the atomic ensemble. The intensity and phase of the two are correlated. In experiment, after the generation of the Stokes photon, the atomic spin wave can be retrieved back to an anti-Stokes photon by a strong read beam within the coherence time. The Stokes and anti-Stokes photons are non-classical correlated photon pairs and the anti-Stokes photon is a single-photon source after detecting the Stokes photon.

(5) Phase-sensitive Raman process

Stokes light has a good intensity correlation and phase correlation with atomic spin excitation. The Stokes field and the atomic spin wave can be put together as an atom-light interface. Manipulation of the atom-light interface leads to phase-sensitive Raman scattering [124], which is a new type of Raman scattering. First, atomic spin wave is initially prepared in atomic ensemble. Then a strong pump beam and a weak Stokes beam, P and S, are injected simultaneously into atoms to interact with the atomic spin wave. When the Stokes beam S has stable phase relation with the atomic spin wave, the intensity of the output Stokes S_{out} changes with the phase difference between the input Stokes beam and atomic spin wave. The phase-sensitive property in such Raman scattering is the unique advantage over other Raman scattering processes. The optical phase shift is very sensitive to the change of propagation distance. The phase of atomic spin excitation is sensitive to the change of atomic internal state by magnetic field, electric field and optical field. This phase-sensitive Raman scattering should be have potential applications for precision measurement of optical displacement, weak magnetic fields, and weak electric fields.

2.8 Optical Quantum Correlation Interferometer

In the field of precision measurements, optical interferometers are effective tools. Interferometers like Mach–Zehnder interferometer are today a central part of the gravitational-wave observatory (GWO). Interferometers are often used to measure changes in phase, and the measurement accuracy of phase changes has long been a challenge for scientists. At the input terminal of the interferometer, vacuum noise fluctuation brings a limit to the measurement accuracy of phase measurement called the standard quantum limit, which is also known as shot noise limit. In 1981, C. M. Caves pointed out in his paper [11] that if a non-classical light source like squeezed light is used as the input to the interferometer, the system noise can be reduced and the signal-to-noise ratio (SNR) can be improved, thus allowing the sensitivity of phase measurement to surpass the standard quantum limit (SQL) or even reach the Heisenberg limit (HL). Since then, several experimental groups have injected squeezed state light sources into the interferometer to improve the SNR of the interferometer and break the SQL for phase measurement [125, 126]. Many theoretical efforts have also demonstrated that the application of non-classical light sources, including squeezed states, Fock states, and NOON states, to fill the vacuum channels of interferometers can break the SQL. However, such interferometers are highly susceptible to optical losses, either outside or inside the interferometer.

In addition to the use of the Mach–Zehnder interferometer discussed above to break the SQL, another interferometer is the SU(1, 1) nonlinear interferometer proposed by B. Yurke in 1986 [80]. The basic structure of this interferometer is similar to that of the Mach–Zehnder interferometer. In the Mach–Zehnder interferometer, two conventional beam splitters are used to combine and split the incident light; in the nonlinear interferometer, the two beam splitters are replaced by two four-wave mixing nonlinear parametric amplifiers. The comparative structure of the classical Mach–Zehnder interferometer and nonlinear interferometer is given in Fig. 2.43. Yurke et al. demonstrated theoretically that this nonlinear interferometer can improve the phase measurement accuracy to HL without losses. A similar scheme was theoretically investigated by Plick et al. [127] who by injecting coherent states into the parametric amplification process finally increased the photon number significantly and thus improved the measurement accuracy. To experimentally implement this type of nonlinear interferometer, we must find an ideal four-wave mixing amplifier. Fortunately, it is recently discovered by Paul Lett's group at NIST in the United States through experiments that an ideal four-wave mixing amplifier can be achieved using the four-wave mixing process in the vapor of rubidium (^{85}Rb)—an alkali metal atom [128]. Our research group has also carried out research in this area, experimentally constructed this nonlinear quantum interferometer successfully using a four-wave mixing process instead of the optical beam splitter in a conventional interferometer, achieving an interference fringe contrast of 99% at least [81], and investigated the dependency relationship of the contrast of this interferometer on system parameters as well [129]. After that, we studied the quantum characteristics of this interferometer and experimentally demonstrated the quantum enhancement of the SNR with respect to the conventional interferometer [82].

Before detailing this type of nonlinear interferometer, it is important to introduce its parametric amplifier (PA), which, virtually, is based on a four-wave mixing process in the hot vapor of rubidium atom to generate twin beams [130, 131]. Since 1985, when R. E. Slusher first implemented the squeezed state by a four-wave mixing process in an optical cavity [132], a lot of research groups have used different experimental approaches to obtain high-quality squeezed light sources. In 2010 for example, Mehmet et al. obtained −11.5 dB vacuum squeezed states by means of optical parametric oscillations [133]; with the four-wave mixing in atomic vapor, several international groups have achieved highly squeezed quantum correlated light beams [134, 135]. In addition, the system has been extensively applied in the fields of quantum information and precision measurements, i.e., controlled delay of EPR entangled states [144].

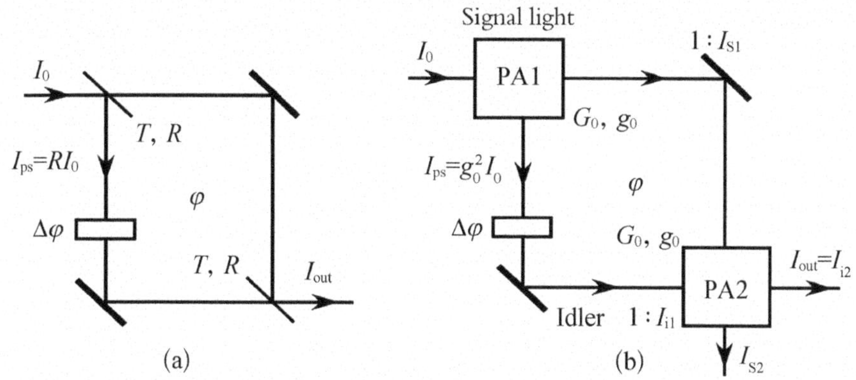

Fig. 2.43 **a** Linear MZ interferometer; **b** Nonlinear interferometer, PA1, PA2 being two four-wave mixing parametric amplifiers

The experimental scheme for implementing a quantum light source by means of four-wave mixing is shown in Fig. 2.44. Both the strong pump light (400 mW) and the weak detection light (20 μW) come from a continuously tunable Ti:sapphire laser. The pump light frequency is located at about 1 GHZ blue detuning (single photon detuning) of Rubidium-85 D1 line ($5S_{1/2}$, $F \rightarrow 5P_{1/2}$, 795 nm). The frequency of the probe light is shifted by an acousto-optic modulator to the red detuning direction of the pump light at 3.04 GHz. Based on the double-λ structure characteristic of the rubidium-85 D1 level structure (see Fig. 2.42), the strong pump light can form a strong coherence with the atomic ensemble, and after probe light injection, it can lead to a four-wave mixing effect and produce an additional beam of conjugate light at the 3.04 GHz blue detuning of the pump light in frequency and on the other side of the pump light in space to meet the phase matching condition. The conjugate light exhibits strong quantum correlations with the probe light, including intensity difference squeezing and quantum entanglement. Experimentally (see Fig. 2.44a), the pump light and probe light are polarization orthogonal and coincide at a 0.4° crossover angle at the center of the rubidium atomic pool where the temperature of the vapor pool is about 120 °C. The intensity of the probe light exiting the atomic pool is amplified, and the amplification is related to experimental parameters such as single-photon detuning, pump light intensity, and pool temperature. The detection light and the conjugate light hit a balanced photodetector (BPD), then the noise spectrum is measured by a spectrometer (SA).

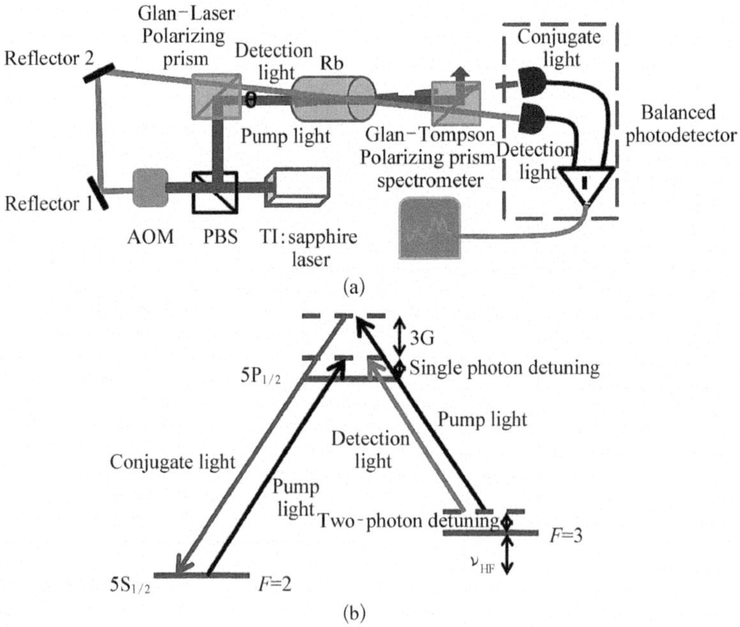

Fig. 2.44 Schematic diagram of four-wave mixing experiment. **a** Experimental setup; **b** Diagram of rubidium-85 D1 line level structure

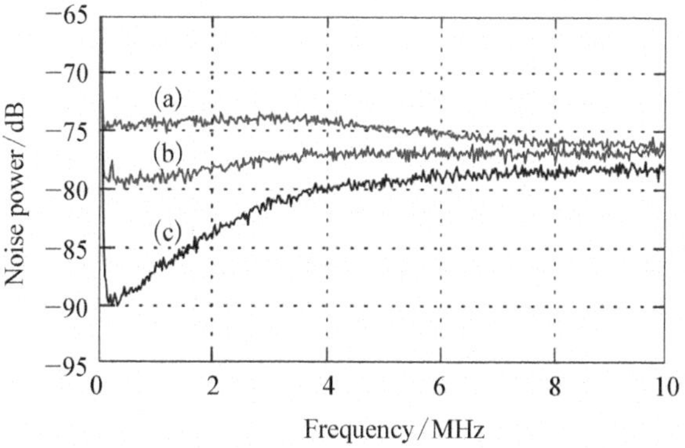

Fig. 2.45 Two-mode compression signal produced by four-wave mixing measured on a apectrometer. **a** Shot noise limit (SNL); **b** Intensity difference squeezing between detection light and conjugate light; **c** Background noise

Experimental results are shown below. We obtain −5 dB intensity difference squeezing between the detection light and the conjugate light, as shown in Fig. 2.45

After introducing the generation of a quantum light source by a parametric amplifier based on four-wave mixing, we return to the nonlinear interferometer. In the following paragraphs, we will explain the basic principles and experimental results of nonlinear interferometer to improve the phase sensitivity. In the MZ interferometer in Fig. 2.43, I_0 is the input signal intensity of the interferometer, T and R are the transmittance and reflectance of the beam splitter; the output intensity of the interferometer is

$$I_{\text{out}} = 2I_0TR(1+\cos\phi) = I_{\text{F}}(1+\cos\phi)/2 \tag{2.101}$$

$I_{\text{F}} = 4I_0TR$ is the light intensity with the strongest interference, and the change in output light intensity is most sensitive to the phase when $\varphi = \pi/2$:

$$\Delta I_{\text{out}} = I_{\text{F}}\frac{\Delta\varphi}{2} = 2TI_{\text{ps}}\Delta\varphi \tag{2.102}$$

where, $I_{\text{ps}} = RI_0$ is called phase-sensitived light intensity. From this, it is known that phase measurement is directly related to the constructive-interference light intensity I_{F}; while for the linear MZ interferometer, the phase change is proportional to the phase-sensitived light intensity I_{ps}.

In view of the nonlinear interferometer in Fig. 2.43, when the input signal intensity is also I_0, PA1 can be regarded as a beam splitter because it amplifies the input signal and produces idler (i.e., conjugate light) at the same time; while PA2 is the amplified signal light (i.e., detection light) coming out of PA1 mixed with the idler light and also undergoes secondary amplification. Thus, PA2 can be considered as a beam combiner. The input and output of the nonlinear parametric process can be expressed by the following relation:

$$A_{\text{s}}^{\text{out}} = GA_{\text{s}}^{\text{in}} + gA_{\text{i}}^{\text{in}x}, A_{\text{i}}^{\text{out}} = GA_{\text{i}}^{\text{in}} + gA_{\text{s}}^{\text{in}x} \tag{2.103}$$

where A represents the amplitude of each light field, s and i denote the signal light and idler, respectively, G indicates amplitude gain and $|G|^2 - |g|^2 = 1$. If there is no input at the idler input terminal of PA1 (i.e., vacuum), the output signal of PA1 is

$$A_{s1} = G_1 A_{s0},\ A_{i1} = g_1 A_{s0}^* \tag{2.104}$$

where A_{s0} is input signal. We can also see that in the case of a large gain, i.e., $G_1 \approx g_1 \geq 1$, the input signal is basically split into two halves in terms of intensity after amplification. Assuming that the idler output from PA1 undergoes phase shift φ, the signal output from PA2 is

$$A_2 = G(\varphi) A_{80},\ A_{i2} = g(\varphi) A_{80}^* \tag{2.105}$$

where

$$G(\varphi) = G_1 G_2 + g_1^* g_2 e^{-ip},\ g(\varphi) = G_1^* g_2 + G_2 g_1 e^{ip} \tag{2.106}$$

If the gains of PA1 and PA2 are equal in magnitude, i.e., $G_1 = G_2 = G_0$ and $g_1 = g_2 = g_0$ (G_0 and g_0 are real numbers), then we have

$$G(\varphi) = 1 + |g_0|^2 \left(1 + e^{-i\varphi}\right),\ g(\varphi) = G_0 g_0 \left(1 + e^{i\varphi}\right) \tag{2.107}$$

In Eq. (2.104), we can see that the intensity of the phase-sensitive light (idler) in the interferometer is

$$I_{i1} = |A_{i1}|^2 = |g_0|^2 I_0 \tag{2.108}$$

where I_0 is the intensity of the input signal light. The output light intensity of the interferometer is also easily derived from

$$I_{s2} = I_0\left[1 + 2|G_0 g_0|^2 (1 + \cos\varphi)\right],\ I_2 = 2I_0 |G_0 g_0|^2 (1 + \cos\varphi) \tag{2.109}$$

From the above equation, we can see that both the output signal light and idle light exhibit interference characteristics, and the interference contrast of the idle light always remains 100%, but the interference contrast of the signal light is

$$V_s = 2|G_0 g_0|^2 / \left(1 + 2|G_0 g_0|^2\right) \tag{2.110}$$

The equation suggests that the interference contrast of the signal light approaches 100% only when $G_0,\ g_0 \geq 1$. From Eq. (2.110), the intensity with the strongest interference of the idler is

$$I_{i2}^{M} = 4I_0 |G_0 g_0|^2 = 4|G_0|^2 I_{ps} = 4\left(I_{ps} + I_0\right) I_{ps} / I_0 \tag{2.111}$$

where $I_{ps} = I_{i1}$ is the intensity of phase sensitive light. In the above equation, it can be seen that when the gain is large as well as the value of the input signal I_0 is determined, there is $I_{ps} \geq I_0$, $I_{i2}^{M} \propto I_{ps}^2$. Equation (2.111) is the biggest difference between nonlinear interferometer and MZ interferometer. The interferometric strongest light intensity is squared dependent on the phase sensitive light intensity, while the constructive-interference light intensity of MZ interferometer is linearly dependent on the phase squared. Meanwhile, the amplification of the constructive-interference light intensity to the phase sensitive light intensity is $4|G_0|^2$. Compared with Eq. (2.102), it can be seen that the output of the nonlinear interferometer is more sensitive to phase change, and thus the phase measurement accuracy can be improved.

The experimental setup for implementing a nonlinear interferometer using a four-wave mixing parametric amplifier is shown in Fig. 2.46, and the two Rb-85s in the figure are PA1 and PA2, which are four-wave mixing quantum light sources as depicted in Fig. 2.44. The signal and idler coming out from PA1 passes through an optical 4F system whose role is to make the spot size of the signal light and idler at the center of the first rubidium actom pool the same as at the center of the second one. Signal light 2 and idler 2 enter two photodetectors (D1, D2) via single-mode coupling (SMF) after exiting PA2, and the detector is connected to a digital scope to detect the interference signal.

Figure 2.47 shows the interferograms of signal light 2 and idler 2 obtained from experiments. Figure (a) is the interferogram of the signal light with an interference contrast of about 94.5%, and figure (b) is the interferogram of the idler with an interference contrast of about 98.6%.

We investigated the dependence of the interference contrast on the gain size of the parametric process, as shown in Fig. 2.48. The figure shows that the interference contrast of the idler (square) always remains at 95% interference efficiency while the interference contrast of the signal light (diamond) relies on the gain size, which agrees well with the results (solid line) simulated by Eq. (2.110).

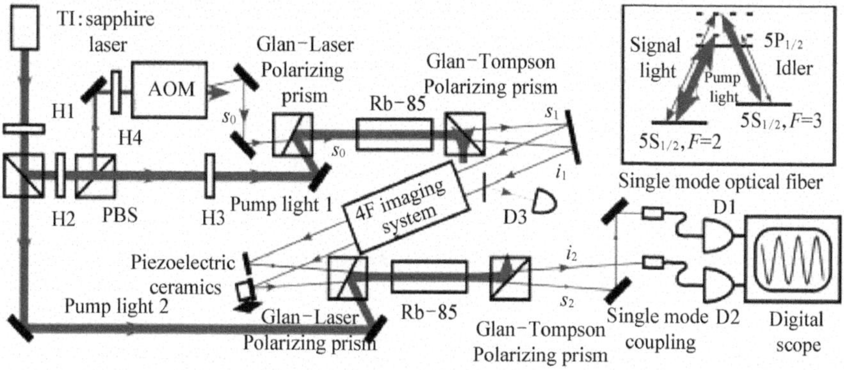

Fig. 2.46 Experimental setup of nonlinear interferometer

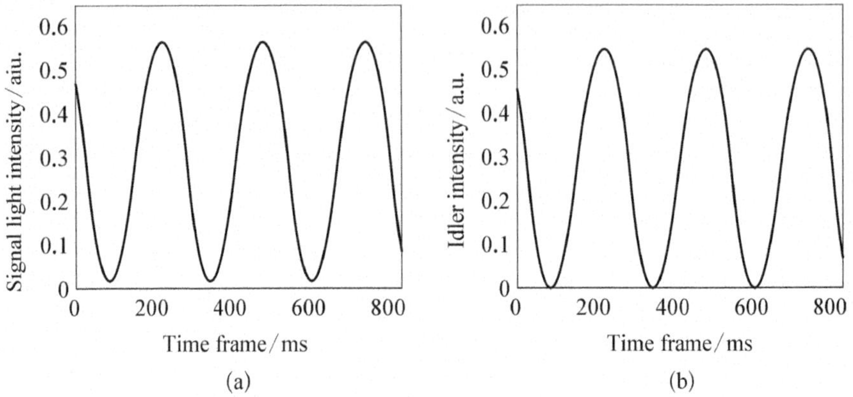

Fig. 2.47 Nonlinear interferometer interference signal

Fig. 2.48 Dependence of interference contrast of nonlinear interferometer output signal on gain. Square: idler; diamond: signal light; solid line: theoretical simulation result of signal light

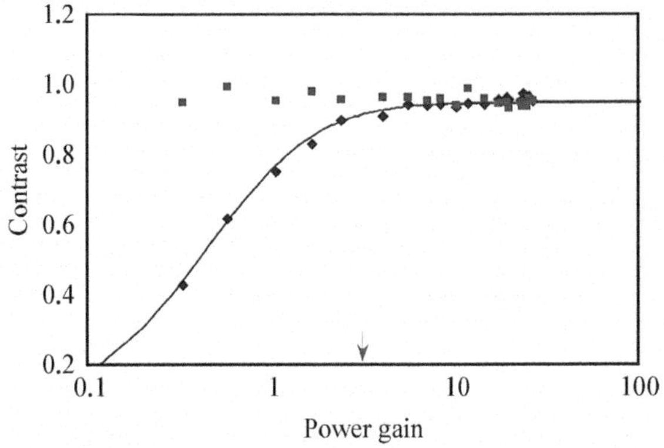

To verify the nonlinear characteristics of the interferometer, we measured the constructive-interference intensity of the idler 2 in relation to the intensity of the phase-sensitived light (idler 1). The change in the intensity of the phase-sensitived light can be achieved by changing the power of the pump light. Taking the logarithm of idler 2 and idler 1 respectively, we get the relationship curve as shown in the diamond-shaped points in Fig. 2.49. The solid line is the theoretical modeling plot obtained according to Eq. (2.110), and it can be seen that the theoretical results are in good agreement with the experimental results. The dashed line is a straight line with a slope of 2. It can be seen that idler 2 and idler 1 are in a quadratic relationship, i.e., a nonlinear relationship between them. The above plot describes a direct intensity relationship curve of idler 2 to idler 1, which is obviously a nonlinear relationship curve.

The sensitivity of phase fluctuation of interferometer measurements is restricted by vacuum noise. To push the measurement accuracy beyond the SQL, a common method is to use a squeezed light source. The noise of the squeezed state is lower than

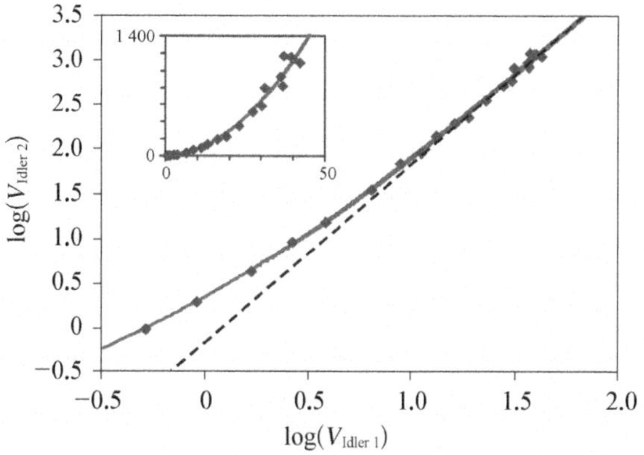

Fig. 2.49 Plot of nonlinear characteristics of nonlinear interferometer

that of the vacuum state. For the same intensity signal, the SNR of the squeezed state is higher than that of the vacuum state. However, the SNR can be enhanced either by reducing the background noise or with the same background noise. Our group used four-wave mixing as a nonlinear parametric amplifier to amplify the signal, and also because four-wave mixing gives rise to twin beams with quantum correlation, the noise of the amplified two beams cancels out each other and maintains at the level of vacuum fluctuation during destructive interference [81]. Therefore, the nonlinear interferometer can improve the SNR of phase measurement. According to Ref. [81], the output signal of the idler end of the nonlinear interferometer is

$$I_{\mathrm{NL}} = 2G^2 I_{\mathrm{ps}}(1 + \cos\phi) \tag{2.112}$$

where G is the amplification of light amplitude of parametric amplification process, $I_{\mathrm{ps}} = g^2 I_0$ is the intensity of phase sensitive light (i.e., idler 1), and g is the intensity of input I_0 light; G and g meet relation $G^2 - g^2 = 1$. For linear interferometer, the interference output signal is

$$I_{\mathrm{L}} = I_{\mathrm{D}\infty}(1 + \cos\phi) \tag{2.113}$$

The interference intensity of the nonlinear interferometer is $2G^2$ times higher than that of the linear interferometer with equal I_{ps}. In the case of destructive interference, assuming a little phase fluctuation δ, $\varphi = \delta \le 1$, we have

$$I_{\mathrm{NL}} \approx 2G^2 I_{\mathrm{ps}}\delta^2,\ I_{\mathrm{L}} \approx I_{\mathrm{ps}}\delta^2 \tag{2.114}$$

In Eq. (2.114), it is known that the signal of nonlinear interferometer has $B_s = 2G^2$ times enhancement to the signal of linear interferometer, and B_s is called enhancement factor of SNR. Therefore, in the ideal case, if the noise of the nonlinear interferometer can be restored to the vacuum noise level, a $2G^2$ times enhancement of the SNR can be obtained.

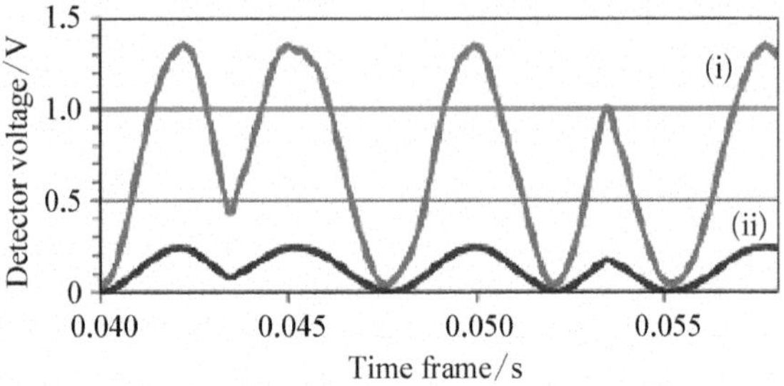

Fig. 2.50 Interference signals of nonlinear interferometer (i) and linear interferometer (ii)

In experiments, we detected the noise of idler 2 by means of balanced homodyne, which could switch between linear interferometer and nonlinear interferometer to measure the noise and signals of nonlinear interferometer and linear interferometer.

Firstly, the interference intensity of two interferometers is measured. Let $I_{ps} = 6\ \mu W$ in both interferometers. In the experiment, the phase of I_{PS} is scanned, and the interference signal obtained is shown in Fig. 2.50. In the figure, the strongest interference intensity of the nonlinear interferometer (i) is about 5.5 times (7.4 dB) that of the linear interferometer (ii), indicating that the nonlinear interferometer has 7.4 dB of signal enhancement than the linear interferometer.

Next, we detected the noise. In the experiment, another set of four-wave mixing method is used to generate new conjugate light as background light (local oscillator) and measure the noise at the output terminal of the nonlinear interferometer via balanced homodyne, as shown in Fig. 2.49, where (i) is vacuum noise (i.e., SQL), equal to the noise of the linear interferometer at destructive interference; (ii) is the noise of the nonlinear interferometer at destructive interference; when scanning phase φ, (iii) is obtained, and (ii) is where the lowest point of (iii) is located.

To illustrate that nonlinear interference can reduce noise, we blocked one of the two arms in the interferometer between pool 1 and pool 2 (signal light 1 or idler 1) to get the noise shown in (iv). It can be seen that the noise in (iv) is much higher than the vacuum noise and the noise produced at the destructive interference of nonlinear interferometer Fig. 2.51. (v) displays the noise of a single vapor pool, and it can be

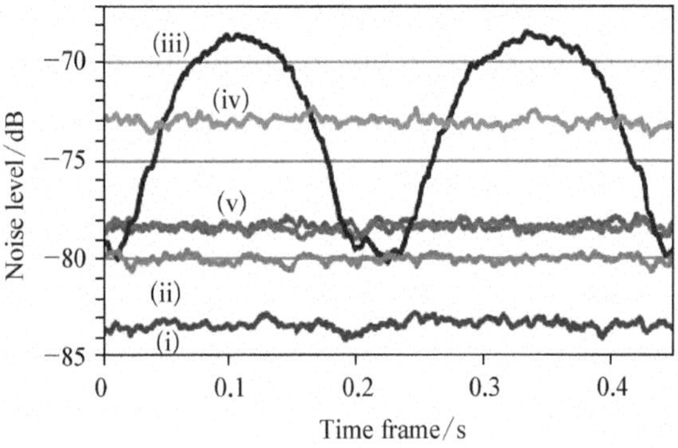

Fig. 2.51 Comparison of noise measurements of nonlinear interferometer and linear interferometer

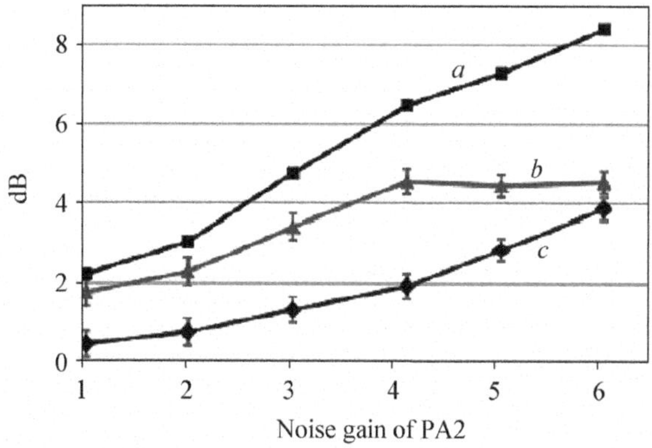

Fig. 2.52 Effect of changing the gain of PA2 on signal amplification, interferometer noise, and SNR enhancement

seen that the noise of the nonlinear interferometer at destructive interference is lower than that of a single amplifier. However, the noise at the interference minimum of the nonlinear interferometer does not reach the level of the vacuum noise, which is still 3.3 dB higher than the vacuum noise. But the results in Fig. 2.50 show a signal enhancement of 7.4 dB relative to the linear interferometer, for which it still improves the SNR by a level of 4.1 dB. This experimentally confirms that the quantum interference occurring in the nonlinear interferometer not only amplifies the signal, but also cancels out most of the correlated noise during destructive interference. The presence of optical losses inside the interferometer brings about additional uncorrelated vacuum noise that cannot be offset, and is a major factor in the noise minimum of the nonlinear interferometer being higher than the SQL.

If the gain of the four-wave mixing in the interferometer increases, the signal gets amplified and the noise of the interferometer increases and is farther away from the vacuum noise level. In Fig. 2.52, the noise gain (dB) of PA2 is changed to get the variation of noise and signals; *c* indicates that the noise at the output terminal of the nonlinear interferometer is higher than the vacuum noise; line *a* represents the enhancement value of the signal; and line b presents the increase of SNR. In Fig. 2.52, it can be seen that the SNR increases with gain provided that the gain is small; when the noise gain reaches 4 dB, the enhancement of SNR basically remains at about 4.1 dB.

References

1. Meystre P. A short walk through quantum optomechanics, Annalen der Physik, 2013, 525: 215–233.
2. Meystre P, Wright E M, McCullen J D, et al. Theory of radiation-pressure-driven interferometers, J . Opt Soc. Am. B, 1985, 2: 1830–1840.
3. Metzger C, Ludwig M, Neuenhahn C, et al. Self-induced oscillations in an optomechanical system driven by bolometric backaction, Phys. Rev. Lett, 2008, 101: 133903(1–4).

4. Carmon T, CrossM, Vahala K. Chaotic quivering of micron-scaled on-chip resonators excited by centrifugal optical pressure. Phys. Rev. Lett., 2007, 98: 167203 (1–4).
5. Dorsel A, McCullen J D, Meystre P, et al. Optical bistability and mirror confinement induced by radiation pressure. Phys. Rev. Lett., 1983, 51: 1550–1553.
6. Marquardt F, Harris J, Girvin S. Dynamical multistability induced by radiation pressure in high-finesse micromechanical optical cavities. Phys. Rev. Lett., 2006, 96: 103901 (1–4).
7. Braginsky V B, Manukin A B. Measurement of weak forces in physics experiments. Chicago: University of Chicago Press, 1977.
8. Drever R, Ford G, Hough J, et al. A gravity-wave detector using optical cavity sensing. Proceedings of the Ninth International Conference on General Relativity and Gravitation. New York, Cambridge University Press, 1983.
9. Braginsky V, Khalili F. Quantum measurement, New York, Cambridge University Press, 1992
10. Milburn G, Woolley M. An introduction to quantum optomechanics. Acta Physica Slovaca, 2011, 61: 483–601.
11. Caves C M. Quantum-echanical noise in an interferometer. Phys. Rev. D., 1981, 23:1693–1708.
12. Kippenberg T, Vahala K J. Cavity optomechanics: back-action at the mesoscale. Science, 2008, 321: 1172–1176.
13. Mancini S, Vitali D, Tombesi P. Optomechanical cooling of a macroscopic oscillator by homodyne feedback. Phys. Rev. Lett., 1998, 80: 688–691.
14. Cohadon P, Heidmann A, Pinard M. Cooling of a mirror by radiation pressure. Phys. Rev. Lett., 1999, 83: 3174–3177.
15. Marquardt F, Chen J, Clerk A, et al. Quantum theory of cavity-assisted sideband cooling of mechanical motion. Phys. Rev. Lett. 2007, 99: 093902(1–4).
16. Schliesser A, Riviere R, Anetsberger G, et al. Resolved-sideband cooling of a micromechanical oscillator. Nat. Phys, 2008, 4: 415–419.
17. Zhang K, Meystre P, Zhang Z. Back-action-free quantum optomechanics with negative-mass Bose-Einstein condensates. Phys. Rev. A, 2013, 88: 043632(1–5).
18. Zhang K, Meystre P, Zhang Z. Role reversal in a bose-condensed optomechanical system. Phys. Rev. Lett., 2012, 108: 240405(1–5).
19. Marshall W, Simon C, Penrose R, et al. Towards quantum superpositions of a mirror. Phys. Rev. Lett., 2003, 91: 130401(1–4).
20. Bose S, Jacobs K, Knight P L. Preparation of nonclassical states in cavities with a moving mirror. Phys. Rev. A, 1997, 56: 4175–4186.
21. Mancini S, Giovannetti V, Vitali D, et al. Entangling macroscopic oscillators exploiting radiation pressure. Phys. Rev. Lett., 2002, 88: 120401(1–4).
22. Paternostro M, Vitali D, Gigan S, et al. Creating and probing multipartite macroscopic entanglement with light. Phys. Rev. Lett., 2007, 99: 250401(1–4).
23. Hartmann M, Plenio M. Steady state entanglement in the mechanical vibrations of two dielectric membranes. Phys. Rev. Lett., 2008, 101: 200503(1–4).
24. Li W, Ates C, Lesanovsky I Nonadiabatic motional effects and dissipative blockade for rydberg atoms excited from optical lattices or microtraps. Phys. Rev. Lett., 2013, 110(5): 213005.
25. Harvey K C, Stoicheff B P. Fine structure of the n2D series in rubidium near the ionization limit. Phys. Rev. Lett., 1977, 38(3): 537–540.
26. Dalgarno A. Rydberg atoms in astrophysics, London, Cambridge University Press, 1983.
27. Vliegen E, Hogan S D, Schmutz H, et al. Stark deceleration and trapping of hydrogen Rydberg atoms. Phys. Rev. A, 2006, 76(8): 023405.
28. Breeden T, Metcalf H. Stark acceleration of Rydberg atoms in inhomogeneous electric fields. Phys. Rev. Lett., 1981, 47(24): 1726–1729.
29. Heidemann R, Raitzsch U, Bendkowsky V, et al. Rydberg excitation of Bose-Einstein condensates. Phys. Rev. Lett., 2008, 100(1), 033601.
30. Anderson W R, Veale J R, Gallagher T F. Resonant dipole-dipole energy transfer in a nearly frozen Rydberg gas. Phys. Rev. Lett., 1998, 80(1): 249–252.

31. Mourachko I, Comparat D, Tomari F, et al. Many-body effects in a frozen Rydberg gas. Phys. Rev. Lett., 1998, 80(1): 253–256.
32. GallagherTF. Rydberg atoms, London: Cambridge University Press, 1994.
33. Reinhard A, Younge K, Liebisch T, et al. Double-resonance spectroscopy of mteracting Rydberg-atom systems. Phys. Rev. Lett., 2008, 100(6): 233201.
34. Carroll T, Claringbould K, Goodsell A, et al. Angular dependence of the dipole-dipole interaction in a nearly one-dimensional sample of Rydberg atoms. Phys. Rev. Lett., 2004, 93(10): 153001.
35. Slichter C P. Principle of magnetic resonance, Berlin: Springer, 1996.
36. Carr C, Ritter R, Wade C G, et al. Nonequilibrium phase transition in a dilute Rydberg ensemble. Phys. Rev. Lett., 2013, 111(9): 113901.
37. Ates C, Pohl T, Pattard T, et al. Antiblockade in Rydberg excitation of an ultracold lattice gas. Phys. Rev. Lett., 2007, 98(1): 023002.
38. Ates C, Pohl T, Pattard T, et al. Many-body theory of excitation dynamics in an ultracold Rydberg gas. Phys. Rev. A, 2007, 76(7), 013413.
39. Amthor T, Giese C, Hofmann C S, et al. Evidence of antiblockade in an ultracold Rydber$_{gg}$as. Phys. Rev. Lett., 2010, 104(1), 013001.
40. Harris S E, Field J E, Imamoglu A. Nonlinear optical processes using electromagnetically induced transparency. Phys. Rev. Lett., 1990, 64(3): 1107–1110.
41. Boiler K J, Imamolu A, Harris S E. Observation of electromagnetically induced transparency. Phys. Rev. Lett., 1991, 66(5): 2593–2596.
42. Ates C, Sevinpll S, Pohl T. Electromagnetically lnduced transparency in strongly interacting Rydberg gases. Phys. Rev. A, 2011, 83(4), 041802(R).
43. Petrosyan D, Otterbach J, Fleischhauer M. Electromagnetically lnduced transparency with Rydberg atoms. Phys. Rev. Lett., 2011, 107(11), 213601.
44. Jaksch D, Cirac J I, Zoller P, et al Fat quantum gates for neutral atoms. Phys. Rev. Lett., 2000, 85(9), 2208.
45. Saffman M, Walker T, Mplmer K. Quantum information with Rydberg atoms. Rev. Mod. Phys., 2010, 82(8): 2313.
46. Comparat D, Pillet P. Dipole blockade in a cold Rydberg atomic sample. J. Opt. Soc. Am. B, 2010, 27(6): A208–A232.
47. Prichard J D, Maxwell D, Gauguet A, et al. Cooperative atom-light interaction in a blockade Rydberg ensemble. Phys. Rev. Lett., 2010, 105(11): 193603.
48. Lukin M D, Fleischhauer M, Cote R. Dipole blockade and quantum information processing in mesoscopic atomic ensembles. Phys. Rev. Lett., 2001, 87 (6); 037901.
49. Viteau M, Huillery P, Bason M G, et al. Cooperative excitation and many-body interactions in a coldRydberg gas. Phys. Rev. Lett., 2012, 109(7): 053002.
50. Pohl T, Sadeghpour H R, Schmelcher P. Cold and ultracold Rydberg atoms in strong magnetic fields. Phys. Rep., 2009, 484: 181–229.
51. Weimer H, MullerM, Lesanovsky I, et al. A Rydberg quantum simulator, Nat. Phys., 2009, 6: 382.
52. Bhaktavatsala R D, Mplmer K. Dark entangled steady states of interacting Rydberg atoms. Phys. Rev. Lett., 2013, 111(7); 033606.
53. Carr A, Saffman M. Preparation of entangled and antiferromagnetic states by dissipative Rydberg pumping. Phys. Rev. Lett., 2013, 111(7): 033607.
54. Isenhower L, Urban E, Zhang X, et al. Demonstration of a neutral atom controlled-NOT quantumgate. Phys. Rev. A, 2010, 104(1); 010503.
55. Saffman M, MØlmer K. Efficient multiparticle entanglement via asymmetric Rydberg blockade. Phys. Rev. Lett., 2009, 102(6): 240502.
56. Fleischhauer M, Imamoglu A, Marangos J. Electromagnetically induced transparency: optics in coherentmedia. Rev. Mod. Phys., 2005, 77(7): 633.
57. Friedler I, Petrosyan D, Fleischhauer M, et al. Long-range interactions and entanglement of slow single-photon pulses. Phys. Rev. A, 2005, 72(10): 043803.

58. Mohapatra A, Bason M, Butscher B, et al. A giant electro-optic effect using polarizable dark states. Nat. Phys., 2008, 4: 890.
59. Gorniaczyk H, Tresp C, Schmidt J, et al. Single-photon transistor mediated by interstate Rydberg interactions. Phys. Rev. Lett., 2014, 113(7): 053601.
60. Tiarks D, Baur S, Schneider K, et al. Single-photon transistor using a Forster resonance. Phys. Rev. Lett., 2014, 113(7); 053602.
61. Baur S, Tiarks D, Rempe G, et al. Single-photon switch based on Rydberg blockade. Phys. Rev. Lett., 2014, 112(2): 073901.
62. Dudin Y, Kuzmich A. Strongyy interacting Rydberg excitations of a cold atomic gas. Science, 2012, 336(5): 887.
63. Bannasch G, Killian T, Pohl T. Strongly coupled plasmas via Rydberg blockade of cold atoms. Phys. Rev. Lett., 2013, 110(6): 253003.
64. Lin Y J, Compton R L, Perry A R, et al. Bose-Einstein condensate in a uniform light-induced vector potential. Phys. Rev. Lett., 2009, 102: 130401.
65. Lin Y J, Compton R L, Jimenez-Garcia K, et al. Synthetic magnetic fields for ultracold neutral atoms. Nature, 2009, 426: 628.
66. Lin Y J, Compton R L, Jimenez-Garcia K, et al. A synthetic electric force acting on neutral atoms. Nature Phys., 2011, 7: 531.
67. Lin Y J, Jimenez-Garcia K, Spielman I B. Spin-orbit coupled Bose-Einstein condensates. Nature, 2011, 471: 83.
68. Wang P, Yu Z Q, Fu K, et al. Spin-orbit coupled degenerate Fermi gases. Phys. Rev. Lett., 2012, 109: 095301.
69. Cheuk L W, Sommer A T, Hadzibabic, et al. Spin-injection spectroscopy of a spin-orbit coupled Fermi gas. Phys. Rev. Lett., 2012, 109:095302.
70. Zhang J Y, Ji S C, Chen Z, et al. Collective dipole oscillations of a spin-orbit coupled Bose-Einstein condensate. Phys. Rev. Lett., 2012, 109: 115301.
71. Hu H, Jiang L, Liu X J, et al. Probing anisotropic superfluidity in atomic Fermi gases with Rashba spin-orbit coupling. Phys. Rev. Lett., 2011, 107:195304.
72. Vyasanakere J P, Shenoy V B. Bound states of two spin-1/2 Fermions in a synthetic non-Abelian gauge field. Phys. Rev. B, 2011, 83:094515.
73. Gong M, Tewari S, Zhang C. BCS-BEC crossover and topological phase transition in 3d spin-orbit coupled degenerate Fermi gases. Phys. Rev. Lett., 2011, 107:195303.
74. Anderson P W. Absence of diffusion in certain random lattices. Phys. Rev., 1958, 109:1492.
75. Zhou L, Pu H, Zhang W. Anderson localization of cold atomic gases with effective spin-orbit interaction in a quasiperiodic optical lattice, Phys. Rev. A, 2013, 87:023625.
76. Giovannetti V, Lloyd S, Maccone L. Quantum-enhanced measurements:beating the standard quantum limb. Science, 2004, 306:1330–1336.
77. Giovannetti V, Lloyd S, Maccone L. Advances in quantum metrology, NatPhotonics, 2011, 5: 222–229.
78. Chun-Hua Yuan, Keye Zhang, and Weiping Zhang. Quantum limits in quantum metrology. Scientia Sinca,2014, 44:343–359.
79. Dowling J P. Quantum optical metrology — the lowdown on high-NOON states. Contemp. Phys., 2008, 49:125–143.
80. Yurke B, McCall S L, Klauder J R. SU(2) and SU(1, 1) interferometers. Phys. Rev. A, 1986, 33(6):4033–4054.
81. Jing J T, Liu C J, Zhou Z F, et al. Realization of a nonlinear interferometer with parametric amplifiers. Appl. Phys. Lett., 2011, 99: 011110.
82. Hudelitt F, Kong J, Liu Cunjin, et al. Quantum metrology with parametric amplifierbased photon correlation interferometers. Nature Commun., 2014, 5: 3049.
83. Li D, Yuan C H, Ou Z Y, et al. The phase sensitivity of an SU (1, 1) interferometer with coherent and squeezed-vacuum light NewJ. Phys., 2014, 16: 073020.
84. Gerry C New, Mimih J. The parity operator in quantum optical metrology,Contemporary Phys., 2010, 51: 497–511.

85. Escher B, Filho R L, Davidovich L. General framework for estimating the ultimate precision limit in noisy quantum-enhanced metrology. Nature Phys., 2011: 406–411.
86. Cooper J Mimi, Hallwood D W, Dunningham J A, et al. Robust sub-shot-noise-limited measurements In a multimode interferometer. Phys. Rev. Lett., 2012, 108: 130402.
87. Ashkin A Acceleration and trapping of particles by radiation pressure. Phys. Rev. Lett., 1970, 24: 156–159
88. Ashkin A Atomic-beam deflection by resonance-radiation pressure. Phys. Rev. Lett., 1970, 25: 1321–1324.
89. Hansch T W, Schawlow A L. Cooling of gases by laser radiation. Opt. Commun., 1975, 13: 68–69.
90. Chu S, Hollberg L, Bjorkholm J E, et al Three-dimensional viscous confinement and cooling of atoms by resonance radiation pressure. Phys. Rev. Lett., 1985, 55: 48–51.
91. Raab E L, Prentiss M, Cable A, et al Trapping of neutral sodium atoms with radiation pressure. Phys. Rev. Lett., 1987, 58: 2631–2634.
92. Lin Y W, Chou H C, Dwivedi PP, et al. Using a pair of rectangular coils in the MOT forthe production of cold atom clouds with large optical density, Optics Express, 2008, 16(6): 3753–3761.
93. Zhang S, Chen J F, Liu C, et al. A dark-line two-dimensional magneto-optical trap of 85Rb atoms with high optical depth. Rev. Sci Instrum., 2012, 83: 073102.
94. Wei D, Chen J F, Loy M M T, et al. Optical precursors with electromagnetically induced transparency in cold atom. Phys. Rev. Lett., 2009, 103; 093602.
95. Steane A M, Chowdhury M, Foot C J. Radiation force in the magneto-optical trap. J. Opt Soc. Am. B, 1992, 9(12): 2142–2158.
96. Radnaev A G, Dudin Y O, Zhao R, et al. A quantum memory with telecom-wavelength conversion. Nature Phys., 2010, 6; 894–899.
97. Imamoglu A, Yamamoto Y. Turnstile device for heralded single photons: coulomb blockade of electron and hole tunneling in quantum confined p-i-n heterojunctions. Phys. Rev. Lett., 1994, 72: 210–213.
98. Kim J, Benson O, Kan, H, et al A single-photon turnstile device. Nature, 1999, 397: 500–503.
99. Brunei C, Lounis B, Tamarat P, et al. Triggered source of single photons based on controlled single molecule Fuorescence. Phys. Rev. Lett., 1999, 83: 2722–2725.
100. Lounis B, Moerner W E. Single photons on demand from a single molecule at room temperature. Nature, 2000, 407: 491–493.
101. Balic V, Braje D A, Kolchin P, et al Generation of paired photons with controllable waveforms. Phys. Rev. Lett., 2005, 94: 183601.
102. Kolchin P, Belthangady C, Du S, et al. Electro-optic modulation of single photons. Phys. Rev. Lett., 2008, 101: 103601.
103. Belthangady C, Du S, Chuu C-S, et al. Modulation and measurement of time-energy entangled photons. Phys. Rev. A, 2009, 80: 031803(R).
104. Du S, Wen J, Belthangady C. Temporally shaping biphoton wave packets with periodically modulated driving fields. Phys. Rev. A, 2009, 79: 043811.
105. Chen J F, Zhang S, Yan H, et al. Shaping biphoton temporal waveform with modulated classical fields. Phys. Rev. Lett., 2010, 104: 183604.
106. Matsukevich D N, Kuzmich A. Quantum state transfer between matter and light. Science, 2004, 306: 663–666.
107. Chaneliere T, Matsukevich D N, Jenkins S D, et al. Storage and retrieval of single photons transmitted between remote quantum memories, Nature, 2005, 438: 833–836.
108. Eisaman M D, Andre A, Massou F, et al. Electromagnetically induced transparency with tunable single-photon pulses. Nature, 2005, 438: 837–841.
109. Liu C, Zhang S, Zhao L, et al. Differential-phase-shift quantum key distribution using heralded narrow-band single photons, Opt. Express, 2013, 21: 9505.
110. Zhang S, Chen J F, Liu C, et al. Optical precursor of a single photon. Phys. Rev. Let 2011, 106: 243602
111. Raman C V. Anew radiation. Indian J. Phys., 1928, 2: 387–398.

112. Shore B W, Bergmann K, Oreg J. Coherent population transfer: Stimulated Raman adiabatic passage and the Landau-Zener picture. Zeitschrift fur Physik D Atoms, Molecules and Clusters, 1992, 23: 33–39.
113. Harris S. Electromagnetically induced transparency. Phys. Today, 1997, 50: 36–42.
114. Raymer M G, Walmsley I A, Mostowski J, et al. Quantum theory of spatial and temporal coherence properties of stimulated Raman scattering. Phys. Rev. A, 1985, 32: 332–344.
115. Mishina O S, Larionov N V, Sheremet A S, et al. Stimulated Raman process in a scattering medium applied to the quantum memory scheme. Phys. Rev. A, 2008, 78: 042313
116. Chen L Q, Zhang G W, Yuan C H, et al. Enhanced Raman scattering by spatially distributed atomic coherence. Appl. Phys. Lett., 2009, 95: 041115.
117. Jain M, Xia H, Yin GY, et al. Efficient nonlinear frequency conversion with maximal atomic coherence. Phys. Rev. Lett., 1996, 77: 4326–4329.
118. Ou Z Y. Efficient conversion between photons and between photon and atom by stimulated emission. Phys. Rev. A, 2008, 78: 023819.
119. Yuan C H, Chen L Q, Ou Z Y, et al. Correlation-enhanced phase-sensitive Raman scattering in atonic vapors. Phys. Rev. A, 2013, 87: 053835.
120. Raymer M G, Mostowski J. Stimulated Raman scattering. Unfed treatment of spontaneous initiation and spatial propagation. Phys. Rev. A, 1981, 24: 1980.
121. Duan L M, Lukin M, Cirac I, et al. Long-distance quantum communication with atomic ensembles and linear optics. Nature, 2001, 414: 413–418.
122. Choi K S, Goban A, Papp S B, et al. Entanglement of spin waves among four quantum memories. Nature, 2010, 468: 412–416.
123. Zhang G W, Bian C L, Chen L Q, et al. Which-way collective atomic spin excitation among atomic ensembles by photon indistinguishability. New Journal of Physics, 2012, 14: 063034.
124. Chen S, Chen Y A, Strassel T, et al. Deterministic and st roable single-photon source based on a quantum memory. Phys. Rev. Lett., 2006, 97: 173004.
125. Goda K, Miyakawa O, Mikhailov E Goad, et al. A quantum-enhanced prototype gravitational-wave detector. Nat. Phys., 2008, 4: 472–476.
126. The LIGO Scientific Collaboration. A gravitational wave observatory operating beyond the quantum shot-noise limb. Nat. Phys., 2011, 7: 962–965.
127. Plick W N, Dowling J P, Agarwal G S. Coherent-light-boosted, sub-shot noise, quantum interferometry. New J. Phys., 2010, 12: 083014.
128. McCormick C F, Boyer V, Arimondo E, et al. Strong relative intensity squeezing by four-wave mixing in rubidium vapor. Opt. Lett., 2007, 32(2): 178–180.
129. Kong J, Jing J T, Wang H L, et al. Experimental investigation of the visibility dependence in a nonlinear interferometer using parametric amplifiersour. Appl. Phys. Lett., 2013, 102: 011130.
130. Liu C J, Jing J T, Zhou Z F, et al. Realization of low frequency and controllable bandwidth squeezing based on a four-wave-mixing amplifier in rubidium vapor. Opt. Lett., 2011, 36: 2979.
131. Qin Z Z, Jing J T, Zhou J, et al. Compact diode-laser-pumped quantum light source based on four-wave mixing in hot rubidium vapor. Opt. Lett., 2012, 37 (15): 3141–3143.
132. Slusher R E, Hollberg L W, Yurke B, et al. Observation of squeezed states generated by four-wave mixing in an optical cavity. Phys. Rev. Lett., 1985, 55(22): 2409.
133. Mehmet M, Vahlbruch H, Lastzka N, et al. Observation of squeezed states with strong photon-number oscillations. Phys. Rev. A, 2010, 81: 013814.
134. Glorieux Q, Guidoni L, Guibal S, et al. Strong quantum correlations m four wave mixing m ^{85} Rb vapor. Proc. SPIE 7727, Quantum Optics, 2010,772703.
135. Marino A M, Pooser R C, Boyer V, et al. Tunable delay of Einstein-Podolsky-Rosen entanglement. Nature, 2009, 457: 859–862.
136. Ho T L, Zhang S. Bose-Einstein condensates with spin-orbit interaction. Phys. Rev. Lett., 2011, 107:150403.
137. Yu Z Q, Zhai H. Spin-orbit coupled Fermi gases across a feshbach resonance, Phys. Rev. Lett., 2011, 107:195305.

138. Sparkes B M. An ultra-high optical depth cold atomic ensemble for quantum memories. Journal of Physics: Conference Series, 2013, 467: 012009.
139. Merriam A J, Sharpe S J, Shverdin M, et al. Efficient nonlinear frequency conversion in an all-resonant double-lambda system. Phys. Rev. Lett., 2000, 84: 5308–5311.
140. Bian C L, Chen L Q, Zhang G W, et al. Retrieval of phase memory in two independent atomic ensembles by Raman process. Europhysics Lett., 2012, 97: 34005.
141. Scully M O, Kattawar G W, Lucht R P, et al. FAST CARS: Engineering a laser spectroscopic technique for rapid identification of bacterial spores, Science, 2002, 99: 10994–11001.
142. Chen L Q, Zhang G W, Bian C L, et al. Observation of the Rabi oscillation of light driven by an atomic spinwave. Phys. Rev. Lett., 2010, 105: 133603.
143. Lawrie B J, Evans P G, Pooser R C. Extraordinary optical transmission of multimode quantum correlations via localized surface plasmons. Phys. Rev. Lett., 2013, 110: 156802.

Index

A
Anti-Stokes photons, 199, 200, 207

B
Beam splitter, 11–14, 18, 24, 27–30, 42–48, 51, 52, 54, 55, 57, 58, 63, 66, 68, 76, 77, 85, 87, 101, 117, 126, 176, 177, 181, 183–186, 197, 198, 206–209, 211

C
Cavity mirror, 136–139, 141, 142, 144–146
Cavity optomechanical system, 136, 138, 141–146
Cavity optomechanics, 132, 134, 136–139, 141–143, 146, 147
Cluster state, 3–5, 19–21, 23–26, 29, 90–97, 100–103, 108–111, 113–115, 119
C-PHASE gate, 11, 100, 113, 120

D
Dipole blocking effect, 159
Dipole-dipole interaction, 150–152, 157, 159, 161–163

E
EIT effect, 157, 158, 163, 199

G
Grover's algorithm, 101, 103

H
High-finesse optical cavities, 142

K
Kerr coefficient, 163
KLM scheme, 10, 13, 16–19, 23, 24, 26, 27

L
Light storage, 28

M
Magneto-optical trap, 134, 190–192, 195, 196, 198

N
Nonlinear optical effects, 5

P
Photonic entanglement, 21

Q
Quality factor, 137, 142, 196
Quantum computing, 2–5, 7, 10, 11, 14, 15, 17, 19, 26, 28–30, 44, 60, 69–72, 76, 78, 79, 89, 90, 100, 101, 104, 106–108, 114, 115, 146, 189
Quantum storage, 4, 5, 28, 29, 101
Quantum teleportation, 3, 8, 9, 14–19, 21, 23, 24, 26

W. Zhang and Z.-B. Chen (eds.), *Special Topics in Quantum Optics*,
https://doi.org/10.1007/978-981-99-8454-1

R
Rabi frequency, 154, 156, 158–161, 164, 166, 169
Raman transition, 165, 166, 169

S
Shor's algorithm, 5–8, 104–108
Sideband cooling principle, 144
Signal-to-noise ratio, 30, 134, 202, 208
Slow light, 194
Spin-orbit coupling, 133, 163, 164, 168, 169, 171–173
Standard quantum limit, 133, 140, 175, 180, 208
Superluminal light, 201

The manufacturer's authorised representative in the EU is Springer Nature Customer Service Centre GmbH, Europaplatz 3, 69115 Heidelberg, Germany. If you have any concerns regarding our products, please contact ProductSafety@springernature.com

Printed and bound by CPI Group (UK) Ltd, Croydon, CR0 4YY
06/10/2026
02228541-0001